ORGANIC SOLAR CELLS

Device Physics, Processing, Degradation, and Prevention

ORGANIC SOLAR CELLS

Device Physics, Processing, Degradation, and Prevention

Pankaj Kumar

CRC Press
Taylor & Francis Group
Boca Raton London New York

CRC Press is an imprint of the
Taylor & Francis Group, an **informa** business

CRC Press
Taylor & Francis Group
6000 Broken Sound Parkway NW, Suite 300
Boca Raton, FL 33487-2742

Printed on acid-free paper
Version Date: 20160323

International Standard Book Number-13: 978-1-4987-2327-5 (Hardback)

Library of Congress Cataloging-in-Publication Data

Names: Kumar, Pankaj, 1981- author.
Title: Organic solar cells : device physics, processing, degradation, and prevention / Pankaj Kumar.
Description: Boca Raton, FL : CRC Press, Taylor & Francis Group, [2017] | ©2017 | Includes bibliographical references and index.
Identifiers: LCCN 2016008260| ISBN 9781498723275 (hardcover) | ISBN 1498723276 (hardcover)
Subjects: LCSH: Solar cells--Materials. | Solar cells--Design and construction. | Solar cells--Deterioration. | Organic electronics.
Classification: LCC TK8322 .K86 2017 | DDC 621.31/244--dc23
LC record available at http://lccn.loc.gov/2016008260

Visit the Taylor & Francis Web site at
http://www.taylorandfrancis.com

and the CRC Press Web site at
http://www.crcpress.com

Printed and bound in the United States of America by Publishers Graphics, LLC on sustainably sourced paper.

Contents

Preface

Clean, long-lasting, and cost-effective means of energy production are urgently required to save the world from an energy crisis, and for that purpose photovoltaic (PV) devices (also known as solar cells) have been identified as important technology. For its maximum benefit, there has always been a quest for efficient, stable, and cost-effective PV technology, which gave birth to its new generations. Organic solar cells (OSCs) have attracted the attention of global scientists, engineers, and industries as they promise to be a potential means of cost-effective energy production. OSCs are quite thin, lightweight, and fully compatible to large-area roll-to-roll (R2R) production on flexible substrates via printing techniques. High throughput through R2R production makes them quite cost effective. Additionally OSCs are supposed to have very short energy payback times, as they neither require expensive processing equipment nor high energy input, and the materials and labor cost are very low.

OSC technology began nearly 25 years ago when the efficiency was hardly 1% and the stability was only few hours. Since then there has been great progress in OSC performance and now the efficiency has gone beyond 11% and the stability of many thousand hours has been reported, but OSCs still miss the figures which can make them commercially viable. Several international research groups are developing new methods and processes to make OSCs more efficient and stable. Several industries have emerged and are contributing to the development of OSCs in different ways like materials synthesis, preparation of substrates, encapsulation materials, equipment manufacturing, and their supply to worldwide customers. Many industries have undertaken the manufacturing and marketing of OSCs. There have been several public demonstrations of prototype OSC modules, but the performance of OSCs is not high enough to compete with the mature traditional PV technologies and more work is needed to make them suitable for replacement of traditional PV technologies. Efficiency, stability, and cost are the main parameters that will decide the success of this technology.

OSCs will not be as stable and efficient as traditional Si solar cells, but they can still be more economic and competitive because of their low cost, and they still need reasonable efficiency and lifetime. OSCs, possessing 8%–10% efficiency with 8–10 years lifetime, can compete with traditional solar cells, but they still have to achieve these figures. Traditional solar cells have struggled for decades to provide an efficient solution to the energy crisis, but the problems with OSCs are different and it is believed that these problems can be easily resolved. This technology would soon serve the society in a better way. Compared to traditional Si solar cells, OSCs will not only be economical but also more fascinating. This book intends to provide students, scientists, and engineers a platform to understand the problems with OSCs and provide them the solutions to develop and manufacture these devices worldwide on a large scale so that everybody has their own technology and their own energy. This book describes in detail the degradation in OSCs and the ways to prevent it. The techniques to measure qualitative and quantitative degradation have also been discussed. This book also covers other important topics like structures, fabrication, device physics, roll-to-roll processing, and cost analysis. This book provides all the information of OSCs from device to technology. I hope this book proves to be quite beneficial for readers and helps them to develop and make OSCs commercially viable. The language of the book is kept very simple for easy understanding. I wish the readers a pleasant reading and hope that they find it a good source of knowledge and helps them to resolve their doubts.

I thank everyone who has helped me in one way or another in successful completion of this book. It is all because of these people that I could write this book tirelessly on time. First, I thank all my collaborators and colleagues who benefited me from their fruitful discussions and brilliant ideas. I thank Dr. R. K. Garg, director, Technology Applications Service, New Delhi, for his encouragement and support all the time. I derived considerable benefits from my stay at University of Newcastle (UON), Australia. Interactions with Professor Paul Dastoor, Dr. Chhinder Bilen, and other colleagues at UON were highly beneficial. I thank all my coworkers and students from CSIR-National Physical Laboratory, New Delhi, including Mukesh Kumar Thakur and Abhishek Kumar Chauhan who supported me directly or indirectly in one way or another during preparation of this book.

I wish to extend my special thanks to Aastha Sharma, commissioning editor, CRC Press, Taylor & Francis Group, for her continuous encouragement and marvelous support during writing of this book. She always showed great consideration to this work that led to its completion on time. I thank Melisa Sedler, project coordinator, CRC Press, Taylor & Francis Group, as well for helping me out at several stages of writing and providing me the necessary information as and when required. I dedicate this book to the late Professor S. C. Jain, former director, Solid State Physics Laboratory (SSPL), New Delhi, Head of Division, CSIR-National Physical Laboratory, New Delhi, and visiting professor at KU Leuven and IMEC, Belgium, who was not only my teacher, mentor, supervisor but also my guardian and a best friend. I learned a lot from Jain and his supervision gave a new direction to my scientific career.

Finally, I sincerely thank my mother, Kusum Lata; father, Mangat Singh; elder brother, Prashant Kumar; younger brother, Dr. Abhishek Bharti; sister, Abhilasha Anand; and brother-in-law, Vikas Anand, who provided me tremendous support and lots of love during preparation of this book. I wish to convey my lots of love and hugs to my little nephew master Avik for entertaining me time to time by his funny activities. I also wish to extend my sincere thanks to uncle Chhinder Bilen and aunty Resham Bilen from Newcastle, Australia, who are like parental figures to me, for their unconditional love, care, and blessings.

Dr. Pankaj Kumar
New Delhi, India

Author

Pankaj Kumar graduated in science and earned a bachelor's degree in 2001 and master's degree in physics (electronics) in 2003 from Ch. Charan Singh University, Meerut. He then joined CSIR-National Physical Laboratory, New Delhi, in 2003 as a research fellow. He has been a recipient of the reputed CSIR-Junior Research Fellowship and CSIR-Senior Research Fellowship and pursued a PhD in science. He has contributed immensely in the field of organic electronics especially in organic solar cells (OSCs), and in 2006 he was appointed as a regular scientist at CSIR-National Physical Laboratory, New Delhi.

Kumar was awarded a PhD in physics in 2010 from University of Delhi, Delhi, for his scientific contributions toward understanding of electrical and optical properties of organic semiconductors. He worked in close association with the late Professor S. C. Jain, former director SSPL, New Delhi, and Professor Vikram Kumar, former director CSIR-NPL, New Delhi, and developed a number of models to explain the characteristics of organic devices.

In 2013, Kumar was honored with the prestigious Indo-Australia early career research fellowship by INSA, New Delhi, to pursue research in Australia and he joined the group of Professor Paul Dastoor, University of Newcastle, Australia, as a visiting fellow for 1 year. During his stay at the University of Newcastle he worked on the degradation issue in OSCs and contributed toward improving their lifetime. Focusing on different issues related to organic materials and devices, Kumar has published more than 45 research articles in reputed national/international research journals and more than 20 papers in national/international conferences, which have garnered worldwide appreciation and citations. For the last 5 years, Kumar has been teaching at the MTech and PhD levels at the Academy of Scientific and Innovative Research-National Physical Laboratory, New Delhi. Kumar is the recipient of several other honors and awards to his credit. He has recently initiated the development of perovskite solar cells as well. Organic and perovskite electronics are the major areas of his current research interest.

1

Solar Cells and Their Generations

1.1 Introduction

Energy plays a very important role in the socioeconomic growth and energy security of a country. The U.S. Energy Information Administration's *Annual Review 2011* identified industrial, residential and commercial, transportation and electric power generation as the prime sectors for energy consumption that account about 21%, 11%, 28%, and 40%, respectively, of total energy consumption [1]. An almost similar report was presented by the European Commission on Energy in 2011, where transportation, industry, household, services, and agriculture were found to share 32%, 25%, 27%, 14%, and 2%, respectively, of total energy consumption. Most energy is drawn from the fossil fuels like coal, oil, and natural gases, which have been buried in the Earth's crust for over millions of years. These fuels are formed from the remains of the dead plants and animals that are trapped and decomposed under the soil. In principle, these fossil fuels should be called renewable sources of energy, but because they take millions of years in generation and are depleting faster than their production, they are called nonrenewable. An estimate by the U.S. Energy Information Administration shows that a large percentage of energy is derived from fossil fuels where they share ~83% of world energy consumption, out of which petroleum, coal, and natural gases account for about 37%, 20%, and 26% shares, respectively. Nuclear energy is another nonrenewable energy that shares 8% of the world's total energy consumption. The rapid depletion in fossil fuels and imbalance in the energy demand and supply is responsible for the hike in inflation rates. According to the 2011 annual report of the European Commission on Energy, to meet the energy demand 35.1% of energy was derived from petroleum and solid fuels accounted for 15.9% of energy. Nuclear power and natural gases accounted for 13.5% and 25.1%, respectively, whereas 9.8% of total energy consumption was derived from renewable sources. The world's energy consumption is increasing very rapidly in every sector of day-to-day life. Fossil fuels are limited in quantity and the population growth is adversely causing rapid reduction in the available energy per capita. Regular consumption of fossil fuels is expected to have severe harmful effects on the climate. Worldwide use of fossil fuels produces a billion tons of CO_2 (a greenhouse gas), which causes an increase in the Earth's surface temperature and contributes to serious concerns like global warming and climate change. Additionally, the combustion of fossil fuels increases environmental pollution, which has very harmful effects on the health of every living being. For example, in 2013 the pollution in the city of Harbin, China, due to coal-powered plants increased to more than 40 times than the set daily target by the World Health Organization [2], which led China to take some serious steps toward minimizing the use of fossil fuels and use of alternate energy sources. In order to prevent pollution and the climate change, the use of alternate energy sources like renewables is now required. China started investing a large sum of money in the development of renewable energy and it has now become the biggest investor in the world for renewable energy production.

The natural sources of energy, which regularly get renewed within the human timescale, are known as renewable sources of energy. Sunlight, wind power, hydropower, geothermal, biomass, and tides are some of the most important renewable sources of energy. They are now promoted and used as much as possible. According to a report from the Renewable Energy Policy Network for the twenty-first century (REN21) in 2014, the renewable energy contributed 19% to the total energy consumption and 22% for electricity generation in 2013. The worldwide investment in renewable energy technologies is now increasing very rapidly. Figure 1.1 shows some of the nonrenewable and renewable sources of energy. Among renewables, solar energy has received special attention and massive promotion over the past few decades. It is the radiant heat and light from the sun, which can be harvested through a number of means like solar thermal, solar photovoltaics, and photosynthesis technologies. Solar thermal includes the solar collectors and concentrators for heating applications, whereas the solar photovoltaics (also known as solar cells) convert solar light into electricity, which can be utilized in all different ways for different needs. The solar energy received by the Earth in 1 h is more than the energy consumed by the world per year. Harvesting of solar energy only from 1% of the land can meet the world's total energy demand. Therefore, solar thermal and solar cell devices have great potential for meeting the world's energy demand, without having any harmful effect on us or the environment. Solar cells could be a godsend for the 2 billion-plus people who do not have access to electricity. Using photovoltaic devices everybody can generate their own energy independently in their homes and offices making them energy secured. Photovoltaic devices can provide long-term power at low operating cost and are virtually free from pollution. Owing to their importance and great potential, solar photovoltaics have become a multibillion industry all over the world. This technology works on the principle of photovoltaic effect.

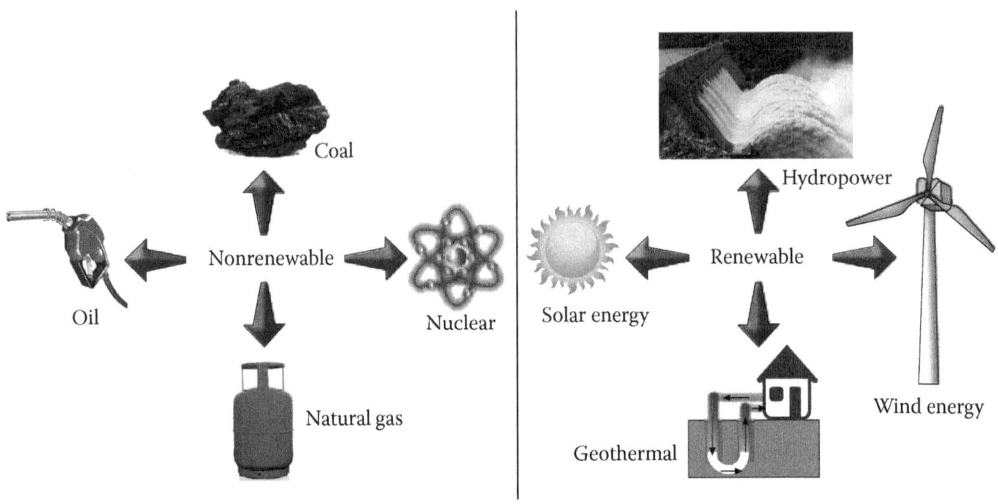

FIGURE 1.1
Some of the nonrenewable and renewable sources of energy.

1.2 Photovoltaic Effect

Generation of a voltage or electric current in a material or a device upon illumination of light is known as the photovoltaic effect. The photovoltaic effect was first observed in 1839 by French physicist A.-H. Becquerel when he found that shining light on an electrode submerged in a conductive solution would create an electric current. Later in 1873, W. Smith discovered photoconductivity in selenium and then this gave birth to photovoltaic technology. The devices that exhibit photovoltaic effect are known as photovoltaic devices or solar cells. In general, this effect takes place in semiconductor devices where exposure to light causes the photons to get absorbed in the semiconducting material that excites electrons from the valance band to the conduction band. Such electrons in the conduction band are known as photogenerated electrons, and they leave behind the corresponding holes in the valance band. These electrons and holes need to be extracted out to get electricity. Due to opposite charge on electrons (–ve) and holes (+ve), they have the tendency of recombination, which should be prevented. The solar cells are designed and prepared in a way that the photogenerated electrons and holes move in opposite directions via drift and diffusion processes and get collected there. A solar cell possesses two electrodes with the light-absorbing medium sandwiched between them. Collection of photogenerated charge carriers on the electrodes in opposite sides of the semiconductor causes an electromotive force, or a photovoltage, to develop across the device. If an electronic circuit is connected to the device, an electric current would pass through the circuit and this way the light energy gets converted into electricity. The drift is provided by the built-in electric field, whereas diffusion is caused by the concentration gradient of the photogenerated charge carriers. If the energy of incident photons is less than the band gap of the semiconductor, the photons will not be absorbed and no photovoltaic effect will be observed. The photovoltaic effect can also be observed in thermoelectric materials where light absorption would cause heating. The heating would increase temperature and due to the temperature gradient in thermoelectric materials, an electromotive force would get generated by the Seebeck effect. If connected to an electric circuit, the electric current would pass through.

1.3 Solar Cells and Their Generations

Solar cells originated more than a 100 years ago, but they were not actually efficient. The first solid-state solar cell was prepared by Charles Fritts in 1883, when he coated a thin film of gold over the selenium semiconductor to form a junction and got around 1% power conversion efficiency (PCE). Some progress was made in selenium solar cells for the next several decades, but nothing appreciable was achieved. Therefore, experiments were performed on other materials and this gave birth to the next generations of solar cells.

1.3.1 First Generation of Solar Cells

Crystalline and polycrystalline Si-based solar cells were the first generation of solar cells. These solar cells are also known as conventional, traditional, or wafer-based solar cells. A brief introduction of them is given next.

1.3.1.1 Crystalline Silicon (c-Si) Solar Cells

In 1940, Russell Ohl, a researcher from Bell Laboratories, had a crystalline silicon (c-Si) sample with a crack in the middle, and the sample exhibited an electric current flow when it was exposed to light. The crack in the sample would have doped one side of the crack as positively charged and the other side as negatively charged, and it would have accidentally made a *p-n* junction. This cell had an efficiency of about 1% [3]. Later the research in diodes and transistors led Daryl Chapin, Calvin Souther Fuller, and Gerald Pearson from the same laboratory to demonstrate the first practical Si solar cell in 1954 that exhibited about 6% efficiency [4]. The efficiency of c-Si solar cells has now improved tremendously to about 25% [5,6]. The progress made in silicon solar cell technology has been discussed in detail by Green [6]. Si is an important elemental semiconductor, which has atomic number 14 with four electrons in the outermost shell. In a Si crystal, four electrons in the outermost shell of each atom make covalent bonds with the neighboring four Si atoms arranged in tetrahedral configuration giving a perfectly diamond-like crystal structure over a long range (see Figure 1.2). It is an indirect band gap material with a band gap of 1.1 eV at room temperature. Si is rarely found in pure form in nature, but after oxygen it is the most abundant element available in the Earth's crust in different forms. The solar cells are processed on a c-Si wafer that usually ranges in diameter from about 25 mm to about 300 mm with thickness ranging from about 200 μm to about 800 μm. Cylindrical ingots of highly pure c-Si are prepared by the Czochralski process from pure molten Si and can be doped into a *n*-type or *p*-type by adding a precise amount of suitable impurity in the molten Si during the growth process. Later these ingots are cut into wafers of suitable size. The wafers are rigid in density (2.33 g/cm^3) but mechanically fragile.

A typical c-Si solar cell possesses a Si *p-n* homojunction with suitable top and bottom electrodes. The top and bottom electrodes are deposited by screen printing. Screen printing is the most mature and simple technique for solar cell fabrication. A typical Si solar cell is prepared on a base substrate (usually *p*-type), which is doped to form an *n*-type emitter on the top. For *n*-type doping, phosphorus is diffused into base substrate at high temperature. Figure 1.3 shows schematically the structure of a typical Si solar cell and the photovoltaic effect through an energy band diagram. A thin antireflection coating (ARC) on the top minimizes the photons loss via back reflection that results in the improved PCE of the cell. The ARC creates a sort of light trap and helps the solar cell to absorb larger fraction of incident light. This is done by the use of a textured surface that reduces the amount of light

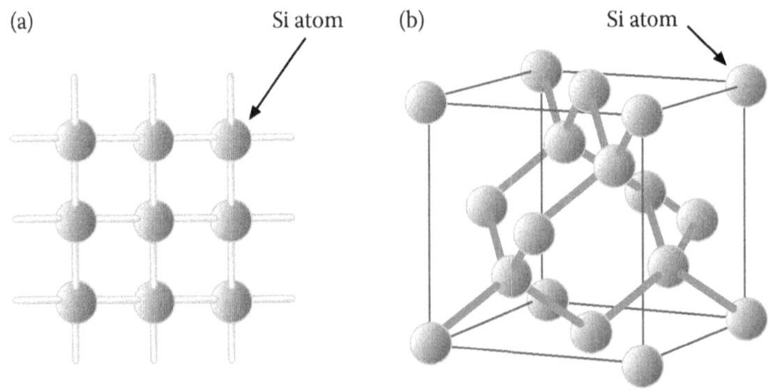

FIGURE 1.2
(a) 2-dimensional (2D) and (b) 3-dimensional (3D) representation of crystal structure of Si.

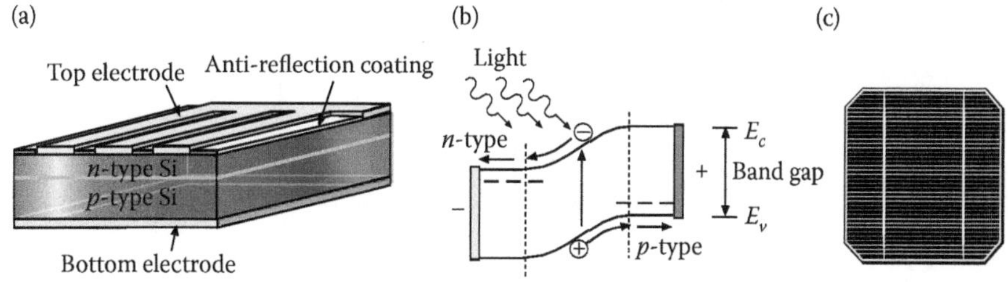

FIGURE 1.3
(a) Schematic structure, (b) photovoltaic effect, and (c) top view of a typical *p-n* junction Si solar cell.

reflected back into the environment, and the light absorbed by the cell is increased resulting in enhanced efficiency. Usually the ARC is done by deposition of silicon nitride (SiN) via the chemical vapor deposition (CVD) technique. For deposition of SiN, the precursor gases of silane (SiH_4) and ammonia (NH_3) are introduced in the CVD chamber where SiH_4 reacts with NH_3 and produces SiN. For the nonstoichiometric film of SiN, a large amount of H_2 is introduced into the chamber to get hydrogenated SiN (SiN:H). TiO_2 is also being used for antireflection coatings, but it does not provide surface passivation. Surface passivation is important to improve the electrical properties of the solar cells. The fingers of top electrodes are formed by screen printing of a metal paste containing cutting agents that allow the metal to go through the antireflection coating and make contact to the underlying Si when fired. For the back contact, a full layer of Al is printed and fired on the back of the substrate that produces a back surface field and improves the cell performance. For solderable contacts, another layer of Al/Ag is printed on the back electrode. The solar cells prepared in the buried contact structure have shown very high efficiency and are better than screen printing solar cells. In the buried contact solar cells, the metal is buried in the groves prepared by laser etching in the solar cells. The buried contact solar cells have less shading losses and allow low reflection giving higher photocurrent. The buried contacts also reduce the parasitic resistance losses due to the high metal aspect ratio. Some of the companies well known for making c-Si solar cells are Alps Technology, ASW, Auxin Solar, and Dmsolar, United States; Orange Solar, Netherlands; Powerwell Solar, China; JS Solar, China; and Motech Industries, Taiwan.

The *p-n* junctions have played a vital role in the development of modern electronic devices and have given foundation to the physics of semiconductor devices. A *p-n* junction is formed by diffusion of suitable impurity atoms in a doped Si substrate. In principle it is like a *p*-type semiconductor has been put in contact with an *n*-type semiconductor. Formation of contact causes the electrons of diffuse from *n*-type to *p*-type semiconductors and holes to diffuse from *p*-type to *n*-type semiconductors across the junction. This movement of charge carriers across the junction leaves the ionized impurity atoms behind generating a space charge, which generates an electric field, known as a built-in electric field. When the electric field becomes sufficient enough, it stops further diffusion of charge carriers. The region around the junction where space charge is formed is known as the depletion region. The built-in electric field helps the charge carriers to drift in the depletion region. Figure 1.4 schematically shows the diffusion lengths of charge carriers around the junction for an *n*-on-*p* solar cell. Only the electron–hole pairs generated within the diffusion lengths and depletion region *W* contribute to the photocurrent and the rest get recombined. The holes (minority carriers) from holes diffusion length (L_{dp}) in the *n*-type region diffuse to the boundary of depletion region and are swiped by the built-in electric

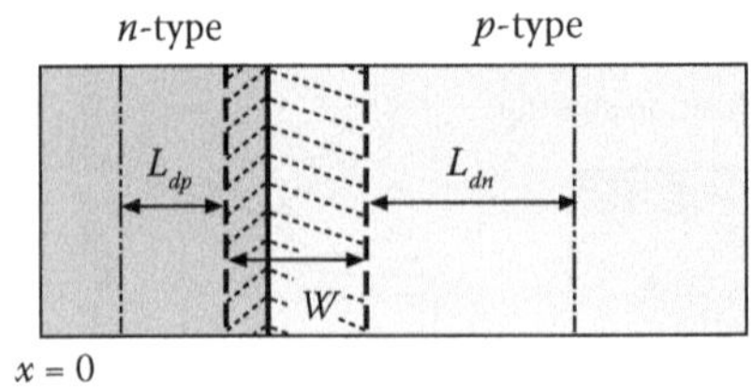

FIGURE 1.4
Diffusion lengths of charge carriers beyond the depletion region (W) of the n-p junction.

field to the p-type region on the other side of the junction. Similarly, the electrons (minority carriers) from electrons diffusion length (L_{dn}) in the p-type region diffuse to the boundary of the depletion region and are swiped to the n-type region on the other side. Due to very high electric field in the depletion region the photogenerated charge carriers are immediately extracted out of the depletion region before they recombine. Photocurrent generation takes place in the depletion region as well and the total photocurrent per unit bandwidth is the sum of these currents.

1.3.1.2 Polycrystalline (Poly-c) Si Solar Cells

Poly-c Si is another form of Si, where it possesses a number of Si crystals of different size and shape, which are oriented in different directions. Each crystalline region is differentiated by a grain boundary and when the grain size is larger than 1 mm ploy-c Si is known as multicrystalline Si (multi-c Si). Grain size plays a very important role in controlling the photovoltaic performance of the solar cells. A larger grain size causes reduced recombination and high efficiency. For preparation of multi-c Si wafers, the broken chunks of poly-c Si rods are mixed and heated to melt. The molten Si is carefully cooled and solidified into ingots, which are cut into wafers and used for solar cell applications. Multi-c Si can be easily recognized visually by the small crystals, which impart the metal flake effect in the material.

Poly-c Si is a potential material for large-scale photovoltaic production and can be prepared in thin film as well via the CVD technique like low pressure CVD (LPCVD) and plasma enhanced CVD (PECVD). Using the CVD technique, the thin films of poly-c Si can be independently deposited on cheaper substrates like glass and plastic. Poly-c Si films are sometime also grown by solid-phase crystallization of amorphous Si (a-Si). For deposition of poly-c Si on plastic substrates via solid-phase crystallization, a-Si is crystallized by the laser crystallization technique. Laser pulse provides the localized heating to a-Si above the melting point of Si and it does not damage the plastic substrate. During the cooling process, the molten Si crystallizes in different grains and the grain size can be controlled by controlling the growth process. Poly-c Si can also be prepared via metal-induced crystallization of a-Si thin films, where a-Si thin film is annealed in contact with some metal like gold, silver, and aluminum. Along with high stability like c-Si, it is comparatively very cheap to produce and requires little material for solar cell fabrication that makes this technology comparatively cost effective.

Like c-Si solar cells, the poly-c Si solar cells also have p-n structure with suitably doped n and p regions. Initial poly-c Si solar cells were prepared in 1970s on cheaper grade Si, graphite, and steel that exhibited PCE of ~1.5% [7,8]. For solar cell fabrication it was very important to make sure that the deposited poly-c Si film did not get contaminated by the substrate material. Therefore application of a suitable barrier layer was important. For example, on the graphite substrate a thin layer of SiC was deposited as a diffusion barrier prior to deposition of poly-c Si. When solar cells were prepared on steel substrates, prior to deposition of poly-c Si films, a thin layer of silica, borosilicate, phosphosilicate, or their mixture was deposited on substrate to prevent contamination by diffusion of iron atoms [7,8]. In these studies the poly-c Si layers were grown by thermal decomposition of SiH_4 and doped suitably via a hot wire CVD (HWCVD) process. The solar cell structure on the graphite substrates was *n*+ -poly-c Si/*p*-poly-c Si/*p*+ -poly-c Si/graphite, which possessed thermally evaporated Ti + Ag + Al grid contact on the top of *n*+ -poly-c Si layer. The *p*+ -poly-c Si layer was deposited by thermal decomposition of SiH_4 in the presence of diborane (B_2H_6). Subsequently the *p*-type and *n*+ -type poly-c Si layers were grown by decomposition of SiH_4 in the presence of B_2H_6 and phosphine (PH_3) as respective dopants. Solar cells were also prepared with an *n*+ base, but no special advantage was observed over a *p*+ base. The highly doped region in the bottom provides a conducting path for the majority carriers. The basic working principle of a poly-c Si solar cell is similar to that of a *p*-*n* junction c-Si solar cell.

The performance of these solar cells is very critical to the processing conditions and should be precisely controlled for high-performance. These solar cells can be prepared in two configurations well known as superstrate and substrate. In superstrate configuration the light enters through the highly transparent supporting substrate, whereas in substrate configuration the light enters from the top of the solar cell. The basic working principle for both types of configuration remains the same but substrate configuration enables us to use a large variety of substrate materials. The structure of a typical poly-c Si solar cell in substrate configuration is shown schematically in Figure 1.5. Though the superstrate configuration has the advantage that it does not have the electrode shadow effect, but its processing is more challenging. To improve the performance of these solar cells, lots of modifications were made in substrates, cell architecture, and processing conditions [9]. In 1990, Green et al. reported a PCE of 17.8% with modified processing conditions and improved device architecture [10]. In 2004, the efficiency of poly-c Si in multi-c form reached a maximum of 20.4% [11]. Some of the well-known companies making multi-c Si solar cells are Luxen Solar Energy, Chinaland, Risen Energy, ECO PV China, and JS Solar, China; Orange Solar, Netherlands; SolarWorld, Sun Perfect Solar, and Zebra Energy, United States; and Qsolar, Canada.

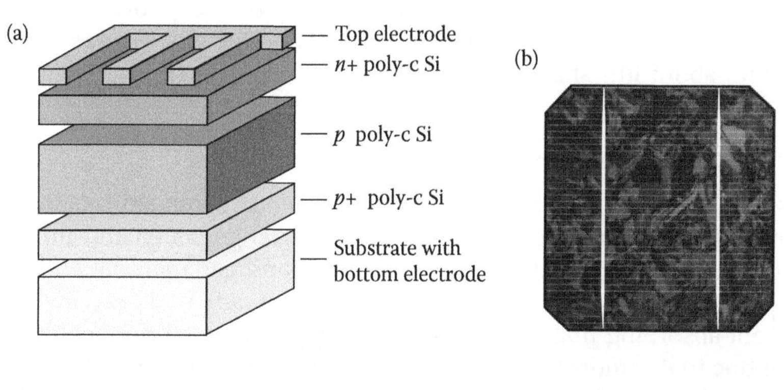

FIGURE 1.5
(a) Schematic representation of stacking of different layers in a typical poly-c Si solar cell. (b) Top view of a multi-c Si solar cell.

Ribbon Si is another form of Si, which is actually poly-c Si grown in flat ribbon from molten poly-c Si. The efficiency of ribbon Si solar cells is low, but they are more cost effective than poly-c Si solar cells because in ribbon formation the Si wastage is reduced. For the growth of ribbon Si, a graphite dye is immersed in the molten Si and a self-supporting 2D sheet of Si is pulled out by the capillary action [12]. The string ribbon method is another way to grow Si ribbons [13]. In 2006, the efficiency of ribbon Si solar cells already crossed 18% [14]. Despite a great hope of becoming a dominating technology in the PV market, ribbon Si only had about 2% share around 2006 and after 2011 they almost lost the interest and presently hold almost no share in global PV production.

c-Si solar cells had reasonably high performance, which made them a reliable source of electricity generation for satellites and space vehicles, but their cost was very high compared to their power output. The price horrified their utilities, but they were still relied on because cheaper alternatives were not available. It was extremely difficult to make them in large area because it multiplies the production cost many times. Though poly-c Si solar cell technology was comparatively cheaper, prices were still too high to be used for cost-effective energy production. Expensive cost and complicated fabrication processes of these solar cells gave birth to their next generations. We now have four generations of solar cells. The second-generation solar cells had lower efficiency but much cheaper production cost than the first ones. Third- and fourth-generation solar cells are supposed to be not only efficient but also highly cost effective and more sophisticated, but they are not yet commercialized and lots of research and development work is ongoing to make them commercially viable. These next generations of solar cells are discussed next in brief.

1.3.2 Second Generation of Solar Cells

For cost-effective solar cell technology, new processes and materials were developed, which included amorphous Si (a-Si), copper indium gallium diselenide (CIGS), and cadmium telluride (CdTe). These materials are processed in thin films and the solar cells made of these materials are also known as thin-film solar cells. These solar cells incorporate a small amount of active materials and are processed on inexpensive substrates like glass and plastic. These solar cells can easily be processed in large area and when processed on plastic substrates, they can be fabricated roll-to-roll (R2R) with ease. These materials are comparatively less efficient than c-Si due to inferior film quality and more recombination losses, but they promise to be highly cost effective. The reduced cost of thin film solar cells comes out from little usage of materials, cheaper substrates, and high throughput in production. However, incorporation of some toxic elements and photoinduced degradation are their main drawbacks. Out of total PV production in the global market, thin-film solar cells account for about 10% share.

1.3.2.1 Amorphous Si (a-Si) Solar Cells

a-Si is a noncrystalline form of Si, where Si atoms are not arranged in a long-range order (an order may exist over nanometer range of atom arrangements) and unlike c-Si not all the atoms are fourfold coordinated. It is an indirect band gap material with a band gap of about 1.7 eV and has high optical absorption coefficient. a-Si had been identified to be an important light-absorbing material for solar cell applications and is relatively cheaper to produce, but due to its amorphous nature some of the atoms possess dangling bonds (see Figure 1.6). The dangling bonds cause defects and charge carrier traps in the films that affect the device performance. a-Si is usually prepared by PECVD at different frequency

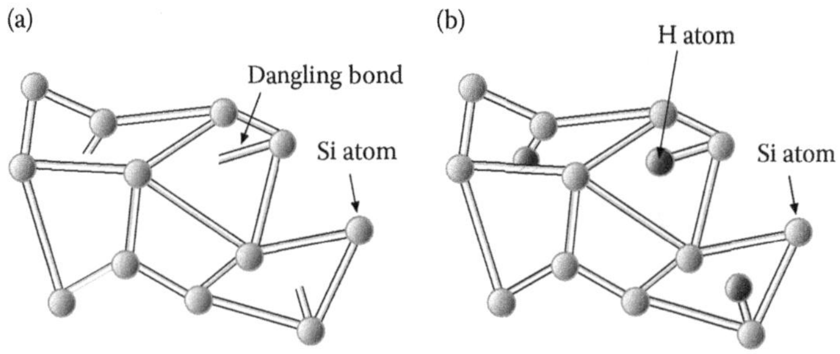

FIGURE 1.6
Schematic representation of chemical bonding in (a) a-Si and (b) a-Si:H.

voltages and HWCVD methods [15–17]. There have been reports on some other methods as well, such as reactive sputter deposition [18], electron beam evaporation [19], photo CVD [20], and pulse laser deposition (PLD) [21]. a-Si was first prepared via the PECVD method using SiH_4 [15,22], and PECVD at 13.56 MHz still remains the widely used method for research and manufacturing of a-Si thin films and devices.

In principle, SiH_4 gas or a mixture of SiH_4 and other gases like H_2 or He is introduced in the reaction chamber where SiH_4 decomposes into radicals, and Si radicals diffuse and deposit on the substrate. The dangling bonds in a-Si could be passivated by hydrogen via bonding with hydrogen atoms. Hydrogen passivated a-Si is known as hydrogenated a-Si (a-Si:H), which had better photovoltaic performance than a-Si but exhibited light-induced degradation, which is commonly known as the Staebler–Wronski effect [23]. Due to better photovoltaic performance with a-Si:H, most solar cells are prepared using a-Si:H but they may be written as a-Si solar cells. For solar cell preparation, a-Si is doped *n*-type and *p*-type like c-Si and a *p-n* junction is prepared for development of internal electric field, which helps in charge extraction. For *n*-type doping PH_3 gas and for *p*-type doping some B_2H_6 gas is introduced along with SiH_4. In *n*-type a-Si, the dopant atoms (phosphorous) make threefold coordination and bond to only three neighboring Si atoms. This way there are two electrons left on the phosphorous atoms in *s* orbital that do not participate in bonding. This configuration is chemically more favorable and happens because of the amorphous nature of Si. This paradox was explained by Street in 1982 as the positively charged fourfold coordinated phosphorous and negatively charged dangling bond seldom occur [24]. Most of the dopant atoms do not contribute to free electrons, and for those that contribute to free electrons, a dangling bond is there to accept it. The dangling bonds induced by doping work as trapping sites for holes. As for the photocurrent, collection of both the electrons and holes is necessary, the photons absorbed in the doped regions do not contribute to the photocurrent therefore solar cells are prepared in *p-i-n* structures.

Both the superstrate and substrate configurations have been employed for the fabrication of a-Si solar cells. Most a-Si solar cells are prepared in superstrate configuration where light enters through the highly transparent supporting substrate. There exists a transparent conducting oxide (TCO) between the transparent substrate and the active a-Si layers deposited in the *p-i-n* structure. The schematic structure of a typical a-Si solar cell in superstrate configuration is shown in Figure 1.7a. The *p*-type, intrinsic, and *n*-type a-Si thin films are deposited sequentially on TCO that are followed by deposition of top metal electrode. The intrinsic layer is relatively thicker than the *p*- and *n*-type layers. For the Fermi level alignment of *n* and *p* regions, electrons diffuse from the *n*-type to the *p*-type layer and make the two sides positively and negatively charged, respectively. This diffusion of electrons

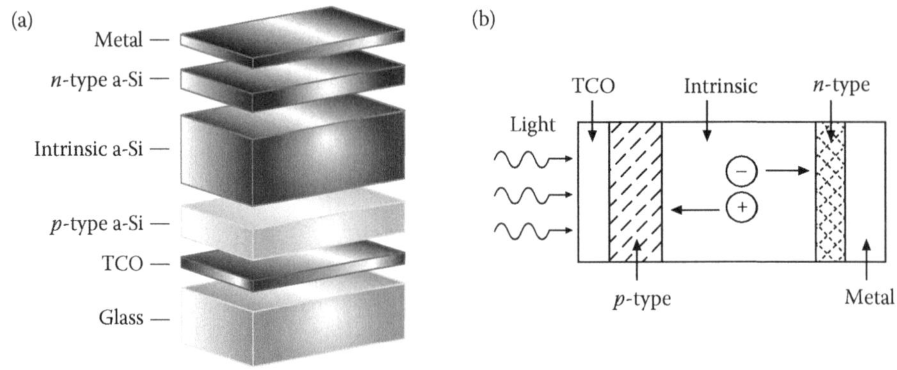

FIGURE 1.7
(a) Schematic stacking of different layers in a typical a-Si solar cell. (b) Schematic representation of photovoltaic effect in an a-Si (*p-i-n*) solar cell.

creates a built-in electric field across the intrinsic layer. The solar cells are illuminated from the TCO side and the photons are absorbed in the intrinsic layer where each photon causes the generation of one electron–hole pair. The built-in electric field sweeps away the photogenerated charge carriers (electrons toward *n*-type and holes toward *p*-type) and a photocurrent is generated. On the other hand, the substrate configuration possesses an *n-i-p* structure on the substrate and the light is illuminated from the top (*p* side). Figure 1.7b schematically shows the working of a superstrate *p-i-n* a-Si solar cell. In the doped regions, due to rapid recombination almost all the photogenerated charge carriers are lost and no photocurrent contribution from the doped layers is obtained, therefore the intrinsic layer is made thick enough to absorb most of the incident light.

D. Charlson and C. Wronski from Radio Corporation of America (RCA), Princeton, fabricated the first a-Si solar cell in 1976 using the PECVD process that showed a PCE of 2.4% [25]. Later, lots of modifications were made in materials and processing conditions to further improve the performance of these solar cells. The properties of grown a-Si thin films depend critically on the deposition conditions. Later in 1986, the HWCVD technique was used to grow a-Si thin films, and then in 1991 Mahan et al. fabricated a-Si thin films with much superior properties [17]. HWCVD proved to be a very promising technique and is similar to the PECVD technique except it possesses a hot wire instead of a radiofrequency (RF) electrode. Using the HWCVD technique for a-Si thin films, Wang et al. achieved the PCE of ~10% in a-Si solar cells [26]. For improved photon harvesting two or more solar cells are stacked one on another to form a multijunction or tandem structure. Usually a subcell absorbs the incident photons that are not absorbed by other subcells. Recently, Kim et al. reported an initial efficiency of 16.1% (in-house) in a triple-junction a-Si:H/a-SiGe:H/μc-Si:H tandem solar cell. The tandem solar cell in the a-Si:H/μc-Si:H/μc-Si:H configuration exhibited a stabilized efficiency of 13.4% that was certified by NREL, United States [27].

Though a-Si exhibits inferior photovoltaic performance compared to c-Si, this technology is relatively very simple and the films can be made much thinner (less than 1 μm) than c-Si that saves the material cost. Also, the films can be deposited at lower temperature on large-area flexible substrates like plastic to give mechanical flexibility. a-Si solar cells are being used for small applications where very little power is required such as pocket calculators, toys, and LCD-based devices. Efficiency is not the main advantage of a-Si solar cells, but it is their lower cost. However, when stacked solar cells are prepared for higher efficiencies, the production cost comes out to be very high and it makes this technology less attractive. Photodegradation is another drawback of this technology and remarkable progress has not yet been made in this direction. Out of 10% share of thin-film technology in world PV production, a-Si solar cell technology accounts for about 1.2%. Some of the well-known companies making a-Si solar cells are DuPont, United States; Onyx Solar, and Gadir Solar, Spain; Apollo New Energy, Japan; Mitsubishi Heavy Industries, Japan; Sungen International, Hong Kong; and FF Solar, Global Solar, GSP, Sunhi Solar, and QS Solar China, China.

1.3.2.2 Copper Indium Gallium Diselenide (CIGS) Solar Cells

CIGS is an I–III–VI_2 compound semiconductor made of copper, indium, gallium, and selenium, which are, respectively, from groups IB, IIIA, IIIA, and VIA of the periodic table. The chemical composition of CIGS is $CuIn_{(1-x)}Ga_xSe_2$, where x varies from 0 to 1. For x being 0 we have pure copper indium diselenide ($CuInSe_2$), whereas for x being 1 we have pure copper gallium diselenide ($CuGaSe_2$) compound semiconductors. CIGS is a direct band gap semiconductor, where band gap varies with variation in x from about 1.0 eV ($CuInSe_2$) to about 1.7 eV ($CuGaSe_2$) [28]. I–III–VI_2 semiconductors are chalcopyrite and have high optical absorption coefficients for the photons with energy higher than their band gaps. Only a few microns thick films can absorb most of the incident photons, making them high potential materials for PV applications. The crystal structure of CIGS is shown in Figure 1.8. It is one of the best materials known so far for efficient light absorption in solar cell applications and is usually prepared by coevaporation or cosputtering of copper, indium, and gallium on the desired substrate followed by thermal annealing in the selenium environment. Alternately, it can also be prepared by coevaporation of copper, indium, gallium, and selenium onto the substrate kept at an elevated temperature. Electroplating and in situ sintering of precursor materials are other nonvacuum-based alternate processes for preparation of CIGS. Preparation of CIGS thin films has also been attempted by CVD and other techniques, which are described by Kemell et al. [29]. Codeposition of copper, indium, and gallium followed by selenization gives better compositional uniformity than coevaporation of all four materials.

Selenization is a very important process that controls the CIGS film quality and its properties. Usually hydrogen selenide (H_2Se) is used as a Se source for selenization, but elemental Se in the vapor phase can also be used for this purpose. Where the complete selenization with elemental Se is achieved at above 500°C, H_2Se imparts faster selenization

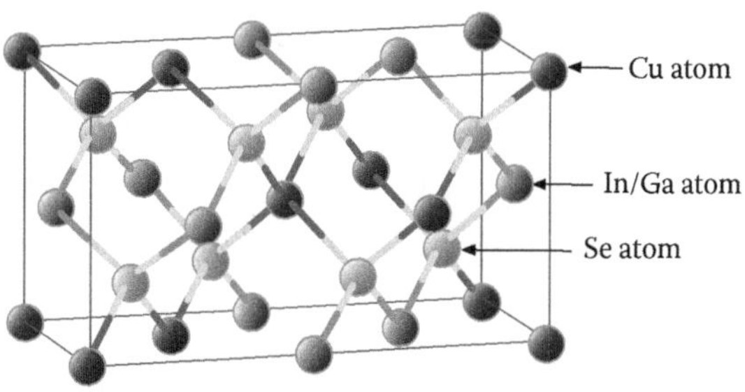

FIGURE 1.8
Crystal structure of CIGS chalcopyrite.

at around 400°C. H_2Se gives better compositional uniformity and larger grain sizes, but it is very toxic and environmentally hazardous. During selenization Se gets adsorbed and diffused into the Cu-In-Ga film and several complex reactions like formation of Cu-In-Ga alloy, metal selenides, and various CIGS compounds take place and form a chalcogenide. Because of complexity of such reactions it is very difficult to prepare good quality CIGS films. If the reaction is not controlled precisely we might end up with separate phases of $CuGaSe_2$ and $CuInSe_2$. Therefore, the reaction time and temperature should be sufficient enough with controlled supply of Se for preparation of good quality CIGS films. Deposition of copper, indium, and gallium in multilayer configuration, for example, Cu/In/Ga/Cu/In/Ga (also known as stacked elemental layer), imparts better crystallinity and smoother surface. Multilayer configuration gives better efficiency, but it is very complicated in preparation.

CIGS solar cells can be prepared on inexpensive substrates like glass, plastic, and metal sheets and the deposited films are polycrystalline in nature. Figure 1.9 shows the basic structure of a typical CIGS solar cell in the substrate configuration and CIGS is the actual light absorber. For glass substrates usually soda-lime glass is preferred as it is less expensive, but, more important, it participates in making the cell more efficient. Soda-lime glass is a good source of alkali impurities, which diffuse into Mo (molybdenum) and CIGS films and create beneficial defect effects. The substrate is coated with Mo, which serves as the back electrode and a thin film of CIGS is deposited on it. Mo also serves as a back reflector to reflect the unabsorbed light back into the device and improves its efficiency. The CIGS film is self doped *p*-type by the intrinsic defects. Cu deficiency in CIGS creates electron acceptor vacancies and makes it *p*-type. For Cu deficient CIGS films, an *n*-type ordered defect compound is formed at the surface that develops a *p*-*n* homojunction near the surface. Before the deposition of the heavily doped *n*+ -type TCO layer, which is usually Al doped ZnO (ZnO:Al), thin films of intrinsic CdS and ZnO are deposited on CIGS film. Intrinsic ZnO is deposited to protect CdS and the absorber layer from the damage

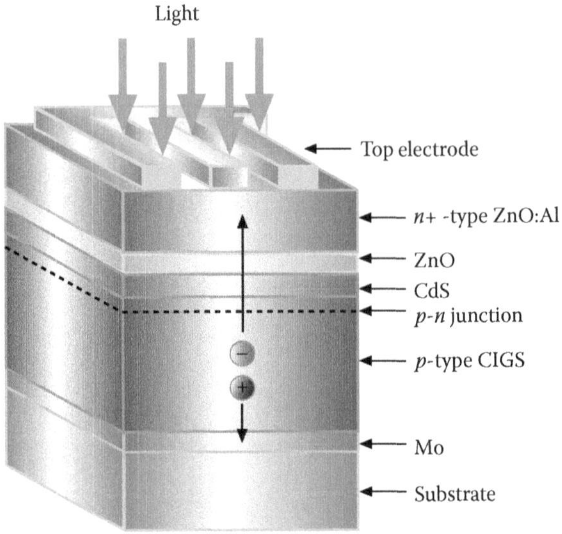

FIGURE 1.9
Schematic structure of a typical CIGS solar cell showing generation and collection of photogenerated charge carriers.

caused by subsequent deposition of the ZnO:Al layer. CdS is used as a buffer layer for electron collection but it is optional (there have been number of reports on CIGS solar cell where no Cd was incorporated at all). However, a heterojunction with CdS gives more stable and efficient solar cells. Fermi level alignment between *p*-type and *n*-type semiconductors builds up an internal electric field, which helps in collection of photogenerated charge carriers. Heavily doped ZnO:Al layer also serves as the front contact. Due to very high band gap, ZnO works as window material and light absorption actually takes place in the CIGS layer. Some people have also used tin doped indium oxide (In_2O_3:Sn), commonly known as indium tin oxide (ITO), in place of ZnO:Al as the front contact. For ZnO contact, Ni/Al grid electrodes are deposited on the top ZnO:Al film.

Though the synthesis and characterization of $CuInSe_2$ was first done in 1953, it was used for solar cell applications at Bell Laboratories in the early 1970s. Deposition of *n*-type CdS on *p*-type single crystal of $CuInSe_2$ gave PCE of around 12% [30], which was a remarkable achievement but due to great difficulty in the growth of single crystal $CuInSe_2$, no further efforts were made on $CuInSe_2$ single crystal solar cells. In 1976, Kazmerski et al. fabricated the first thin-film $CuInSe_2$/CdS solar cell with thermal evaporation of $CuInSe_2$ that exhibited the PCE of about 6% [31]. When Mickelsen et al. demonstrated around 10% efficiency in 1981 [32], research in CIS solar cells got accelerated. Partial substitution of In by Ga resulted in CIGS, a semiconductor with better optical and electrical properties with wider band gap and improved film crystallinity. Too much introduction of Ga resulted in highly resistive films, which were not good for solar cell applications. CIGS with band gap of around 1.25 eV resulted in optimum photovoltaic performance, therefore most CIGS solar cells are prepared with such band gaps. More interesting, the behavior of CIGS/CdS solar cells was insensitive to junction defects that led to highly efficient solar cells. Among the chalcopyrites, CIGS proved to be the most efficient light absorber. Additionally, introduction of a little amount of Na in CIGS results in high-performance solar cells. Usually it is done by the soda-lime glass, otherwise Na is deliberately added. It increases the hole conductivity, passivates the grain boundaries, and controls the grain size. Further development in device engineering made them the most efficient among all thin-film solar cell technologies and the efficiency has already taken over multi-c Si solar cells with 21.7% PCE. Copper indium disulfide ($CuInS_2$) is another important chalcopyrite with energy band gap of 1.5 eV. It also has great potential to be a candidate for producing efficient solar cells [33]. Among all PV technologies, CIGS has the widest spectral response and perform well in low light conditions as well. CIGS solar cell technology is quite reliable, cost effective, and highly efficient. Out of the 10% share of the thin-film global PV market, CIGS solar cells account for about 3.8%. Some of the well-known companies making CIGS solar cells are Ascent Solar Technologies, Solo Power, and Global Solar Energy, United States; AVANCIS, Germany; Hulk Energy Technology, Taiwan; Japan Solar, Japan; LBS, UK; and Shurjo Energy Private, India.

1.3.2.3 Cadmium Telluride (CdTe) Solar Cells

CdTe is a II–VI compound semiconductor with a zinc blend crystal structure having a direct band gap of 1.5 eV. It has a very high optical absorption coefficient and because the film is only a few microns thick it is sufficient enough to absorb most of the incident light. It is an excellent light-absorbing material for solar cell applications. Photoconductivity in CdTe was first observed by Bube in 1955 [34], but the solar cells were prepared by Cusano in 1963 [35]. The *n*-type Cu_2Te was deposited on *p*-type CdTe to form a *p-n* heterojunction, but these cells were very unstable. Replacement of Cu_2Te by *n*-type CdS resulted in

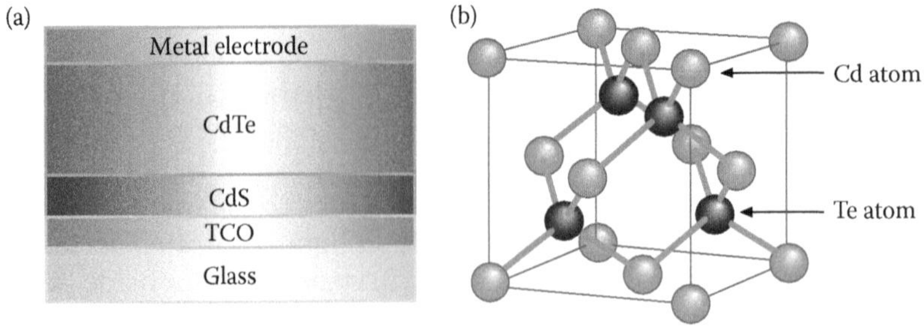

FIGURE 1.10
(a) Schematic cross-sectional view of a typical CdTe solar cell. (b) Crystal structure (zinc blend cubic) of CdTe.

better performance and in 1972 Bonnet et al. fabricated CdTe/CdS solar cells of around 5% efficiency [36]. Later in 1981 Tyan demonstrated 10% efficient CdTe/CdS solar cells using the closed space sublimation (CSS) technique (a thermal evaporation technique where source and substrate are placed very close to each other) [37,38]. Like other thin-film solar cells, these solar cells could also be prepared on inexpensive substrates like glass, plastic, and metal foils. Figure 1.10 shows the schematic cross-sectional view of a typical CdTe solar cell in superstrate configuration on glass substrate. The light is illuminated from the transparent glass side that is coated with a low resistive TCO, which works as the bottom electrode. TCO is usually ITO or fluorine doped tin oxide (SnO_2:F) (FTO). A thin layer of CdS, which works as a window layer for the incident light, is followed by the CdTe absorber layer. The CdS layer is usually deposited by a chemical bath process [39], where cadmium acetate ($Cd(CH_3CO)_2$) is used as Cd source, thiourea ($CS(NH_3)_2$) is used as S source, ammonium acetate (NH_4CH_3CO) and ammonia (NH_4OH) are used as buffer and complexing agents, respectively. CdTe is usually deposited by CSS technique, which is based on the principle of dissociation of CdTe at high temperatures. The two gases of Cd and Te then diffuse and recombine on the substrates, placed about 2 mm away from the source [40]. After deposition of CdTe layer, a $CdCl_2$ annealing step is performed, which had proved to be an important step for efficient energy conversion [41–45]. $CdCl_2$ has been proposed to eliminate the recombination centers in CdTe film and at the CdTe/CdS junction [46,47]. Usually for $CdCl_2$ treatment the substrates are dipped in $CdCl_2$ solution followed by annealing [43], but several people have also used vapor-based $CdCl_2$ treatment in place of solution-based treatment [44,45,48]. A dry fabrication process is most suitable for in-line production to give cost effective, high throughput PV production. After $CdCl_2$ treatment an electrode is coated on the top of CdTe layer for the top contact. CdTe possesses a high work function and no known metal makes an ohmic contact with CdTe, therefore a small amount of Cu in addition to other materials like HgTe and graphite has been used as top contact for CdTe [49,50]. Such a low resistive Cu paste resulted in improved performance of solar cells. Finally, Ag paste is applied on the top electrode to reduce its lateral resistivity. Kuribayashi et al. used Cu-added graphite paste as the top electrode with final deposition of Ag via screen printing and got a PCE of 12.8% [49]. Some people also approached the deposition of a heavily doped high work function *p*-type interlayer like Cu doped *p*-ZnTe films, before metallization on the top of CdTe layer [51–55]. Sputtering, spray pyrolysis, electrodeposition, and screen printing are some other techniques that have also been used for the production of CdTe/CdS solar cells.

In 1992, introduction of a thin resistive transparent oxide (SnO_2) buffer layer between low resistive TOC and CdS film, and thinning down of the CdS buffer layer improved the cell efficiency up to 15%. The efficiency of these solar cells could be improved further by doing device engineering, pre- and postdeposition treatments and controlling the CdTe growth and its quality [41,56]. Considerable efforts have been made to improve the cell performance and very recently in February 2015, First Solar announced the record efficiency of 21.5% for CdTe solar cells [57]. CdTe solar cell technology shares about 5% of PV production in the global market. Toxicity of CdTe is the main drawback of CdTe solar cells. But safe use and encapsulation of devices may render it harmless and at the end of life of the cell it could be recycled and the concerns of environmental harm can be resolved. Improved efficiency and recycling of CdTe solar cells has the potential to make CdTe PV technology a leading technology in the PV market. Some well-known companies making CdTe solar cells are First Solar, Dmsolar, and Willard & Kelsey (WK) Solar, United States; Advanced Solar Power, China; Calyxo, and Antec Solar, Germany; and LBS, UK.

1.3.3 Third Generation of Solar Cells

The third generation of solar cells includes multijunction solar cells and emerging PV technologies like dye-sensitized solar cells (DSSCs), organic solar cells (OSCs), and quantum dot solar cells (QDSCs). A brief introduction to each of these is given next.

1.3.3.1 Multijunction Solar Cells Based on III–V Compound Semiconductors

III–V compound semiconductors like gallium arsenide (GaAs), gallium antimonide (GaSb), and indium phosphide (InP) were also observed to have excellent optical and electrical properties making them potential candidates for solar cell applications. They possessed very high optical absorption coefficients and absorbed most of the incident light within few micrometers. GaAs has a direct band gap of 1.43 eV and the atoms are arranged in zinc blend cubic crystal structure. First, reasonably high efficiency from GaAs-based *p-n* homojunction solar cells was reported in early 1960s that was about 11% [58]. Later in the early 1970s, Alferov et al. first reported AlGaAs/GaAs-heterojunction-based *p-n* junction solar cells [59]. In the 1980s, GaAs-based solar cells surpassed Si solar cells efficiency and in 1990s took them over for satellite applications. GaAs-based single-crystal single-junction thin-film solar cells hold the record maximum efficiency of 28.8%, which has remained unchanged since early 1990s [60]. GaAs homojunction solar cells were more superior than c-Si-based solar cells both in terms of efficiency and resistant to damages caused by space radiation [61]. Their superior properties made them the most suitable candidate for power generation in space. So, whenever very high power or small size solar arrays are required in spacecrafts, III–V compound multijunction solar cells are used instead of or in combination with c-Si solar cells. From the radiation resistance point of view, InP-based solar cells have been found to be more superior compared to those based on both c-Si and GaAs [62]. The only drawback of GaAs solar cells was their high production cost. Initially single crystals of GaAs were produced via Czockralski method or Bridgmann method. The cheaper processes used for material growth did not work as those resulted in imperfect crystals and the efficiencies came out to be very low. Additionally, the single crystal wafers of GaAs were dense but mechanically very fragile. To overcome the issues related to cost and fragility, the solar cells needed to be grown on lighter and less expensive substrates. Efforts were made in this direction by growing thin films of the semiconductors, but the lattice mismatch, caused by the substrate, remained the biggest problem. For the growth of high-quality epitaxial thin films, liquid phase epitaxy (LPE) and metalorganic CVD (MOCVD) techniques were used that resulted in highly efficient solar cells, but these techniques were quite expensive. However, these techniques proved to be cost effective when monolithic multijunction solar cells were fabricated. Therefore, over the last 20 years, instead in single-junction solar cells, the focus has been on the development of multijunction solar cells.

It is important to note that a single-junction solar cell possesses lots of energy losses, where transmission of photons with energy less than the band gap of the semiconductor and loss of excess energy as heat for the photons with energy more than the band gap of the semiconductor are some very important examples. These losses limit the performance of a solar cell. Detailed calculation by Shockley and Queisser showed that the efficiency of a single-junction solar cell cannot exceed 31% [63]. The maximum efficiency of 31% is achievable with a semiconductor of band gap around 1.3 eV. As per the same calculation, Si (band gap of 1.1 eV) can have 29% efficiency as the limit for a single *p-n* junction solar cell. However, the efficiency can cross these limits if two or more semiconductors having different band gaps are incorporated in the same cell. Multijunction solar cells emerged from the idea of minimizing the energy losses by stacking multiple *p-n* junction solar cells one on another, from previous generations that absorb different portions of solar light. Where Loferski et al. proposed the maximum theoretical limit of efficiency to be 60% in multijunction solar cells [64], Henry predicted 72% efficiency with 32 component cells [65]. Accounting the thermodynamic loss mechanisms, an infinite-layer solar cell would have a theoretical limit of 86% efficiency [66]. Multijunction structures indeed increased the PCE significantly [67,68]. The stacking of different cells should possess the sequential reduction or increment in the band gaps of constituting subcells and the light should enter from the side of the subcell with highest band gap. The concept is that the photons that are not absorbed by one cell would be absorbed by the following cell. The multijunction concept includes both the monolithic growth and mechanical stacking of individual cells one on other. In monolithic growth, the discrete *p-n* junctions forming different cells of different band-gap materials are grown one on other, and they are connected to each other physically, electrically, and optically. But in mechanical stacking the individual *p-n* junction cells are stacked one on other such that the cathode of one individual cell faces the anode of other. Mechanical stacking is represented by the symbol "//" between the connecting cells and it gives great flexibility in connecting the cells, which are completely incompatible, but this configuration remained less efficient than monolithic stacking. Figure 1.11 shows the schematic cross-sectional view of mechanically and monolithically stacked multijunction solar cells. Hot carrier cells and spectrum conversion have been some other ideas, which have been implemented to minimize the energy losses for improved performance.

In monolithic growth, the growth techniques and intercompatibility of constituent materials are very important for fabrication of efficient solar cells. For this purpose, MOCVD is the best technique for film growth and ternary III–V compounds have shown the best intercompatibility giving most efficient multijunction solar cells. MOCVD enabled growth of high-quality multilayer complex structures of different III–V compound materials with an ease. In 1988, Olsen et al. fabricated a two-junction $GaInP_2$/GaAs solar cell that had an efficiency of about 22% [69]. This paved a path to develop a GaInP/GaAs/Ge triple-junction solar cell, where the $GaInP_2$/GaAs junction was grown on a Ge substrate and the device exhibited 34.1% efficiency [67]. Then in 2012, Agui et al. achieved 36.9% efficiency in a GaInP/GaAs/GaInAs triple-junction solar cell [68]. The triple-junction solar cells hold the current maximum efficiency of 37.9% without any concentrator. The concentrated triple-junction solar cells crossed 40% efficiency in 2007 [70,71] and the efficiency has now gone beyond 44% [72]. King et al. predicted that even five- or six-junction solar cells with concentrators can give more than 50% efficiency [73,74].

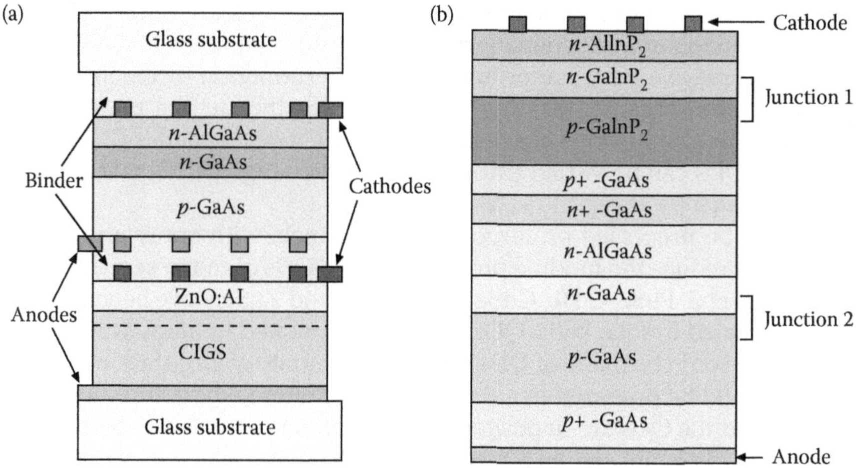

FIGURE 1.11
Schematic cross-sectional view of the typical (a) mechanically stacked GaAs//CIGS and (b) monolithically stacked $GaInP_2$/GaAs, multijunction solar cells.

Multijunction solar cells are the only third-generation solar cells that are commercially available. They are highly efficient and hold the world records in performance, but these solar cells are very expensive and are being used in satellites and other terrestrial applications only. Multijunction solar cells cost ~\$40,000/m^2. Hopefully with solar concentrators these cells would become less expensive. The ultimate goal for multijunction solar cells is to reduce their cost so that they could also participate in fulfilling the world's energy demand. Some well-known companies making multijunction solar cells are Akhter Solar, Solargy, UK; ArimaEco, Delta Electronics, and Everphoton Energy, Taiwan; Daido, Japan; AZUR Space, Germany; HT Sun, Italy; Morgan Solar, Canada; Solar System, Australia; ZSun Technology, China; Zytech Solar, Spain; and Semprius, United States.

1.3.3.2 Quantum Dot Solar Cells (QDSCs)

QDSCs is another solar cell technology based on the concept of minimization of energy loss using semiconductors of different band gaps like planer multijunction solar cells but with better tunability across the solar spectrum and ease in production. The idea of using quantum dots (QDs) in solar cells was introduced by Burnham and Duggan in 1990 [75]. QDs are basically the tiny particles (a few nanometers) of a semiconductor that have a size of exciton Bohr radius. QDs incorporate only a few atoms/molecules of the semiconductor. Due to such a small size the electron energies in the particle become finite and discrete like in an atom. These energy levels are tunable with the size of the particle and so the band gap varies. The band gap of bulk semiconductor is fixed but the band gap of a QD of that semiconductor can be tuned easily by changing its size. Ease in tunability of band gaps makes QDs potential candidates for multijunction solar cells, where materials of different band gaps are required for efficient photon harvesting from different regions of solar spectrum. Initially, MOCVD was used for preparation of QDs, but later they could be prepared with less expensive wet chemical processing. Wet chemical synthesis and processing of QDs is quite easy and can be deposited on any desired substrate by the spin coating technique from their colloidal suspension in liquids. For long-term sustainability of the concentrated solutions, some long-chain hydrocarbons are used as ligands [76]. A ligand is usually an ion or a functional group that is attached to a QD to form a coordination complex and plays a very important role in determination of its optical properties. A ligand can be both organic and inorganic, however, the surface passivation through inorganic ligands like halide ions has resulted in efficient performance [77]. Even the band edge energies of QDs can be varied through ligand exchange [78,79] and can be used for different functions [80].

QDs can easily be processed on large-area R2R panels with spray and printing techniques that would reduce the module production cost. QDs of many semiconductors, such as Si, GaAs, GaSb, PbS, PbSe, CdTe, CdSe, CdS, CIGS, and ZnSe, have been grown and successfully incorporated in solar cells. QDs can be incorporated in many ways in a solar cell. A QDSC solar cell could be made of QDs only in superstrate configuration on glass/plastic substrates or it could be prepared in substrate configuration where the substrate is a bulk semiconductor and the QDs are deposited on it. In multijunction solar cells, QDs could be used in a series of different sized QDs to absorb different regions of the solar spectrum. The structure of a typical multijunction all-QDs tandem solar cell on a glass substrate is shown in Figure 1.12. For ease in production of a cell, usually different sized QDs of the same semiconductor are used. The colloidal solutions of different sized QDs are spin cast one on other with interconnecting layers to formulate a solar cell. Chuang et al. prepared single-junction PbS QD solar cells [80]. The solar cells were prepared on ITO-coated glass substrates. Before deposition of PbS QDs, a thin layer of ZnO was deposited on ITO via the spin coating technique. PbS QDs capped with oalic acid were solution processed on ZnO film with two ligands, tetrabutylammonium iodide (TBAI) and 1,2-ethanedithiol (EDT), for solid-state ligand exchange. Finally, an Au anode was deposited by thermal evaporation in vacuum. Light absorption takes place in QDs and the photogenerated charge carriers get collected at respective electrodes with the help of an internal electric field developed from the Fermi level alignment of two electrodes. The device consisting of

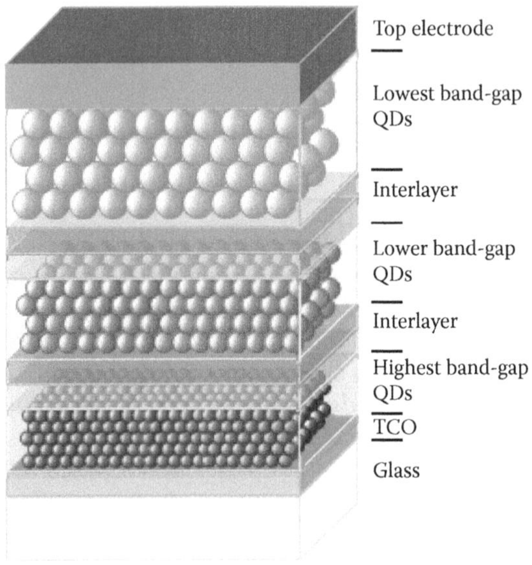

FIGURE 1.12
Schematic structure of an all-QDs tandem solar cell.

12 films of TBAI-treated QDs exhibited around 6% efficiency, and replacing the two topmost layers of PbS-TABI with two PbS-EDT layers improved the efficiency to about 9% [80]. The top layers of PbS-EDT served as the hole transport and electron blocking layer that resulted in enhanced efficiency. This is the maximum efficiency reported so far for single-junction QDSCs [80,81]. It is not a remarkable figure compared to conventional solar cells, but theoretically they promise to have a surprising efficiency limit of 45%. This is possible because absorption of a photon in QDs generates more than one electron–hole pair and the efficiency can be more than double of a single-junction solar cell [82,83]. It makes the QDs very efficient at converting light into electricity. Some companies making QD solar cells are Voxtel, United States; Solterra Renewable, United States; and Shenzhen YinXuanShen Technology, China.

1.3.3.3 Dye-Sensitized Solar Cells (DSSCs)

Dye-sensitized solar cells (DSSCs) is another class of thin-film solar cells that is based on organic dyes. DSSCs are quite thin, lightweight, and can be processed at low temperatures on nonexpensive substrates like glass, plastic, and metal foils. Low cost processing and inexpensive materials make these cells very cost effective. Figure 1.13a shows the schematic structure of a modern DSSC, where the light-absorbing medium is an organic dye. A typical DSSC possesses a TCO-coated substrate, which is subsequently coated with a thin porous layer of large band gap semiconductor like TiO_2. The porous structure of TiO_2 film (TiO_2 nanoparticles connected with each other to form a porous network) is the basic element for DSSCs. Due to the large band gap of TiO_2, it absorb very little light in UV range and works as a window material. Sometime ZnO is also used for this purpose, but TiO_2 has shown superior properties. The porous TiO_2 layer is coated with an organic dye, which works as a light absorber and sensitizer for TiO_2. The top electrode, also known as the counter electrode, is again a TCO coated on other substrate. Usually FTO is used as TCO in DSSCs. The counter electrode is coated with a thin layer of catalytic platinum or lead, and to formulate a solar cell, the counter electrode is coupled with a dye-coated electrode with an electrolyte medium (usually nitrile derivatives) between them. Usually the electrolyte medium possesses an I^-/I^-_3 redox couple, which is achieved by addition of I_2 to some metal iodide. I_2 gets converted into I_3^- according to the chemical equilibrium $I_2 + I^- \leftrightarrow I_3^-$. The I^-/I^-_3 redox couple in electrolyte medium helps in transportation of charge in the solar cell.

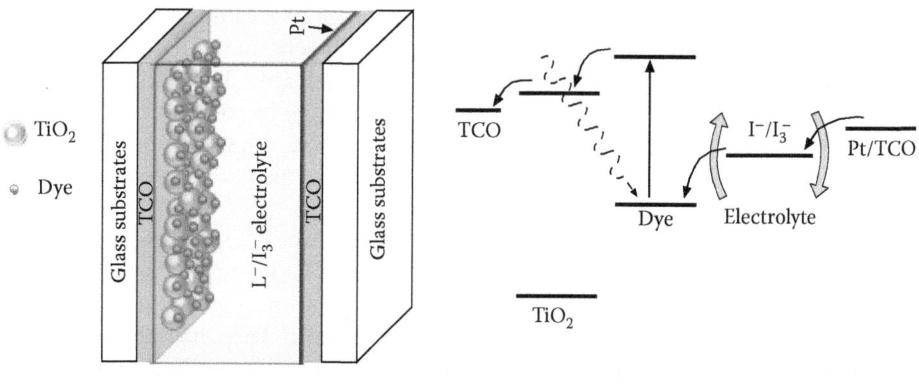

FIGURE 1.13
(a) Schematic structure of a DSSC and (b) process of photocurrent generation on light absorption by the cell.

The concept of DSSCs originated from the photosynthesis phenomenon in the plants, where chlorophyll, an organic compound, absorbs the sunlight and initiates photosynthesis process. The photovoltaic effect in DSSCs was first demonstrated in the early 1970s, but instability was their biggest problem [84,85]. Though the efficiency improved in subsequent years but poor stability remained a big problem. In 1991, O'Regan and Gratzel presented the modern version of DSSCs that incorporated TiO_2 nanoparticles deposited on a TCO-coated substrate. The nanoparticles were sensitized by the Ru-based organic dye and the cell exhibited a PCE of about 7%–8% [86]. The structure presented by Gratzel was a breakthrough in DSSCs and further development in electrolyte solution and dye molecules led to higher efficiencies; therefore, these solar cells are also known as Gratzel solar cells. Later in 1996, the record efficiency reached a plateau at 10%–11% until ~12% efficiency was reported in 2011 [87]. Very recently a PCE of 13% was achieved through molecular engineering of porphyrin sensitizers [88]. Though DSSCs are in the early stage of development, they promise to be a cost-effective alternate for solar energy harvesting. Professor Michael Gratzel was awarded the 2010 Millennium Technology Prize for making this technology an affordable way to harvest solar energy on a large scale.

The fabrication of DSSCs is extremely simple and does not require elaborate equipment for manufacturing. A 1–2 μm thin porous layer of TiO_2 is deposited on the TCO-coated substrates by the doctor-blading technique from a nanocrystalline paste of TiO_2. Then, the substrates are immersed in a solution of the organic dye that results in infiltration and adsorption of dye molecules onto TiO_2 surface. The adsorption of dye molecules on the TiO_2 semiconductor forms a heterojunction, which helps in generation of the photovoltaic effect. For the counter electrode, a thin layer of Pt is coated by sputtering on another TCO substrate. The two substrates are then joined together with coating sides facing each other and a redox couple electrolyte is filled in between. The two substrates are then sealed together to prevent leakage of the electrolyte. The operation of a DSSC can be described as artificial photosynthesis, where light absorption takes place in organic dye, and electrons and holes are generated. Due to low dielectric coefficients of organic molecules, the electrons and holes are tightly bound together and form excitons. The exciton dissociation is achieved at the heterojunction of dye and the TiO_2 nanocrystals. The electrons are transferred to the conduction band of TiO_2 within femtoseconds, leaving the holes in the dye. The electrons transport through TiO_2 via diffusion and reach the TCO electrode. The positive charge on the dye molecules is neutralized by oxidation of I^- ion to I_3^- ion in the I^-/I_3^- redox couple electrolyte. This process regenerates the dye molecule. The I_3^- ions diffuse to the Pt-coated TCO electrode and reduce back to I^- ions by releasing the charge to counter electrode. The reduction of I_3^- ions requires the presence of a catalyst, and the purpose is served by Pt coated on the TCO electrode. Figure 1.13b schematically shows the mechanism of photocurrent generation on light absorption in a DSSC. The photocurrent generation in DSSCs is completely regenerative and no net change in composition of constituent materials is witnessed. In principle, the efficiency of a DSSC depends on the energy levels of dye molecules, conduction/valance band, or Fermi level of the TiO_2 semiconductor, and the redox potential of the electrolyte medium.

Electrical and morphological properties of nanocrystalline TiO_2 are also very important for high performance solar cells. The relative positions of TiO_2 energy bands, energy levels of the dye, and redox couple are very crucial in determination of photovoltaic parameters. A reduction in potential difference between Fermi level of TiO_2 and the energy level of redox couple would reduce the open-circuit voltage (V_{oc}). On the other hand, an increase in Fermi level or conduction band of TiO_2 might cause less efficient electron transfer from dye to TiO_2 resulting in reduced photocurrent. For efficient light absorption, the TiO_2 film should be porous enough to give high surface area for maximum adsorption of dye molecules. The liquid electrolyte fills all the pores in the film and ensures maximum charge extraction. The fraction of the dye molecules that are not strongly attached to the TiO_2 surface do not allow fast transfer of electrons and are less photostable. These molecules are dissolved in electrolyte or form higher molecular aggregates. Presence of liquid electrolyte is the main drawback for this technology. Liquid electrolyte makes DSSCs less stable and not suitable for all weather conditions. The temperature variation causes variation in viscosity and volume of the electrolyte, which hinders the charge extraction and affects sealing of the cell. Also the electrolyte possesses volatile organic compounds, which are environmentally hazardous and require high-quality sealing. Use of expensive Pt as the catalyst and ruthenium dyes as the light absorber in highly efficient solar cells is also a drawback for cost-effectiveness. The most efficient devices have remained unchanged from the original concept of DSSCs. It took almost two decades for DSSCs to go beyond 11% PCE and there remains the issue of instability. Attempts have been made to replace the liquid electrolyte medium with some suitable solid-state medium and the solar cells were named solid-state dye-sensitized solar cells (SSDSSCs). Though the SSDSSCs have shown better stability, the efficiencies are low. More important, the concept of SSDSSCs recently gave birth to a new emerging solar cell technology known as perovskite solar cell technology. Perovskite solar cells have shown great potential and promise to be more cost-effective and efficient than any other solar cell technology known so far. These solar cells are discussed in detail in Section 1.3.4.

Just for the information to readers, there have been many integrations of DSSCs on a small scale, but only the Swiss Tech Convention Center of Ecole Polytechnique Fédérale de Lausanne (EPFL), Switzerland, the best equipped conference center in the world that was opened in April 2014, has installed 1400 DSSCs solar modules each of 35×50 cm^2 on the west façade, covering a total area of about 300 m^2. This is the world's first exterior architectural integration of DSSCs. The modules possess five different shades including red, green, and orange giving the ensemble a warm and dynamic look. They not only produce the electricity but also provide shade to the building from direct sunlight. Such an integration will boost the large-scale production and utilization of DSSCs. Some of well-known companies that have secured the license for manufacturing and marketing of DSSCs are G42i Power, UK; AmoSolar, Korea; 3GSolar Photovoltaics, Israel; Samsung Electronics, Korea; Sharp, Japan; Exeger, Sweden; and Solaris Nanoscience, United States.

1.3.3.4 Organic Solar Cells (OSCs)

OSCs are solid-state thin-film solar cells based on organic semiconductors. Organic semiconductors is a special class of aromatic hydrocarbons where carbon atoms are sp^2 hybridized and have alternate single and double bonds. Due to alteration in single and double bonds among C atoms these materials are also known as conjugated semiconductors. In principle these materials are insulators in pure form, but they are called semiconductors because their electrical conductivity increases exponentially with temperature and some other properties like band gap are similar to those of the inorganic semiconductors. Alteration of single and double bonds is responsible for the conductivity in these materials. The overlapping of p_z orbitals in two sp^2-hybridized C atoms form two π molecular orbitals named as π and π^* molecular orbitals. The π orbitals, also known as bonding orbitals, exist lower in energy, and the π^* orbitals, also known as antibonding orbitals, exist higher in energy. The electrons in p_z orbitals occupy the lower energy π orbital. The energy difference between π and π^* orbitals corresponds to the band gap of the semiconductor.

Most of the organic semiconductors are amorphous in nature, which gives mechanical flexibility to devices. OSCs are prepared in superstrate configuration where TCO-coated glass or plastic is used as the substrate. Depending on the molecular weight, the organic semiconductors can be classified into small molecules (low molecular weight materials with no repeating units), oligomers (low molecular weight materials with some repeating units), and conjugated polymers (high molecular weight materials with repeating units in the polymer chain). Usually thin films of small molecules and oligomers are prepared by thermal evaporation in vacuum, whereas the thin films of conjugated polymers are prepared by spin coating from their solutions. Therefore, there are two subcategories of OSCs: those incorporating small molecules are known as small molecular solar cells (SMSCs), whereas those based on conjugated polymers are known as polymer solar cells (PSCs).

The first OSC was fabricated in 1978 with a single material (merocyanine) sandwiched between Ag and Al electrodes that had an efficiency only almost half a percent [89]. Due to low dielectric constant of organic semiconductors, the photon absorption creates excitons instead of free electrons and holes. The thermal energy at room temperature is not sufficient enough to break the excitons into free electrons and holes. For dissociation of excitons an additional force was required, and in 1986 Tang introduced an electron donor–acceptor concept that resulted in an efficiency of ~1.0% [90]. This solar cell had a donor–acceptor bilayer structure, where a thin film of electron acceptor material was coated on the light-absorbing material, which worked as an electron donor. Electron acceptor material provided the necessary force to break the excitons into free charge carriers at the donor–acceptor interface. The donor–acceptor concept was the most important finding for OSCs and led to more efficient solar cells when the donor–acceptor materials were mixed together [91]. Mixing of donor–acceptor materials forms lots of junctions in a single layer and it is known as the bulk-heterojunction (BHJ) configuration. Figure 1.14 shows the

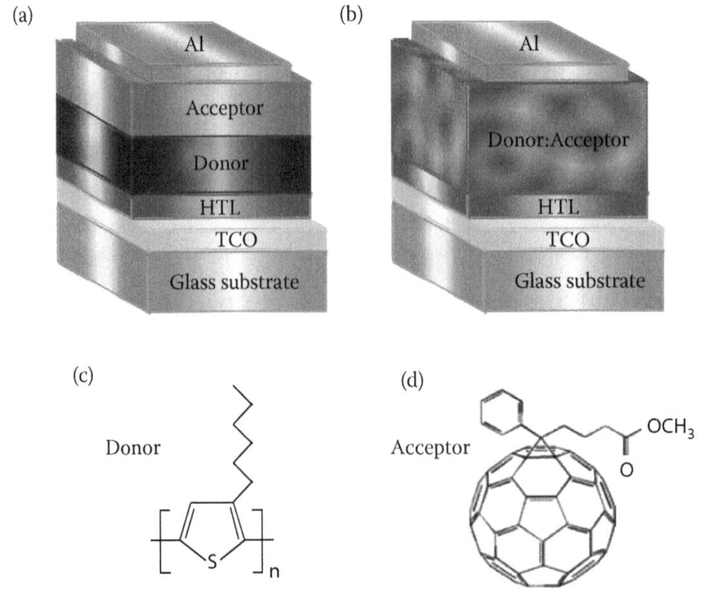

FIGURE 1.14
Schematic structures of (a) bilayer and (b) BHJ OSCs. Molecular structures of (c) poly(3-hexylthiophene) (P3HT) and (d) phenyl C_{61} butyric acid methyl ester (PCBM).

schematic structures of a typical bilayer and BHJ OSCs. Some of the most commonly used donor and acceptor organic materials are also shown in Figure 1.14.

BHJ configuration proved to be the most efficient design and revolutionized OSC technology. Further developments in materials and device architectures led to highly efficient solar cells and the efficiency has now gone beyond 11% [92,93]. Though there has been great progress in the performance of bilayer solar cells, the most efficient solar cells are based on BHJ structure. OSCs can be fabricated R2R on plastic sheets at room temperature. Due to ease in production of large-area solar cells and very low material and processing cost, OSCs are supposed to be highly cost-effective solar cell technology. Though these cells will not be as efficient as those based on c-Si, but lower efficiency can be compensated by lower cost. OSCs having 8%–10% efficiency with 5–10-year lifetime can compete with c-Si solar cells [94]. Though the efficiency of OSCs has improved tremendously, the lifetime remains too low to be commercially viable. The basic properties, limitations, difficulties, and progress made toward the market feasibility of OSCs are discussed in detail in the subsequent chapters. Some well-known companies making OSCs are Haliatek, Germany; LBS, UK; Solar Press, UK; and Plextronics, Solarmer Energy, and Solar Ivy, United States.

1.3.4 Fourth Generation of Solar Cells

The fourth generation of solar cells is quite futuristic and is expected to have very high PCE. They can be prepared in large area with great ease like DSSCs and OSCs; in addition, high PCE would make them potential candidates for future power generation. The fourth generation of solar cells includes the organic–inorganic hybrid and organometal halide perovskite solar cells. A brief introduction to these solar cells is given next.

1.3.4.1 Organic–Inorganic Hybrid Solar Cells

High PCE in OSCs required the donor/acceptor materials of broad absorption spectra and high charge carrier mobilities. On the contrary, the low charge carrier mobilities in organic semiconductors had always been a serious concern. As the inorganic semiconductors have high charge carrier mobilities, high thermal stability, and high photoconductivity, they drew the attention of researchers toward their possible use in OSCs. Interfacial charge separation in OSCs is very critical and it requires large interfacial area for efficient charge separation. Therefore, for the use of inorganic semiconductors along with conjugated polymers, they have been grown in nanocrystalline structures and incorporated as electron acceptors to formulate the solar cells. The solar cells having both organic and inorganic semiconductors as electron donors and acceptors are known as organic–inorganic hybrid solar cells. The organic–inorganic hybrid solar cells utilize high electron mobility of inorganic nanocrystals and wide-range absorption and solution processability of conjugated polymers. Additionally, the combined absorption spectrum of organic–inorganic semiconductors in the composite improves photon harvesting. The relative energy levels of donor/acceptor materials in hybrid solar cells are of high importance as the photocurrent of the solar cells depends on the energy level alignment at both the donor–acceptor and electrode interfaces. Therefore, in hybrid solar cells the selection of organic–inorganic semiconductors along with suitable electrodes is very crucial.

To deal with the issues of charge separation and transportation in organic–inorganic hybrid solar cells, some of the important inorganic semiconducting nanomaterials that have been used as electron acceptors are CdSe, CdTe, ZnO, CdS, ZnS, TiO_2, PbS, and PbSe [95–109]. The structure and working principle of organic–inorganic hybrid solar cells are

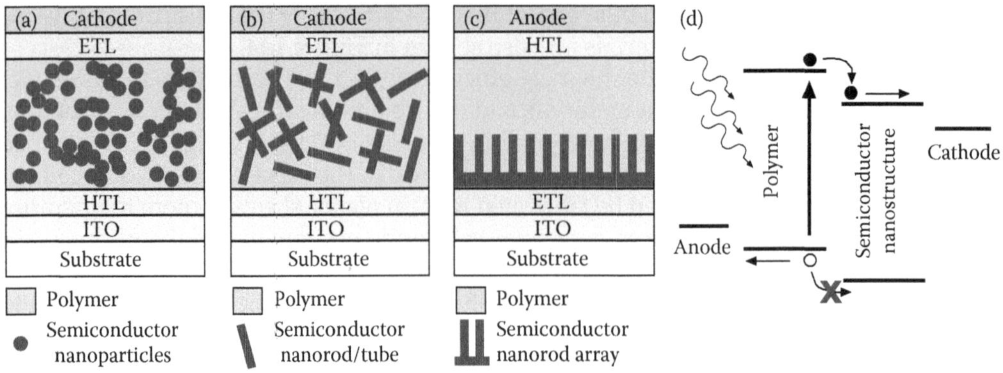

FIGURE 1.15
Schematic representation of some of the organic–inorganic hybrid solar cells with different inorganic semiconducting nanostructures like (a) nanoparticles, (b) nanorods or nanotubes, and (c) nanorod or nanotube arrays. (d) Schematic representation of working principle of a typical organic–inorganic hybrid solar cell.

quite similar to those of OSCs except the electron acceptors in hybrid solar cells are crystalline nanostructures of inorganic semiconductors. The excitons generated in the polymers on light absorption get dissociated at the polymer–inorganic nanocrystal interfaces where electrons get transferred to inorganic nanocrystals and the holes remain in the polymer. Electrons transport through nanocrystalline inorganic semiconductor channels and holes transport through polymer channels to collect at the respective electrodes. The diagrams shown in Figure 1.15a–c represent some of the hybrid solar cells incorporating different nanocrystalline structure of inorganic semiconductors like quantum dots, nanoparticles, nanorods, nanotubes, needles, and arrays of nanorods or nanotubes, and Figure 1.15d represents the working principle of a typical organic–inorganic hybrid solar cell. Researchers have also tried ternary blends where both the inorganic nanocrystalline structures and fullerenes were used as acceptors along with conjugated polymers [110,111].

Quantum dots and nanoparticles have the beauty of band gap tuning through variation in their size. Therefore, use of quantum dots and nanoparticles with conjugated polymers would have additional advantages of energy level matching with other constituents and harvesting the solar energy from the regions where the polymer does not absorb. However, the blending of quantum dots and nanostructures with conjugated polymers for efficient photon harvesting is not simple. Physical contact of nanostructures and conjugated polymers is essential for charge separation and transportation. The charge separation and migration of charge carriers is strongly affected by the interfacial contacts. Therefore, the interface and interaction between polymers and nanostructures are of high importance. To avoid agglomeration of quantum dots they are capped with some capping agents and these capping agents hinder the contact of quantum dots with polymers. Different procedures adopted to blend quantum dots with polymers influenced the polymer-quantum dot interfaces and thus the photovoltaic response. To overcome the issues related to capping agents, a number of other strategies such as in situ growth of quantum dots [112], ligand exchange [113], thermally cleavable ligands [114], novel surfactants [115], and polymerization of polymers on a quantum dot surface [116] have also been employed. These methods led to better PCE in hybrid solar cells. For example, the in situ growth of CdS quantum dots in P3HT resulted in direct interconnection and good contact between P3HT and CdS quantum dots, giving much higher V_{oc} and short-circuit current density (J_{sc}) compared to normal blending of oleic-acid-capped CdS quantum dots with P3HT [117].

Morphology, interfaces, and relative energy levels of the organic–inorganic hybrid systems are very important for efficient photon harvesting, and unfortunately due to unavailability of suitable hybrid combinations and due to unoptimum interfacial/bulk morphologies, the organic–inorganic solar cells have not yet utilized the benefits of organic–inorganic semiconductors to their fullest. The maximum PCE achieved so far in organic–inorganic hybrid solar cells is 5.5% that employed a low band gap polymer poly(2,6-(*N*-(1-octylnonyl)dithieno[3,2-b:20,30-d]pyrrole)-alt-4,7-(2,1,3-benzothiadiazole)) (PDTPBT) as the donor and nanocrystals of PbS_xSe_{1-x} as the acceptor [118]. Efforts are still in the works to improve their performance. A detailed description of organic–inorganic hybrid solar cells and the issues related to their performance can be found in a number of review articles [119–122].

1.3.4.2 Perovskite Solar Cells

The perovskite solar cell is a recently emerged technology, which has shown great potential and promise to be highly cost effective and more efficient than any other thin-film solar cell technology known so far. Perovskite materials do not incorporate any rare elements that keep their cost very low. These devices can easily be prepared on flexible substrates at room temperature with the spin coating process on a small scale, whereas via the printing process on a large scale. Due to its high efficiency and high cost effectiveness, in 2013 *Science* magazine declared this technology to be a breakthrough in the field of solar cells [122]. Figure 1.16 compares the yearly progress in efficiency of perovskite solar cells with that of other cost-effective solar cell technologies. The DSSCs fabricated in 1991 exhibited the PCE of ~7%–8% [86]. Later the record efficiency reached a plateau at 10%–11% until ~12% efficiency was reported in 2011 [87]. The most efficient devices have remained unchanged from the original concept of DSSCs. It took almost two decades for DSSCs to go beyond 11% PCE and they are still struggling with the instability. Similarly, the OSCs fabricated in 1995 exhibited a PCE of ~1.5% [91] and it took around six years to reach 2.1% in 2001 [124]. Further the improvement in efficiency of OSCs accelerated but still took more than one decade to go beyond 10%, and the maximum efficiency reported so far in OSCs is ~11% [92,93,125]. On the contrary, the first perovskite solar cell was prepared in

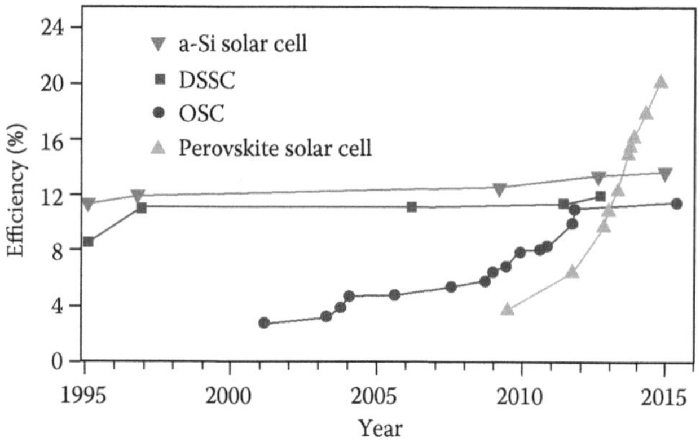

FIGURE 1.16
Yearly progress in efficiency of some of the cost-effective thin-film solar cell technologies. (Modified from National Renewable Energy Laboratory, Best research-cell efficiencies, Graphic, http://www.nrel.gov/ncpv/images/efficiency_chart.jpg.)

2009 that had an efficiency of ~3.8% [126]; since then there has been remarkable improvement in their performance and the efficiency has gone beyond 20% [127] in just 6 years. The efficiency is still expected to boost further and it may easily go beyond 30% in the next few years.

Perovskites are not new materials. They are known since 1839 when the first perovskite material, calcium titanium oxide ($CaTiO_3$), was discovered and its structure was named after Russian mineralogist L. A. Perovski. $CaTiO_3$ has a cubic crystal structure where the Ca atoms are located at the corners of the unit cell and the Ti atoms are body centered and the O atoms are face centered. Since then, all the materials having chemical formula ABX_3 (where A is the bigger cation, B is the smaller cation, and X is an anion) and a crystal structure like $CaTiO_3$ are called perovskites. The anions can be, for example, oxygen, halogens, or transition metals. In cubic symmetry structure the B cation has six fold coordination, surrounded by an octahedron of anions, and the A cation has twelve fold octahedral coordination. The relative size difference of the two cations is quite stringent. A slight buckling and distortion can produce several lower symmetry distorted versions, where coordination numbers of A and B cations are reduced and the orthorhombic and tetragonal phases are their most common noncubic variants.

Perovskite materials exhibit many interesting properties (e.g., colossal magnetoresistance, ferroelectricity, superconductivity, and spin-dependent transport) and have been used frequently for fuel cells, superconductors, memory, and spintronic devices. Nobody thought of using perovskite materials for solar cell applications until in 2009 when Miyasaka et al. used synthetic organometal halide perovskite semiconductors as light absorbers in place of organic dye in DSSCs [126]. They used the two organometal halides, namely, methyl-ammonium lead iodide ($CH_3NH_3PbI_3$) and methyl-ammonium lead bromide ($CH_3NH_3PbBr_3$), as TiO_2 sensitizers, and for the respective sensitizers 0.15 M LiI + 0.075 M I_2 in methyloxyacetonitrile and 0.4 M LiBr + 0.4 M Br_2 in acetonitrile were used as the electrolyte mediums. The solar cells based on $CH_3NH_3PbI_3$ and $CH_3NH_3PbBr_3$ exhibited PCE of 3.81% and 3.13%, respectively. Figure 1.17 shows the schematic structure of the first solar cell based on $CH_3NH_3PbI_3$ perovskite sensitizer and the crystal structure of $CH_3NH_3PbI_3$. These solar cells were not promising as the efficiency was quite low and devices exhibited very poor stability because the electrolyte medium dissolved

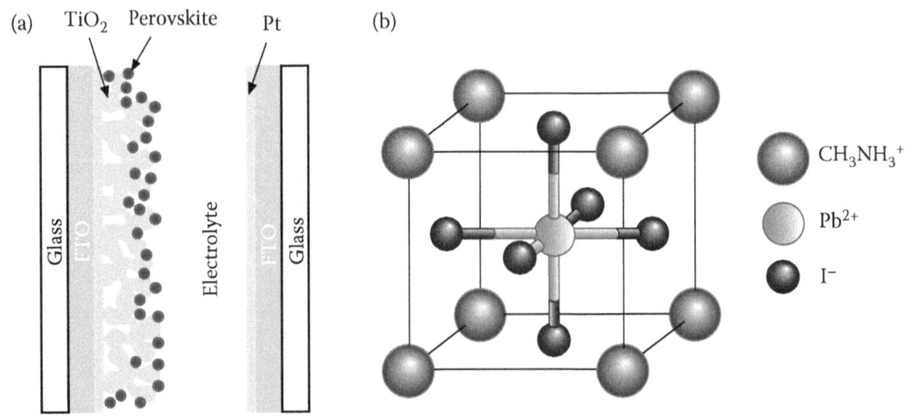

FIGURE 1.17
(a) Schematic structure of perovskite solar cell. (Prepared by A. Kojima et al., 2009. *J. Am. Chem. Soc.* 131: 6050.) (b) Crystal structure of organometal halide perovskite $CH_3NH_3PbI_3$.

perovskite materials. However, $CH_3NH_3PbI_3$ was found to have a broad absorption spectrum, high charge carrier mobilities, long charge carrier lifetime, very long charge carrier diffusion lengths, and does not incorporate any rare element. It has got a band gap of 1.5 eV with a conduction band minimum at Pb atom and valance band maximum at I atom. According to the Shockley–Queisser limit, a semiconductor with the band gap of 1.5 eV can lead to a PCE of ~30%. Additionally these perovskites were highly soluble in organic solvents, and thin film could easily be prepared on any substrate with spin coating and printing processes. Therefore, this demonstration of synthetic perovskite semiconductors as light absorbers in solar cells generated the hope of getting cost-effective and efficient solar cells.

In 2012, Gratzel et al. prepared the first solid-state $CH_3NH_3PbI_3$ perovskite-sensitized solar cells, where 2,2′,7,7′-tetrakis-(*N*,*N*-di-*p*-methoxyphenyl-amine)-9,9′-spirobifluorene (spiro-OMeTAD) was used as a solid-state medium for hole transportation [128]. This solar cell exhibited a PCE of 9.7%, which stimulated the interest in perovskite materials and gave birth to a new kind of solar cell technology known as perovskite solar cells. These solar cells were prepared on FTO-coated glass substrates where FTO was coated with a thin compact layer of TiO_2 prepared by spin coating of titanium diisopropoxide bis(acetylacetonate) and subsequent annealing at 500°C for 15 min. The compact layer of TiO_2 serves as the electron transport and hole blocking layer and also prevents direct contact of perovskite with the FTO electrode. On the prepared compact TiO_2 film a nanocrystalline TiO_2 paste was deposited by the doctor-blading method and sintered at 500°C for 1 h to form a mesoporous TiO_2 film. Mesoporous nanocrystalline TiO_2 film was treated with $TiCl_4$ and then coated with $CH_3NH_3PbI_3$ perovskite precursor and heated at 100°C for 15 min to get perovskite nanocrystals on TiO_2 mesoporous film. The perovskite precursor consisted of 1:3 wt% ratio of CH_3NH_3I and PbI_2 in γ-butyrolactone. The hole transport layer of spiro-OMeTAD doped with bis(trifluoromethane)sulfonamide lithium (LiTFSI) and 4-tert-butylpyridine (TBP) was deposited by spin coating. To complete the cell a 60 nm of Au was deposited by thermal evaporation as the counter electrode. Figure 1.18 shows the architecture and cross-sectional SEM (scanning electron microscope) images of the solar cell prepared by Gratzel et al.

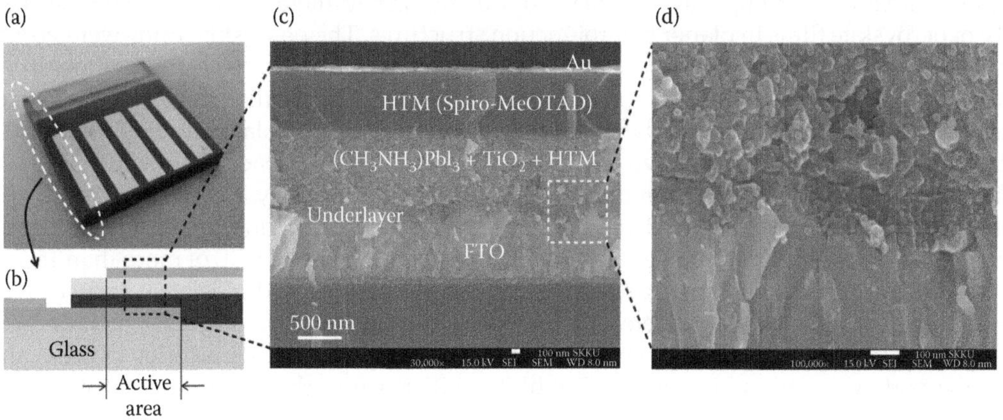

FIGURE 1.18
(a) Photograph of the actual solid-state perovskite solar cell. (b) Schematic cross-sectional structure of the solar cell. (c) Cross-sectional SEM image of the entire solar cell. (d) Cross-sectional SEM image of the FTO-under layer-active layer junctions interfaces. (Reproduced from H. S. Kim et al., 2012. *Sci. Rep.* 2: 591. With permission.)

Later in 2013 Gratzel et al. modified the process of deposition of the perovskite active layer [129]. The growth of $CH_3NH_3PbI_3$ perovskite on TiO_2 nanocrystals was done by sequential deposition method where, first, the PbI_2 solution in di-methyl formamide (DMF) was allowed to infiltrate into mesoporous TiO_2 film and then dried. After that the PbI_2-coated TiO_2 films were dipped in a CH_3NH_3I solution in 2-propanol for 20 s and then rinsed with propanol and dried. This resulted in the conversion of PbI_2 into $CH_3NH_3PbI_3$. The transformation of PbI_2 into $CH_3NH_3PbI_3$ was monitored by ultraviolet–visible and photoluminescence measurements. This new method allowed the in situ growth of $CH_3NH_3PbI_3$ nanocrystals on the TiO_2 mesoporous film. The in situ growth of the perovskite material resulted in better and controlled morphology of perovskite film. Even the material was able to be in the deep of the pores, which was difficult before. The HTM layer was grown by the solution casting of spiro-OMeTAD from its chlorobenzene solution. On the top of spiro-OMeTAD the Au electrodes were deposited by thermal evaporation. Better morphology and availability of the perovskite material in the deep of the pores resulted in a strikingly very high efficiency of 15%.

Henry Snaith, of Oxford University, thought of the necessity of TiO_2 mesoporous film and replaced it with insulating Al_2O_3 mesoporous film. Though Al_2O_3 is an insulator, the researchers were not expecting any photovoltaic response, but it was surprising that the solar cells not only exhibited the photovoltaic response but also the efficiency was more than that with TiO_2 [130]. The solar cells with TiO_2 mesoporous film showed a PCE of 7.9%, whereas those with Al_2O_3 scaffold showed a PCE of 10.9%. They used a mixed halide perovskite methyl-ammonium lead iodide-chloride ($CH_3NH_3PbI_2Cl$) as the light absorber. It is worth mentioning that TiO_2 is an electron-transporting semiconductor and Al_2O_3 is an insulator. High efficiency even with an insulator suggested that the electron transport happened within the perovskite itself and electron mobility in the perovskite was much more than that in TiO_2. As Al_2O_3 is an insulator and was just providing a scaffold for the perovskite, Snaith thought of eliminating the mesoporous structure and prepared a planer heterojunction solar cell with a structure FTO/compact TiO_2/perovskite/spiro-OMeTAD/Au and got an efficiency of ~2%. Though the efficiency was not as high as that with the mesoporous structures, the important thing they found was that the perovskite itself could work as the electron transporter and the mesoporous structures were not necessary. Later in 2013 Snaith and his group played with the optimization parameters and improved the quality of perovskite films in planer heterojunction structures. The perovskite films were grown by spin coating of the precursor solutions and then the thickness, annealing temperature, and annealing time were optimized to improve the quality of the films and performance of the solar cells. The optimized planer heterojunction perovskite solar cells exhibited a PCE of 11.4% [131]. Studies have shown that the morphology and compactness of the perovskite films play a very important role in determination of cell efficiency. Keeping the same idea in mind, Snaith et al. employed the thermal evaporation technique to get more uniform and compact layers of perovskites and interestingly they achieved PCE of more than 15% in the planer heterojunction structure itself [132]. This was the record efficiency of perovskite solar cells prepared in planer heterojunction structure by thermal evaporation. There was absolutely no difference in the perovskite structures deposited from solution process and by thermal evaporation in vacuum. The higher efficiency in these devices was attributed to more uniform and compact films of perovskite semiconductors. To make the perovskite solar cells more cost effective, the researchers are trying to find out a replacement of expensive materials like spiro-OMeTAD, the hole transport material. Prashant Kamat and his group replaced spiro-OMeTAD with highly cost-effective CuI and got a reasonably good efficiency. Where the efficiency with spiro-OMeTAD was 7.9%, the replacement of spiro-OMeTAD with CuI resulted in a PCE of 6.0% [133]. Though the efficiency with CuI was relatively low, it is cheap and would result in very cost-effective perovskite solar cells.

A typical perovskite solar cell possesses several hundred nanometer-thick perovskite film sandwiched between two electrodes along with electron and hole transport layers with or without mesoporous scaffold. The incident light is absorbed in the perovskite semiconductor that generates electrons and holes, which transport through transport layers and collect at the respective electrodes. Recently, Yang et al. doped TiO_2 with yttrium to improve electron transportation and used polyethyleneimine ethoxilated (PEIE) to reduce the work function of ITO, and by decreasing carrier recombination via controlling the humidity conditions they achieved 19.3% PCE with mixed halide $CH_3NH_3PbI_{3-x}Cl_x$ perovskite [134]. It is quite amazing that just within a couple of years the efficiency improved more than sixfold. Why the perovskite materials are so efficient for solar cell application is still a mystery. Scientists are doing basic studies and trying to find the reasons behind their high performance. Some of the reasons discovered very recently are their high charge carrier mobilities and long charge carrier diffusion lengths. Snaith's group tried to find out the charge carrier lifetime and their diffusion lengths in organometal iodide and organometal mixed halides [135]. Quite surprisingly the $CH_3NH_3PbI_3$ exhibited diffusion lengths more than 100 nm for both types of charge carriers, whereas for mixed halides the diffusion lengths were more than one micron [135]. Diffusion lengths of around 10 times more than the absorption lengths (~100 nm) is a quite unique phenomena in perovskites. Large diffusion lengths lead to thicker films and thicker films absorb more light causing generation of more photocurrent and high efficiencies.

Efforts are being made to improve further the performance of these solar cells. New materials and new growth methods are being developed. The replacement of I^- with other halogens and replacement of Pb^{2+} and $CH_3NH_3^+$ with other suitable cations would give birth to new perovskite semiconductors that could be more promising for solar cell applications. Some researchers have already synthesized some other organometal halides, which have shown quite high performance [130,132,136,137]. The perovskite solar cells have shown V_{oc} more than 1 V with the semiconductors of band gap ~1.5 eV. High V_{oc} in perovskite solar cells is quite remarkable, but the origin is still not very clear. There is lots of information that we are still missing about such perovskite materials and solar cells, therefore intense research is being carried out.

1.4 Performance Analysis of a Solar Cell

1.4.1 Solar Spectrum and Irradiance

The universe is made of several galaxies having numerous stars. Each star with all objects traveling in orbits around it forms a solar system. Our solar system comprises the sun, nine planets and their natural satellites, dwarf planets, asteroids, and comets. The sun is the star of our solar system and is situated in the center with all other objects traveling around it. It is composed of mainly H_2 (~73%) and He (~25%). The remaining 2% is covered by O_2, C, Ne, and Fe. Sun is an abundant source of energy which originates from nuclear fusion reaction through a series of proton–proton chains where H_2 converts into He. In proton–proton chains, every second ~6.2×10^{11} kg of H_2 is converted into He with emission of about 4×10^{26} J of energy. The total mass of the sun is about 1.989×10^{30} kg, and with the same rate of energy emission it is predicted that it will continue to emit this much energy for over 10 billion years. This reaction takes place in the inner core of the sun where the temperature is about 15.7 million kelvins. In nuclear fusion super high energy γ rays are produced that are absorbed by the outer core, and relatively lower energy photons are emitted. These lower energy photons are transmitted into space as electromagnetic radiation. The temperature of the outer surface of the sun is about 5778 K. The electromagnetic radiation from the sun possesses x-rays, ultraviolet, visible, and infrared waves.

The x-ray content is very little and the solar radiation entering the Earth's atmosphere span from 100 nm to about 1 mm. The radiation from 100 to 280 nm is designated as the ultraviolet C region, 280–315 nm is designated as ultraviolet B region, and 315–400 nm is designated as the ultraviolet A region. The visible radiation is visible to human eyes and spans from 380 to 780 nm. It is important that visible radiation has the maximum intensity over the entire solar spectrum. The radiation from 700 nm to 1 mm is designated as infrared radiation and can be divided into three regions: near infrared (700–1500 nm), mid-infrared (1500–5000 nm), and far infrared (5000 nm–1 mm). In terms of energy the solar radiation at the Earth's surface possess about 52%–55% infrared, 42%–43% visible, and 3%–5% ultraviolet radiation. The solar intensity at the mean distance from the Earth is about 1361 W/m^2 and is known as solar constant [138,139]. It is the amount of incoming electromagnetic radiation per unit area from the sun. When solar radiation enters into the Earth's atmosphere, some of its fraction is absorbed and scattered by the atmospheric gases. The absorption and scattering by atmospheric gases alters the solar spectrum and its intensity, which is different at different solar elevation angles. That is why we have the blue dominating sunlight before sunrise and after sunset, and red dominating sunlight during sunrise and sunset. Due to atmospheric effects, the solar radiation varies season to season. The amount of alteration in the solar spectrum is defined by the air mass (AM), which is secant of the angle between the sun and the zenith (sec θ). Air mass defines the optical path length traveled through the Earth's atmosphere. If L is the optical path length of solar radiation through the Earth's atmosphere, Z is the zenith path length (normal to the Earth's surface), and θ is the angle between sun and zenith, then AM is defined as $AM = L/Z = \sec\theta$ (see Figure 1.19). When solar radiation is not blocked by clouds and it comes directly to the Earth's surface, the radiation is commonly known as sunshine. And the sunshine duration is the time when the direct radiation on the Earth is at least 120 W/m^2.

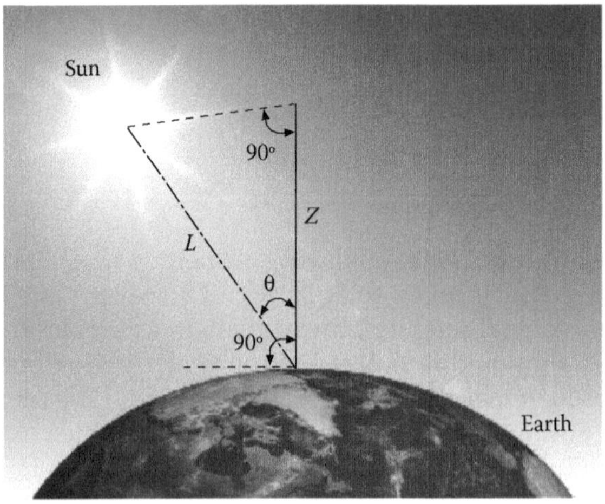

FIGURE 1.19
Position of the sun at solar zenith angle θ.

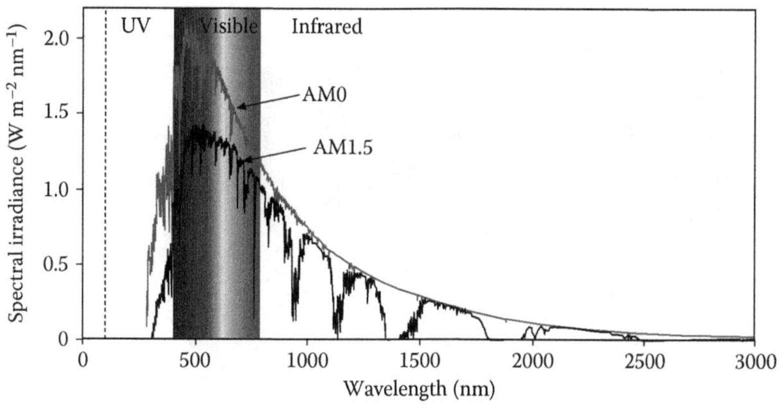

FIGURE 1.20
Solar spectral irradiance corresponding to AM0 and AM1.5 conditions.

The solar radiation outside the Earth's atmosphere is referred as air mass zero (AM0), meaning no atmosphere and no deflection from the original solar spectrum. The solar spectrum corresponding to AM0 conditions is shown in Figure 1.20. This spectrum can be equated to that of a black body at 5800 K. This spectrum is important for the performance analysis of satellite and space-vehicle solar cells. When the sun is at the zenith ($\theta = 0$), the solar spectrum at the Earth's surface is referred as AM1, and the solar irradiance at the Earth's surface is about 1050 W/m^2. This spectrum is important for analysis of solar cells placed at equatorial and topical regions. For an angle of 60° or 70° the solar spectrums are represented by AM2 and AM3, respectively. AM2 and AM3 spectrums are useful for estimation of performance of solar cells placed at high altitudes. AM38 corresponds to the air mass in horizontal direction ($\theta = 90$). The sun is not always at the same angle and the air mass varies. For the mid-latitudes, an AM number is required that could represent the overall yearly average condition and for that purpose AM1.5 is quite reasonable. The solar spectrum corresponding to the AM1.5 condition is also shown in Figure 1.20. AM1.5 corresponds to a solar zenith angle of 48.2°. Solar industries have been using AM1.5 as the standard air mass condition for testing of different solar cells and modules. It is important to note that for accurate comparison of performance of different solar cell technologies, there should be some standard test conditions for performance evaluation. In view of this many international standards, including the American Society for Testing & Materials (ASTM), International Electrotechnical Commission (IEC), and International Organization for Standardization (ISO), were developed for different testing purposes [140]. In general for testing of a solar cell or module, there are three standard test conditions (STCs) used in industries for testing of solar cells: cell temperature, cell illumination intensity, and air mass values. For STCs, the temperature should be 25°C, solar irradiance should be 1000 W/m^2, and air mass value should be AM1.5G. Here G stands for Global.

For energy generation from solar power, it is very important to know the worldwide distribution of solar energy. The solar radiation is unevenly distributed on the Earth and the intensity varies from one location to another depending on the season, latitude, and day time. To be very simple and convenient, the distribution of solar intensity on the Earth can be divided into four broad belts designated as most favorable, moderately favorable, less favorable, and least favorable for solar applications. Figure 1.21 shows the distribution of four belts in the northern hemisphere. A similar distribution would apply to corresponding

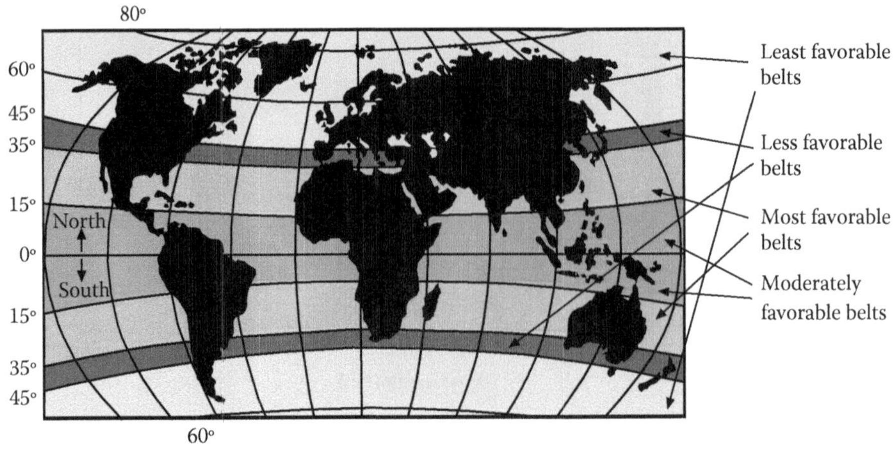

FIGURE 1.21
Worldwide distribution of solar energy in terms of four belts, which define the most and least favorable regions for solar energy applications. (Adapted and modified from Google Search.)

belts in the southern hemisphere. The most favorable belt lies between latitudes 15°N and 35°N, and is most suitable for solar energy applications. Due to limited cloud coverage and less rainfall, the regions falling in this belt have the largest amount of solar radiation. The sunshine hours per year in these regions are more than 3000 h. The moderately favorable belt lies between latitude 15°N and the equator (latitude 0°N). Due to high humidity and regular clouds, the solar intensity in the regions falling on this belt is less compared to those in the most favorable belt. There are usually about 2500 h of sunshine per year. There are only slight seasonal variations and the solar intensity remains almost uniform throughout of the year. The less favorable belt lies between latitude 35°N and 45°N. In the regions falling in this belt, the radiation intensity and daylight hours varies due to frequent seasonal variations. The average solar intensity is almost similar to most and moderately favorable belts, but solar radiation in winter is very low compared to summer. The fourth belt, which is least favorable for solar energy application, lies below latitude 45°N. The regions falling in this belt encounter widespread and quite regular cloud coverage, and most part of the year is covered by winter. The average solar intensity in these regions is comparatively very low. Most developing countries fall in the most favorable region, therefore the solar energy could be a solution of their energy problems and it can play a huge role in their socioeconomic growth.

1.4.2 Power Conversion Efficiency (PCE)

PCE of a solar cell defines its capability to convert incident light into electricity. It is given by the ratio of electrical power generated by the cell over the optical power incident on it. In general, when a solar cell is exposed to light, not all the incident photons are absorbed by the cell. Some of the photons are transmitted through without absorption. The photons with energy less than the band gap energy of the semiconductor are not absorbed. Therefore, the band gap of the semiconductor places a limit to the PCE of the solar cell. The absorbed photons excite electrons from the ground state (valance band) to the excited state (conduction band) of the semiconductor giving photogenerated electrons in the conduction band and corresponding holes in valance band. Irrespective of the binding energy of photoexcited electrons and holes, the photon absorption generates an electron–hole pair, which should be separated and transported in the opposite direction to constitute an electric current. The process of charge separation and collection may be different in different solar cells, but the basic principle of electric current generation remains the same. After separation, the electrons and holes transport and collect, respectively, at the cathode and anode. The direction of transportation is decided by the drift and diffusion processes. Diffusion comes from the concentration gradients of the charge carriers, whereas drift comes from the internal electric field of the device. The collection of charge carriers at respective electrodes develops a potential difference between them. If a load is connected across the two electrodes, an electric current would pass through it. Not all photogenerated charge carriers reach the electrodes, as some of them get recombined. Recombination of charge carriers reduces the cell efficiency. There are lots of loss mechanisms and limiting factors in a cell that control its efficiency. It is almost impossible to overcome all the loss mechanisms and get 100% efficiency. The different loss mechanisms and factors limiting the cell efficiency will be discussed later.

If the cell is not connected to the load or the cell is in open-circuit condition, the potential difference developed across the two electrodes of the cell under illumination is known as open-circuit voltage (V_{oc}). In the open-circuit condition, no current is derived from the cell. If the two electrodes of the cell are short circuited, an electric current would pass through the short and this would be the maximum possible current that could be derived from the cell for given light intensity. This current is known as the short-circuit current (I_{sc}). Under short-circuit conditions the voltage across the cell becomes zero. Under the load condition the current and voltage across the load would be less than I_{sc} and V_{oc}. To calculate the PCE, it is important to know the maximum electric power given to the load, and for this purpose current–voltage (*I–V*) characteristics of the cell are measured under illumination. To measure the *I–V* characteristics a source-measure unit is used as an alternate of the load, where a range of voltage from negative to positive values is applied and the corresponding current through the cell is measured and recorded. The schematic *I–V* characteristic of a solar cell under illumination is shown in Figure 1.22. The I_{sc} and V_{oc} are directly measured from the *I–V* characteristics as the current at zero voltage and voltage at zero current, respectively. In the reverse bias ($V < 0$), the current through the cell is negative and

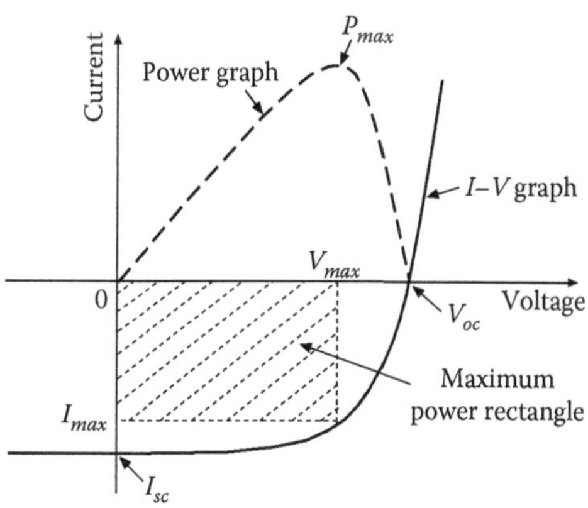

FIGURE 1.22
Schematic *I–V* characteristics of a solar cell under illumination (solid curve). Dashed curve represents the corresponding inverted power graph.

remains negative even in forward bias ($V > 0$) until $V = V_{oc}$. For $V > V_{oc}$ the current magnitude becomes positive and the current increases rapidly with the applied voltage.

The region between I_{sc} and V_{oc} on the I–V characteristics represents the power derived from the cell for different load conditions. The schematic inverted power graph between I_{sc} and V_{oc} is also shown in Figure 1.22. The point corresponding to the maximum value on the power graph represents the maximum electrical power (P_{max}) derived from the cell. The current and voltage corresponding to P_{max} are written as I_{max} and V_{max}, respectively on the I–V graph. Now the PCE (η) of a cell is obtained from

$$\eta = \frac{P_{max}}{P_{in}} = \frac{I_{max}V_{max}}{P_{in}}, \tag{1.1}$$

where P_{in} represents the incident optical power in W. The shaded area in Figure 1.22 represents the P_{max} rectangle. The ratio of P_{max} over $I_{sc}V_{oc}$ is known as the fill factor (FF), that is,

$$FF = \frac{I_{max}V_{max}}{I_{sc}V_{oc}}. \tag{1.2}$$

FF is the measure of cell quality and defines the shape of the characteristics. For very low FF the shape of the characteristics would be concave, whereas for high FF the shape would be convex. High FF corresponds to fewer recombination losses and high efficiency of the cell. Ideally the maximum value of FF can be 1, but due to recombination losses it is always less than 1. For maximum output of the cell, a load should be chosen such a way that the power given to the load is P_{max}. From Equations 1.1 and 1.2, the expression for η can also be written as

$$\eta = \frac{I_{sc}V_{oc}FF}{P_{in}}. \tag{1.3}$$

The percentage efficiency of a solar cell is calculated from

$$\eta = \frac{I_{sc}V_{oc}FF}{P_{in}} \times 100. \tag{1.4}$$

If $P(\lambda)$ is the spectral power density of the incident light, then the incident optical power (P_{in}) can be calculated from

$$P_{in} = \int_0^{\infty} P(\lambda)\, d\lambda. \tag{1.5}$$

If $\Phi(\lambda)$ is the photon flux density, then

$$P_{in} = \int_0^{\infty} \Phi(\lambda) \frac{hc}{\lambda}\, d\lambda \tag{1.6}$$

Please note that the efficiency of a solar cell depends on its temperature and the intensity of incident light. It will vary with variation in testing conditions. Therefore while quoting the cell performance, the testing condition should always be mentioned.

Due to different device architectures and material properties of different types of solar cells, it is not necessary for the physics to be same for all types of devices. It should therefore be studied and understood independently for different types of solar cells. The physics of OSCs will be discussed later in detail, but the basic understanding of a conventional *p-n* junction solar cell is discussed next in brief. It is important to note that a *p-n* junction solar cell is actually a *p-n* junction diode, and application of a voltage would cause a current to flow through the junction even when the cell is not illuminated. This current is called dark current (I_D). Therefore, during *I–V* measurement under illumination, along with the photogenerated current a dark current also flows that should always be taken into account while calculating the actual photogenerated current. For an ideal *p-n* junction diode, I_D is given by [141]

$$I_D = I_s(\exp^{(qV/k_BT)} - 1), \tag{1.7}$$

where I_s is the reverse saturation current of the diode, q is the elementary charge, k_B is the Boltzmann constant, and T is the absolute temperature of the diode. The photogenerated electrons and holes move toward the cathode and anode, respectively, and constitute a photocurrent (I_L), which flows from *n*-type to *p*-type across the junction, therefore I_L is negative. For an ideal *p-n* junction there are no recombination losses and I_{sc} is equal to I_L. In reverse bias both the I_D and I_L are negative, but in forward bias I_D becomes positive and I_L remains negative. As a result, both currents in reverse bias add up and constitute a total negative current; but in forward bias, the two currents are in opposite directions, which causes the total current to decrease until it becomes zero at V_{oc} where both I_D and I_L are equal. For $V > V_{oc}$ the positive I_D dominates the negative I_L and the total current becomes positive, which increases with applied voltage due to increase in I_D. The total current (I_{Total}) through the cell under illumination can be written as

$$I_{Total} = I_s(\exp^{qV/k_BT} - 1) - I_L. \tag{1.8}$$

In the short-circuit condition ($V = 0$), the total current through the cell would be

$$I_{Total} = I_{sc} = -I_L. \tag{1.9}$$

In the open-circuit condition $I_{Total} = 0$ the voltage across the cell would be V_{oc}, which will be given by

$$V_{oc} = \frac{k_BT}{q}\ln\left(\frac{I_L}{I_s} + 1\right) \approx \frac{k_BT}{q}\ln\left(\frac{I_L}{I_s}\right). \tag{1.10}$$

Hence for a given temperature and light intensity the V_{oc} of a *p-n* junction solar cell depends logarithmically on the inverse of I_s.

In a practical solar cell there are lots of parameters that inhibit charge extraction and cause recombination of charge carriers. All the semiconductors inevitably have some bulk resistivity that limits the charge extraction. Also the defects in the film and energy barriers at contacts would inhibit the charge extraction. All the factors causing hindrance to charge extraction are represented by a resistance connected in series (R_s) with the diode. The recombination of charge carriers is represented by shunting the diode with a resistance connected in parallel (R_p). Being an electronic device, a solar cell can be represented by an equivalent circuit. In general, the equivalent circuit of a *p-n* junction solar cell in dark is

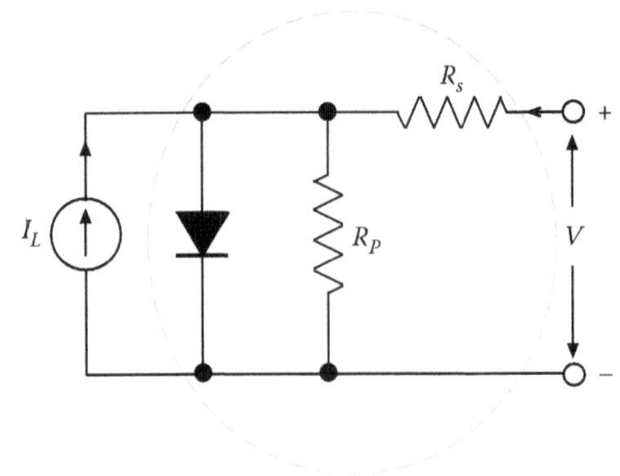

FIGURE 1.23
Equivalent circuit of a *p-n* junction solar cell under illumination. Encircled portion represents equivalent circuit in dark.

represented by a diode with R_s and R_p. Under illumination the photovoltaic effect is represented by a current source generating I_L connected to the diode. In a conventional *p-n* junction solar cell I_L is constant for a given light intensity. Figure 1.23 shows the equivalent circuit of a *p-n* junction solar cell. Considering the Shockley–Read–Hall recombination in a practical solar cell, Equation 1.8 changes to

$$I_{Total} = I_s\left(\exp^{(qV - IRs/nk_BT)} - 1\right) + \frac{V - IR_s}{R_p} - I_L \tag{1.11}$$

where n is ideality factor of the diode.

A solar cell possesses both the optical and electrical losses that are crucial in determination of its efficiency. Optical losses include the reflection and transmission of incident light. Some of the light incident on the cell is reflected back by the top surface of the cell. Out of the light entering in the cell, that with energy less than the band gap of the semiconductor is not absorbed and gets transmitted. Electrical losses include loss of excess energy through thermal relaxation and recombination of charge carriers. Recombination can take place in many ways during separation and transportation of the charge carriers. Usually the electron and hole mobilities in the semiconductor are not the same and one moves slower than the other. In this case the slower one recombines at the atom and loses its energy. Recombination of charge carriers in a solar cell is detrimental and should be minimum. The materials with high charge carrier mobilities possess low recombination and high PCE. The imperfections and grain boundaries in the semiconductor film create charge carrier traps, which work as recombination sites. The series resistance causes hindrance in collection of charge carriers, whereas shunt resistance provides the path for recombination. The series and shunt resistances in an ideal solar cell are zero and infinitely large, respectively.

Considering the physics of an ideal *p-n* junction Si solar cell, thermal relaxation of excess energy of the photon, and that each absorbed photon generates one electron–hole pair in the semiconductor, Shockley and Queisser calculated the ideal PCE for different semiconductors using the method of detailed balance limit [142]. Figure 1.24 shows the Shockley–Queisser ideal single-junction solar cell efficiency graph for different band gaps of the semiconductors. The maximum efficiency under one sun illumination was found to be 33.7% for the semiconductor having a band gap of 1.34 eV and the materials parameters of III–V compound semiconductors. For most studied Si having a band gap of 1.1 eV, the

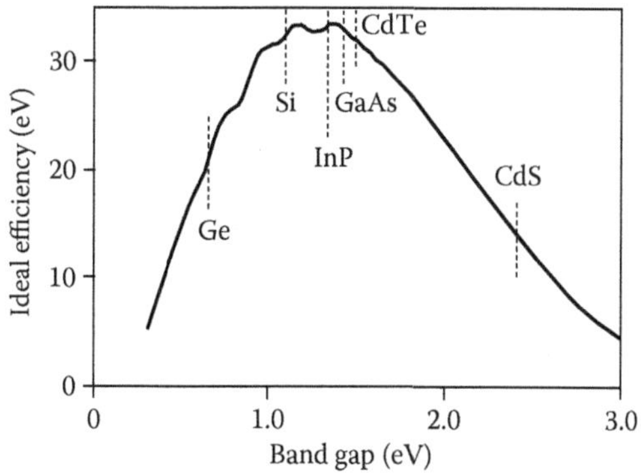

FIGURE 1.24
Ideal single junction solar cell efficiency as a function of band gap of the semiconductors.

efficiency limit was calculated to be 29%. The ideal efficiencies corresponding to other semiconductors are also indicated on the graph. The slight swinging in the curve is caused by IR absorption in the atmosphere. The actual Shockley–Queisser graph was smoother and the efficiency values were different because the solar spectrum was approximated by a blackbody radiation at 6000 K. This graphs shows that the efficiency does not critically depend on the band gap and has broad maximum. Reduction in efficiency for larger band gaps is because of greater transmission losses and that for smaller band gaps is because of greater thermalization losses.

1.4.3 Incident Photon to Converted Electron (IPCE) Efficiency

The incident photon to converted electron (IPCE) efficiency of a solar cell is also known as the external quantum efficiency (EQE). IPCE is related to the spectral response of the solar cell that depends upon the absorption coefficient of the semiconductor. The absorption coefficient of a material is the reciprocal of the distance traveled by the photons before absorption (1/e of incident intensity) in the material. Absorption coefficient of a material depends strongly on the wavelength of the incident light. When a solar cell is exposed to light, not all the photons incident on the solar cell are absorbed by the semiconductor. The photons with energy less than the band gap of the semiconductor are not absorbed and do not contribute to the output of the cell. Only the photons with energy larger than the band gap of the semiconductor are absorbed and generate electron–hole pairs in the cell. The generation rate of electron–hole pairs in the cell depends upon the absorption coefficient of the semiconductor and decreases exponentially from the illuminated surface. When a light of wavelength λ is incident of the cell, the generation rate of electron–holes at a distance x from the illuminating surface is given by

$$G(\lambda, x) = \alpha(\lambda)F(\lambda)(1 - R(\lambda))\exp(-\alpha(\lambda)x), \quad (1.12)$$

where $\alpha(\lambda)$, $F(\lambda)$, and $R(\lambda)$ are, respectively, the absorption coefficient, the number of incident photons per unit area per unit time per unit bandwidth, and fraction of incident photons reflected from the surface [141].

Usually for a semiconductor $\alpha(\lambda)$ is large for short wavelengths and small for long wavelengths. For example, Figure 1.25a shows the absorption coefficients of Si and P3HT:PCBM. For better understanding, the electron–hole generation rates in a semiconductor for short

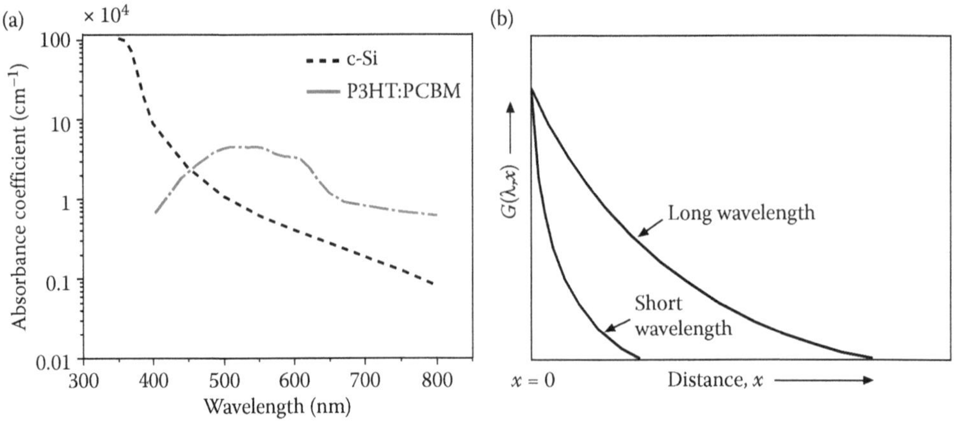

FIGURE 1.25
(a) Absorption coefficient of c-Si and P3HT:PCBM as a function of wavelength. (b) Schematic representation of electron–hole generation with distance for short and long wavelengths.

and long wavelengths are shown schematically in Figure 1.25b. At low photon energies most carriers are generated far from the front surface due to their low absorption coefficient and the front region takes over at high photon energies because of their high absorption coefficient. The spectral response of a solar cell is defined by the ratio of photocurrent over the photon flux and is given by

$$S(\lambda) = \frac{J_{sc}(\lambda)}{\Phi(\lambda)}, \tag{1.13}$$

where $\Phi(\lambda)$ is the photon flux density at wavelength λ and $J_{sc}(\lambda)$ is the corresponding short-circuit current density generated by the cell. EQE is described by the number of electrons leaving the device per unit time and per unit area over the number of photons incident per unit time per unit area. The spectral response is directly related to EQE as

$$EQE(\lambda) = \frac{hc}{q}\frac{J_{sc}(\lambda)}{P(\lambda)\lambda}. \tag{1.14}$$

or

$$EQE(\%) = 1240\frac{J_{sc}(\lambda)}{P(\lambda)\lambda}, \tag{1.15}$$

where $P(\lambda)$ is the incident optical power, h is Planck's constant, c is speed of light in air, and q is the elementary charge. If the loss factors are taken into account, the effective carrier generation after correction for optical losses is called the internal quantum efficiency (IQE). The spectral response measurements can also be used to calculate the J_{sc} of a solar cell as

$$J_{sc} = \frac{q}{hc}\int_{\lambda 1}^{\lambda 2} P(\lambda)EQE(\lambda)\lambda\, d\lambda, \tag{1.16}$$

where λ_1 and λ_2 are the wavelength limits of the absorbed spectrum. The J_{sc} calculated from the spectral response measurements is more accurate compared to that measured from the illuminated current density–voltage (J–V) characteristics as there might be some mismatch between the spectrum of solar simulator and that of the sun.

1.5 Photovoltaic Modules

An individual solar cell produces a little electrical power, and for large power production a number of solar cells are interconnected together to formulate a solar module or solar panel. For different types of solar cells the module structures are different. In a solar module based on wafer solar cells, the individual solar cells are connected together by soldering of wires, whereas in a thin-film solar module the solar cells are grown monolithically on the same substrate and are fabricated in a way that they are interconnected monolithically. Though the monolithic growth of solar cells is a bit complicated but it does not require soldering of individual cells that makes the solar module sturdier. Figure 1.26a shows a typical thin-film polymer solar module from Konarka, United States, whereas the monolithic connection of solar cells in the module is shown schematically in Figure 1.26b. All the layers including electrodes are deposited in a predefined patterned. For example, for the series connection of solar cells the bottom electrodes are deposited or etched in a pattern of parallel strips. It provides separate bottom electrodes for different solar cells. Laser scribing and chemical etching are the most frequently used techniques for

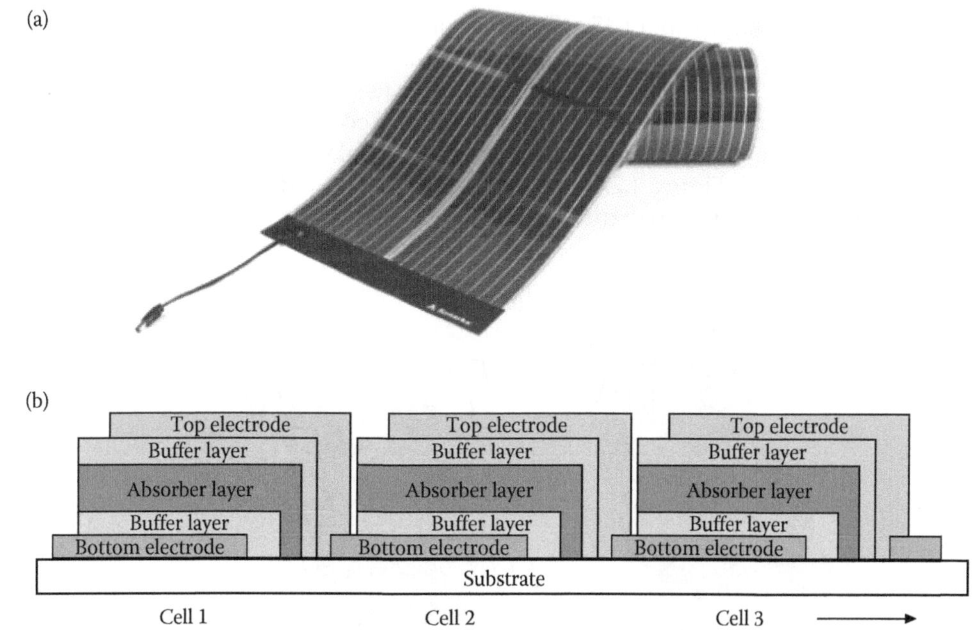

FIGURE 1.26
(a) Thin-film polymer solar module by Konarka, United States, on a plastic base. (b) Schematic representation of monolithic connection of solar cells in a thin-film solar module.

patterning of bottom electrodes. Subsequently the buffer and light-absorbing layers are deposited in the desired sequence in the same pattern but with a slight lateral shift from the bottom electrode pattern. For patterning of the buffer and absorber layers, either the layers are mechanically scribed or printed by suitable means in desired pattern. Finally, the top electrode is deposited in the same pattern but with a slight lateral shift compared to the buffer layer pattern such that it makes intimate contact with the bottom electrode of the next adjacent cell. For robustness of the modules they are packaged by suitable means. Packaging is very important to avoid mechanical damage to the cells and wire connections. In addition the packaging is important to avoid corrosion of the top surface of solar cells and interconnections by rain or water vapors in the environment. The modules are packaged properly to prevent them from environmental damages and to protect the users from electrical shocks.

Apart from solar cells and their connection, front surface material, encapsulant, rear surface material, and framing are other main components of a solar module. The front surface material should necessarily be transparent for the light to enter and be absorbed by the solar cells. Additionally it should be highly stable to high temperatures, high UV exposures, and high barriers for water vapor entrance. Glass, acrylic, and high barrier plastic sheets are usually used for this purpose. The thin-film solar cells are encapsulated in flexible array, whereas wafer solar cells are packaged in rigid array with glass front surface. The top surface of the solar module should have low reflection, which could be achieved by texturing the top surface, but it would cause soiling of the top surface. For rigid panels, glass is the most commonly used material and it also possesses self-cleaning properties. The encapsulant is used to bind the front and rear surface materials to the solar cells and therefore should be optically transparent and highly stable at high temperatures and high UV exposures. Ethyl vinyl acetate (EVA) is the most commonly used encapsulant for solar modules. For encapsulation EVA sheets are inserted between solar cells and the top and rear surfaces and heated to about 150°C. At elevated temperature EVA gets polymerized and binds the modules together. Depending upon the applications of the solar module the rear material could be transparent or opaque, rigid, or flexible, but it should provide a high barrier to water vapor ingress. Usually a thin sheet of Tedlar, polyvinyl fluoride (PVF), is used as a rear surface material. For framing of the module a light weight, corrosion free material like Al is used. Figure 1.27 shows a typical Si PV module with schematic representation of its layer by layer structure.

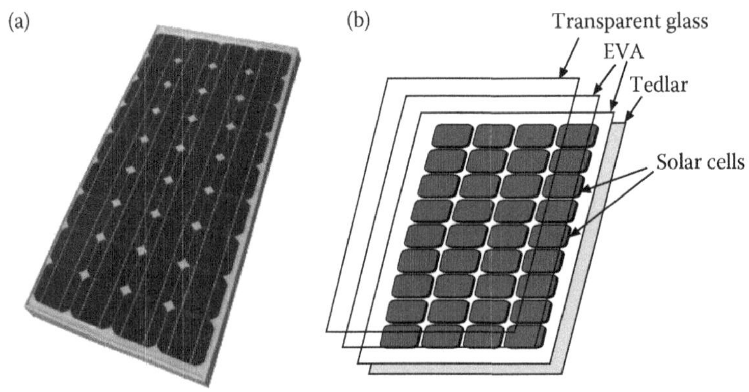

FIGURE 1.27
(a) A typical Si PV module and (b) schematic representation of its layer by layer structure.

Depending on the applications and power requirement, the solar cells in a module could be series connected, parallel connected, or both. To ensure maximum power output, proper design of a solar module is very important. The packing factor is another important parameter that defines the area covered by the solar cells in the module. It is given by the ratio of the total area of the cells in the module over the total area of the module. It directly affects the power output of the module and depends on the shape of the solar cells and distance between them. For a larger packing factor the power output of the module would be more. The packing factor can be increased by placing the solar cells with maximum possible proximity in the module. Even with the maximum proximity the packing factor is minimum for round solar cells and maximum for square solar cells. It is important to note that a higher packing factor also increases the module temperature by increasing the heat generated by the module per unit area that may not be good for module performance.

Ideally when all the solar cells in a module have identical characteristics with identical operating current and voltage, the *I–V* characteristics of the module can be expressed by

$$I_T = YI_s\left(\exp\left(\frac{qV_T}{Xnk_BT}\right)-1\right)-YI_L, \tag{1.17}$$

where Y is the number of solar cells connected in parallel, X is the number of solar cells connected in series, I_T is the total current from the module, and V_T is total voltage from the module. The short-circuit current of the module would be YI_{sc}, open-circuit voltage of the module would be XV_{oc}, where I_{sc} and V_{oc} are short-circuit current and open-circuit voltage of the individual cell. In a series connection of the cells the total current remains the same as that of the individual cell, but the voltage is given by the sum of voltages of the cells in series. In parallel connection of the cells, the total voltage remains equal to that of the individual cell, but the current is given by the sum of currents of the cells in parallel.

Usually the module efficiency is less than the efficiency of an individual cell and there could be lots of reasons behind this difference. Optical losses due to reflection and absorption by front surface material, electrical losses due to nonohmic contacts, the packing factor, the difference in cell and module temperatures, and the mismatch losses due to nonidentical photovoltaic performance of solar cells in the module are some of the reasons behind the difference in cell and module efficiencies. When the photovoltaic parameters of individual solar cells in a module do not match, mismatch losses take place. Mismatch losses are very crucial for maximum output and stable performance of a module. Mismatch losses may cause irreversible damage to the module and should be as minimum as possible. While integrating the solar cells in a module, the mismatch factors should be taken into account. Mismatch in I_{sc} and V_{oc} are the two most commonly observed mismatches in solar cells. For the cells connected in series, the mismatch in V_{oc} is not a problem as the voltages of the cells accumulate and the total module voltage would be the sum of the corresponding voltages of the cells at a particular current. But a mismatch in I_{sc} may cause a serious problem for the cells connected in series. The current through the series cannot exceed that of the cell-producing minimum current. The extra current generated by good cells is dissipated in poor cells and cause local hot-spot heating. Therefore, the series connection is recommended only for solar cells having no mismatch in I_{sc}. For the cells connected in parallel, the mismatch in I_{sc} is not a problem as the currents from individual cells accumulate and build a total current more than that of an individual cell (around the sum of the currents of all individual cells), but a mismatch in V_{oc} of the cells would cause a serious problem. Parallel connection of the cells with V_{oc} mismatch would cause the total voltage to be less than the voltage of the cell having maximum voltage. The extra power generated by the good cell would dissipate across the cell-producing less voltage. The power dissipation would cause local heating and may damage the module. Parallel connection is recommended only for the solar cells having same V_{oc}. For more information on the mismatch effect, see Section 2.3.2.

Usually the solar cells in a module are connected in series. A typical Si PV module possesses about 36 solar cells connected in series. For charging of a 12 V battery using a solar module the output voltage of the module should be its compatible. The output voltage of an individual Si solar cell is around 0.6 V, but it will be slightly less during operation, therefore typically 36 solar cells are connected in series to give an operating voltage of around 17 V. As some of the voltage drops across the module elements during operation, the open-circuit voltage of the module would be around 21 V. The current from the module would be around that from the individual cell, which is usually around 3.5 amp (35 mA/cm^2 × 100 cm^2). For higher power requirements the large PV arrays are prepared where individual modules are connected in series and parallel. The rated output of a solar module depends on its efficiency. For the two solar modules having the same area the rated output power of the module with higher efficiency will be more than that having lower efficiency. Module temperature is an important parameter that decides the module performance. A fraction of light incident on the module generates heat and increases the module temperature. The light not contributing to electrical power is absorbed by the cells or by the module and contributes to heat generation. If the solar cell or module does not operate at maximum power point, a fraction of the generated electrical energy will also convert into heat via recombination of charge carriers. A very high temperature would cause deterioration in the module performance. Therefore, for maximum output the module temperature should be maintained at lower values.

1.6 Utilization of Solar Modules for Different Applications

Power output of a solar cell or the PV module is in direct current (DC) form, and for the applications where DC power is required the output power can directly be fed to the load. But for alternate current (AC) appliances the output of the PV array is fed to an inverter, which converts the DC power into AC power. Depending on the applications, the solar inverters vary in size and specification from residential to solar power stations. In solar power stations, the electric power produced by PV arrays is fed to the grid tie inverter, and the output AC power is transported from one place to another through grid lines. The grid tie inverter synchronizes the AC frequency and voltage to those of the grid. Where grid lines are difficult to reach like in remote areas or the grid connection comes out to be quite expensive, the solar power is used as stand-alone solar power systems or off-grid power systems where DC power generated by the PV system is stored in batteries and converted to AC using an inverter. The output of the inverter is used for domestic and commercial purposes. During the day the output of solar array is used to charge the batteries as well as for direct utilization in appliances. During the night the power stored in the batteries is used for suitable applications. Figure 1.28 schematically shows the utilization of solar power through stand-alone PV systems and its transportation through grid lines.

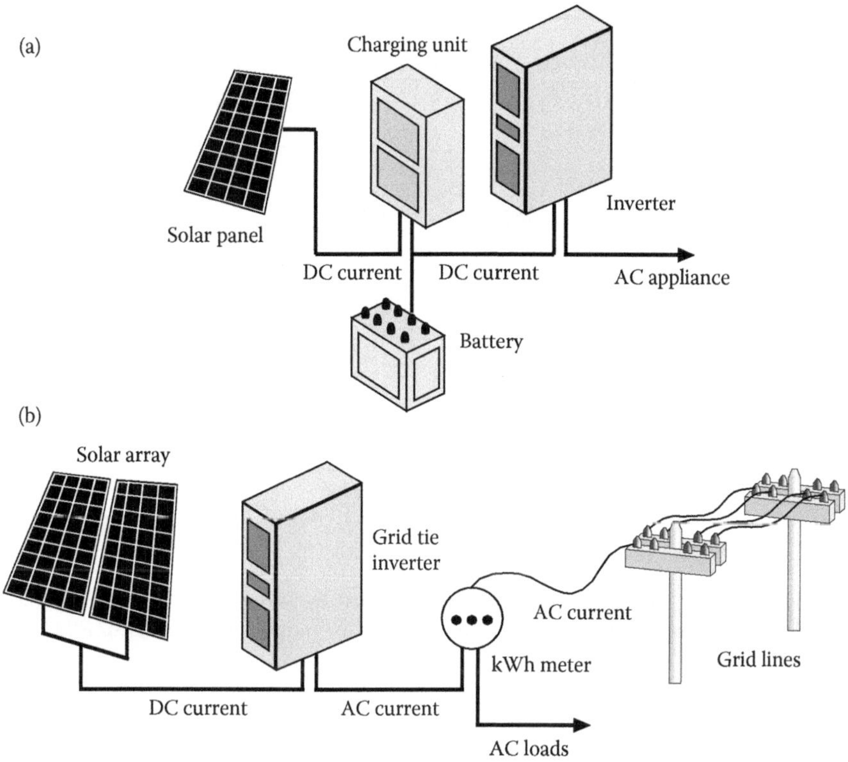

FIGURE 1.28
Schematic representation of (a) stand-alone solar power system and (b) grid tie solar power system.

1.7 Energy Payback Time (EPBT) and Carbon Footprint

Energy payback time (EPBT) is an important parameter for the economic and cost-effective analysis of a PV technology. EPBT is the time span required to generate the same amount of energy that was used in production and installation of the PV technology and is defined by the ratio of energy invested on the module during its life cycle over the energy produced by the module. For example, most of the PV modules in the market are based on crystalline and multicrystalline Si, and their wafers use purified Si. Purification and crystallization of Si are the most energy-consuming processes. Achievement of metallurgical-grade Si from raw materials, cutting of Si ingots into wafer, processing of wafers into cells, assembling of cells into modules, and installation of modules are other processes that require lots of energy. Apart from them, other overhead energy uses in manufacturing of solar cells further increase the overall energy consumption. In case of thin-film solar cells the major energy cost for manufacturing includes the substrates and methods of film deposition. The EPBT of a solar PV module depends on a number of parameters such as manufacturing materials and processing, lifetime of solar module, module size and efficiency, encapsulation and framing, the application type, and the weather conditions of the place where it is installed. Calculation of accurate EPBT is bit difficult, as not each and every process involved is done in the same industry. Therefore,

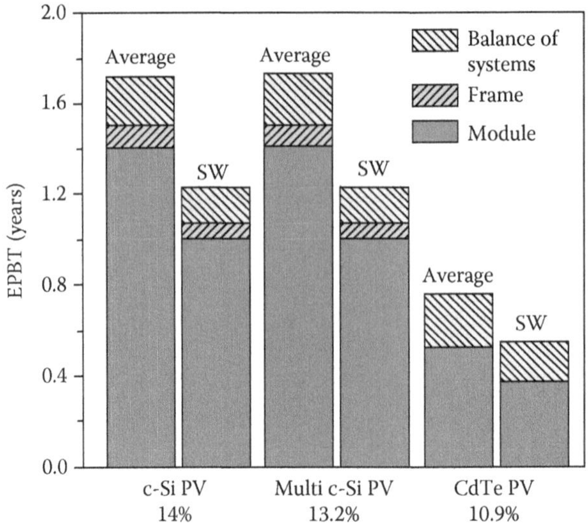

FIGURE 1.29
EPBT of c-Si, multi c-Si and CdTe PV modules installed in U.S. average and Southwest (SW) conditions. (Modified from V. Fthenakis, 2012. How long does it take for photovoltaics to produce the energy used? *PE Magazine*, 16–17.)

calculation of EPBT of a PV technology requires some assumptions. However, the EPBT of current PV technologies are from 6 months to 2 years. Depending upon the PV technology and place of installation, with expected lifetime of around 30 years the PV modules returns 15–60 times more energy than invested. For example, Figure 1.29 shows the EPBT of three major commercial PV technologies installed in U.S. average (1700 kWh/m^2 year) and Southwest (SW) (2380 kWh/m^2 year) solar conditions. CdTe-based PV technology has the shortest EPBT among all commercial PV technologies [143]. Several other estimated EPBT studies on different PV modules in different weather conditions can be found in the literature. EPBT does not tell us about the energy gained from the module for the rest of the economic lifetime. Therefore, another parameter was established that is known as energy return factor (ERF) and represents the total energy gained from the module per unit invested energy.

The production of PV modules may also have issues related to climate change and environmental toxicity and acidification. Some indicators are required for comparison of usefulness and environmental issues of different PV technologies. Similar to EPBT, carbon footprint is another parameter that indicates the effect of a particular PV technology on the climate change. Usually PV modules do not use much energy after manufacturing or installation, but a certain amount of energy is required in their production and installation. The energy required in manufacturing of PV modules comes from burning of fossil fuels that emit CO_2 and other greenhouse gases. Carbon footprint of a PV technology corresponds to the level of greenhouse gas emission responsible for over the life cycle of the PV module. It is defined in terms of carbon dioxide equivalent (CO_2-eq) per kWh of electricity generation (gCO_2e/kWh). The carbon footprint of a PV module depends strongly on the module type and manufacturing processes. For example, Figure 1.30 shows the total CO_2-eq load contribution of different steps in global production of 1 m^2 of c-Si module. Similar to Si PV technology, other PV technologies too have significant carbon footprints and the PV technologies are not really as green as we think. However, among all the PV

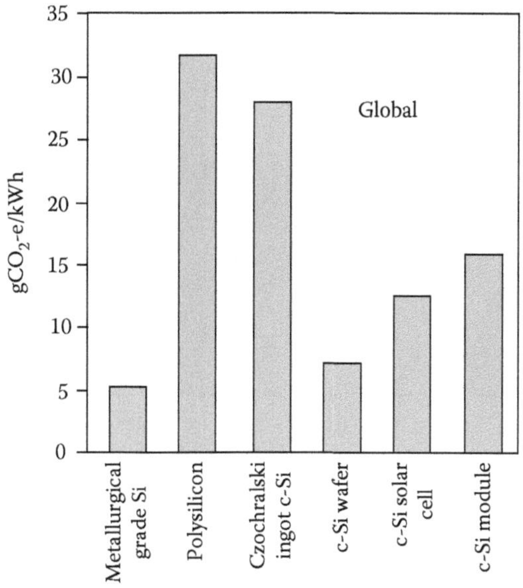

FIGURE 1.30
Global CO_2-eq contribution of different steps in production of 1 m^2 of c-Si PV module. (Modified from A. D. Ramos et al., 2010. Carbon footprint assessment of photovoltaic modules manufacture scenario, *20th European Symposium on Computer Aided Process Engineering – ESCAPE20.*)

technologies, organic and perovskite PV technologies are supposed to have minimum EPBT and carbon footprint values. Short payback time would lead to faster reduction in carbon content in the environment without energy deficits.

1.8 Current State of the Art

Since the early realization of solar cells there have been continuous efforts to improve their efficiency and stability. Cost effectiveness of energy generation by solar cells has been another decisive parameter for their success. This gave birth to new generations of solar cells and many photovoltaic industries were established for their manufacturing and development. Still, many new photovoltaic industries are growing rapidly to ensure maximum utilization of this technology. According to the 2014 "Photovoltaics Report" from Fraunhofer Institute for Solar Energy Systems, about 90% of the global photovoltaic market is shared by crystalline and multicrystalline Si solar cells only, where multicrystalline Si solar cells account for about 55% share and about 35% of the market is covered by c-Si solar cells. The remaining 10% is covered by thin-film a-Si, CIGS, and CdTe solar cell technologies [146]. Figure 1.31 shows the yearly development in the shares of different PV technologies in annual PV production of global market. The recorded cell efficiency in the laboratory is highest for c-Si, which is 25%, and is followed by multi-c Si (20.8%), CIGS (21.7%), CdTe (21.5%), and a-Si (13.6%). The best performing c-Si module in laboratory has shown maximum efficiency of about 23%, but commercial c-Si PV modules have efficiency of about 16% only. Where lab CdTe modules have shown efficiency of 17.5%,

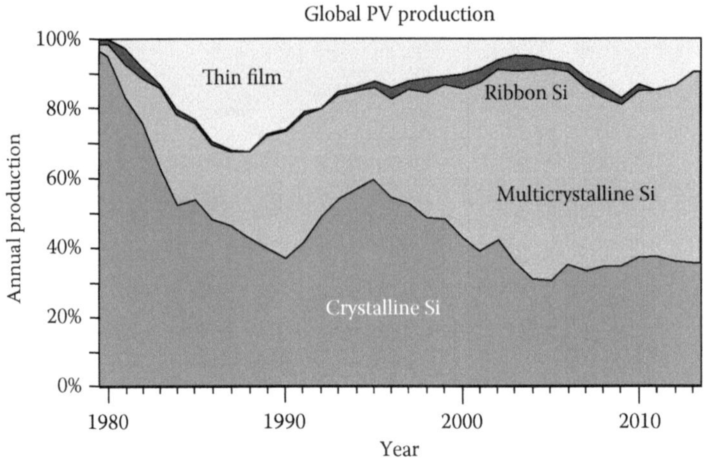

FIGURE 1.31
Yearly progress in production contribution of different PV technologies in global PV market. (Modified from Fraunhofer Institute for Solar Energy System ISE, *Photovoltaics Report*, October 2014.)

the commercial CdTe modules have shown around 13% efficiency. The concentrated multijunction solar cells have reached up to 46% efficiency in the laboratory, whereas the module efficiency is around 38.9%.

China/Taiwan hold around 69%, Europe contributes to around 6%, whereas Japan and United States/Canada each contribute to around 4% of global PV module production. In 2014, Germany installed about 1.5 million PV systems, accounting for 21% (38 GWp) of total worldwide PV capacity installed (183 GWp). It covered almost 7% of Germany's electricity demand. By production of around 34.9 TWh electrical energy from PV systems in 2014, Germany avoided about 25 Mio.t CO_2 emission. Thin-film technologies are less mature compared to first-generation solar cells, and therefore share the market modestly. They are struggling with the long-term stability, toxicity, material availability, and price issues. Unfortunately the overall price of electricity from solar cells is still very high compared to that from other sources. Because of this reason most research and development efforts are being devoted to potentially less cost-effective solar cell technologies based on organic and perovskite materials. These technologies are in the research and development phase and have yet to be developed to the level of commercialization.

References

1. U.S. Energy Information Administration (EIA), *Annual Energy Review 2011*, September 2012.
2. C. Nunez, Harbin smog crisis highlights china's coal problem, *National Geographic*, October 22, 2013.
3. R. S. Ohl, Light sensitive electric device, US Patent 240252, filed March 27, 1941.
4. D. M. Chapin, C. S. Fuller, and G. L. Pearson, 1954. A new silicon *p-n* junction photocell for converting solar radiation into electrical power, *J. Appl. Phys.* 25: 676.
5. National Renewable Energy Laboratory, www.nrel.gov.

6. M. A. Green, 2009. The path to 25% silicon solar cell efficiency: History of silicon cell evolution, *Prog. Photovolt.: Res. Appl.* 17: 183.
7. T. L. Chu, June 1976. Fabrication of polycrystalline solar cells on low cost substrates, US patent 3961997.
8. T. L. Chu, J. C. Lien, H. C. Mollenkopf, S. C. Chu, K. W. Heizer, F. W. Voltmer, and G. F. Wakefield, 1975. Polycrystalline silicon solar cells on foreign substrates, *Sol. Energy* 17: 229.
9. C. Becker, D. Amkreutz, T. Sontheimer, V. Preidel, D. Lockau, J. Haschke, L. Jogschies et al., 2013. Polycrystalline silicon thin film solar cells: Status and prospective, *Sol. Energy Mater. Sol. Cells* 119: 112–123.
10. S. Narayanan, S. R. Wenham, and M. A. Green, 1990. 17.8-percent efficiency polycrystalline silicon solar cells, *IEEE Trans. Electron. Devices* 37: 382.
11. O. Schultz, S. W. Glunz, and G. P. Willeke, 2004. Multicrystalline silicon solar cells exceeding 20% efficiency, *Prog. Photovolt.: Res. Appl.* 12: 553–558.
12. A. Eyer, A. Rauber, and A. Goetzberger, 1990. Silicon sheet materials for solar cells, *Optoelectron. Devices Technol.* 5: 239–257.
13. R. L. Wallace, J. I. Hanoka, A. Rohatgi, and G. Crotty, 1997. Thin silicon string ribbon, *Sol. Energy Mater. Sol. Cells* 48: 179–186.
14. D. S. Kim, V. Yelundur, K. Nakayashiki, B. Rounsaville, V. Meemongkolkiat, A. M. Gabor, and A. Rohatgi, 2006. Ribbon Si solar cells with efficiencies over 18% by hydrogenation of defects, *Sol. Energy Mater. Sol. Cells* 90: 1227–1240.
15. R. Chittick, J. Alexander, and H. Sterling, 1969. The preparation and properties of amorphous silicon, *J. Electrochem. Soc.* 116: 77–81.
16. H. Matsumura, 1986. Catalytic chemical vapor deposition (CTC–CVD) method producing high quality hydrogenated amorphous silicon, *Jpn. J. Appl. Phys.* 25: L949–L951.
17. A. Mahan, J. Carapella, B. Nelson, R. Crandall, and I. Balberg, 1991. Deposition of device quality, low *H* content amorphous silicon, *J. Appl. Phys.* 69: 6728–6730.
18. T. Moustakas, H. Maruska, and R. Friedman, 1985. Properties and photovoltaic applications of microcrystalline silicon films prepared by rf reactive sputtering, *J. Appl. Phys.* 58: 983.
19. D. Miller, H. Lutz, H. Weismann, E. Rock, A. Ghosh, S. Ramamoorthy, and M. Strongin, 1978. Some properties of evaporated amorphous silicon made with atomic hydrogen, *J. Appl. Phys.* 49: 6192.
20. R. Rocheleau, S. Hegedus, W. Buchanan, and S. Jackson, 1987. Novel photochemical vapor deposition reactor for amorphous silicon solar cell deposition, *Appl. Phys. Lett.* 51: 133–135.
21. M. Hanabusa and M. Suzuki, 1981. Reactive laser-evaporation for hydrogenated amorphous silicon, *Appl. Phys. Lett.* 39: 431–432.
22. W. Spear and P. LeComber, 1972. Investigation of the localised state distribution in amorphous Si films, *J. Non-Cryst. Solids* 8–10: 727–738.
23. C. R. Wronski, J. M. Pearce, J. Deng, V. Vlahos, and R. W. Collins, 2004. Intrinsic and light induced gap states in a-Si:H materials and solar cells-effect of microstructure, *Thin Solid Films* 451–452: 470–475.
24. R. Street, 1991. *Hydrogenated Amorphous Silicon*, Cambridge University Press, Cambridge.
25. D. Carlson and C. Wronski, 1976. Amorphous silicon solar cell, *Appl. Phys. Lett.* 28: 671.
26. Q. Wang, E. Iwaniczko, J. Yang, K. Lord, S. Guha, K. Wang, and D. Han, 2002. Amorphous silicon films and solar cells deposited by HWCVD at ultra-high deposition rates, *J. Non-Cryst. Solids* 299–302: 2–8.
27. S. Kim, J. W. Chung, H. Lee, J. Park, Y. Heo, and M. Lee, 2013. Remarkable progress in thin film silicon solar cells using high efficiency triple junction technology, *Sol. Energy Mater. Sol. Cells* 119: 26–35.
28. T. Tinoco, C. Rincon, M. Quintero, and G. S. Perez, 1991. Phase diagram and optical energy gaps for CuInyGa1-ySe2 alloys, *Physica Status Solidi (a)* 124: 427.
29. M. Kemell, M. Ritala, and M. Leskela, 2005. Thin film deposition methods for $CuInSe_2$ solar cells, *Crit. Rev. Sol. State Mater. Sci.* 30: 1–31.

30. S. Wagner, J. Shay, P. Migliorato, and H. M. Kasper, 1974. $CuInSe_2$/CdS heterojunction photovoltaic detectors, *Appl. Phys. Lett.* 25: 434.
31. H. Yamamoto, H. Ishiwara, and S. Furukawa, 1985. Enhancement of lateral solid phase epitaxial growth in evaporated amorphous Si films by phosphorus implantation, *Appl. Phys. Lett.* 46: 268.
32. R. A. Mickelsen and W. S. Chen, 1981. Development of a 9.4% efficient thin film $CuInSe_2$/CdS solar cell, *Proceedings of 15th IEEE Photovoltaics Specialist Conference*, 800–803.
33. U. Rau and H. W. Schock, 2001. Cu(In,Ga)Se2 solar cells, in M. D. Archer and R. Hill (Eds.), *Clean Electricity from Photovoltaics*, Imperial College Press, pp. 277–332.
34. R. H. Bube, 1955. Photoconductivity of the sulfide, selenide, and telluride of zinc or cadmium, *Proceedings of the IRE* 43: 1836–1850.
35. D. A. Cusano, 1963. CdTe solar cells and photovoltaic heterojunctions in II–VI compounds, *Sol. State Electron.* 6: 217.
36. D. Bonnet and H. Rabenhorst, 1972. New results on the development of a thin film p-CdTe/n-CdS heterojunction solar cell, *Proceedings of the 9th IEEE Photovoltaics Specialists Conference*, 129–131.
37. Y. S. Tyan, 1978. Polycrystalline thin film CdS/CdTe photovoltaic cell, Kodak, US Patent 4207119.
38. Y. S. Tyan and E. A. Perez-Albuerne, 1982. Integrated array of photovoltaic cells having minimized shorting losses, Kodak, US Patent 4315096.
39. T. L. Chu, S. S. Chu, C. Ferekides, C. Q. Wu, J. Britt, and C. Wang, 1991. High efficiency CdS/CdTe solar cells from solution-grown CdS films, *Proceedings of the Twenty-Second IEEE Photovoltaic Specialists Conference* 2: 952–956.
40. T. C. Anthony, A. L. Fahrenbruch, and R. H. Bube, 1984. Growth of CdTe films by close-spaced vapor transport, *J. Vacuum Sci. Technol. A*, 2: 1296–1302.
41. D. H. Rose, F. S. Hasoon, R. G. Dhere, D. S. Albin, R. M. Ribelin, X. Li, Y. Mahathongdy, T. A. Gessert, and P. Sheldon, 1999. Fabrication procedures and process sensitivities for CdS:CdTe solar cells, *Prog. Photovolt.: Res. Appl.* 7: 331–340.
42. H. Uda, H. Matsumoto, and Y. Komatsu, 1983. All screen printed CdS:CdTe solar cell, *Proceedings of the 14th Photovoltaic Specialists Conference*, New York, 1982, IEEE, 801–804.
43. P. V. Meyers, C. H. Liu, and M. Doty, 1989. Methods of making photovoltaic cell with chloride dip, US Patent 4,873,198.
44. T. X. Zhou, N. Reiter, R. C. Powell, R. Sasala, and P. V. Meyers, 1995. Vapor chloride treatment of polycrystalline CdTe:Cds films, *Proceedings of the First World Conference on Photovoltaic Energy Conversion*, 103–106.
45. Y. Mahathongdy, D. S. Albin, C. A. Wolden, and R. M. Baldwin, 1999. Vapour $CdCl_2$ optimization and screening experiments for an all dry chloride treatment of CdS:CdTe solar cells, *NCPV Photovoltaics Program Review, AIP Conference Proceedings* 462: 236–241.
46. H. Moutinho, M. Al-Jassim, D. Levi, P. Dippo, and L. Kazmerski, 1998. Effects of $CdCl_2$ treatment on the recrystallization and electro-optical properties of CdTe thin films, *J. Vacuum Sci. Technol. A*, 16: 1251–1257.
47. D. Bonnet, B. Henrichs, and H. Richter, 1992. Some phenomena in CdTe/CdS thin film solar cells made by close-spaced sublimation, *Int. J. Sol. Energy* 12: 133–136.
48. B. E. McCandless, H. Hichiri, G. Hanket, and R. W. Birkmire, 1997. Vapour phase treatment of CdTe:CdS thin films with $CdCl_2$:O_2, *Proceedings of the 14th Photovoltaics Specialists Conference*, 781–784.
49. K. Kuribayashi, H. Matsumoto, H. Uda, Y. Komatsu, A. Nakano, and S. Ikegami, 1983. Preparation of low resistance contact electrode in screen printed CdS/CdTe solar cell, *Jpn. J. Appl. Phys.* 22: 1828.
50. D. Levi, D. Albin, and D. King, 2000. Influence of surface composition on back-contact performance in CdTe:CdS PV devices, *Prog. Photovolt.: Res. Appl.* 8: 591.
51. T. A. Gessert, X. Li, T. J. Coutts, A. R. Mason, and R. J. Matson, 1994. Dependence of material properties of radio-frequency magnetron-sputtered, Cu-doped, ZnTe thin films on deposition conditions, *J. Vac. Sci. Technol.* 12: 1501.

52. M. Neumann-Spallart and C. Konigstein, 1995. Electrodeposition of zinc telluride, *Thin Solid Films* 265: 33–39.
53. J. Tang, D. Mao, L. Feng, W. Song, and J. U. Trefny, 1996. The properties and optimization of ZnTe:Cu back contacts on CdTe/CdS thin film solar cells, *Proceedings of the 25th IEEE PVSC*, 925–928.
54. J. Tang, L. Feng, D. Mao, W. Song, Y. Zhu, and J. U. Trefny, 1996. Study of ZnTe:Cu back contacts on CdTe/CdS thin film solar cells, *Proceedings of Mater. Res. Soc. Symp.* 426: 355.
55. L. Feng, D. Mao, J. Tang, R. T. Collins, and J. U. Trefny, 1996. The structural, optical, and electrical properties of vacuum evaporated Cu-doped ZnTe polycrystalline thin films, *J. Electron. Mater.* 25: 1422–1427.
56. R. W. Birkmire and P. V. Meyers, 1995. Processing issues for thin film CdTe cells and modules, *Proceedings of the First World Conference on Photovoltaic Energy Conversion*, 76–82.
57. First Solar, www.firstsolar.com.
58. A. R. Gobat, M. F. Lamorte, and G. W. McIver, 1962. Characteristics of high-conversion-efficiency gallium-arsenide solar cells, *IRE Trans. Military Electron.* 6: 20.
59. Z. I. Alferov, V. M. Andreev, M. B. Kagan, I. I. Prostasov, and V. G. Trofim, 1971. Solar-energy converters based on *p-n* $Al_xGa_{1-x}As$-GaAs heterojunctions, *Sov. Phys. Semicond.* 4: 2047.
60. M. A. Green, K. Emery, Y. Hishikawa, W. Warta, and E. D. Dunlop, 2012. Solar cell efficiency tables (version 39), *Prog. Photovolt.: Res. Appl.* 20: 12–20.
61. I. Weinberg, C. K. Swartz, and R. E. Hart Jr., 1985. *Proceedings of the 18th IEEE Photovoltaics Specialists Conference*, 5344–5349.
62. K. Ando, M. Yamaguchi, A. Yamamoto, and C. Umera, 1985. *Proceedings of the 18th IEEE Photovoltaics Specialists Conference*: 770–775.
63. W. Shockley and H. J. Queisser, 1961. Detailed balance limit of efficiency of p-n junction solar cells, *J. Appl. Phys.* 32: 510.
64. N. A. Gokcen and J. J. Loferski, 1979. Efficiency of tandem solar cell systems as a function of temperature and solar energy concentration ratio, *Sol. Energy Mater. Sol. Cells* 1: 271–286.
65. C. H. Henry, 1980. Limiting efficiencies of ideal single and multiple energy gap terrestrial solar cells, *J. Appl. Phys.* 51: 4494.
66. A. Brown and M. Green, 2002. Detailed balance limit for the series constrained two terminal tandem solar cell, *Physica E* 14: 96.
67. A. W. Bett, F. Dimroth, W. Guter, R. Hoheisen, E. Oliva, S. P. Philipps, J. Schöne et al., 2009. Highest efficiency multi-junction solar cell for terrestrial and space applications, *Proceedings of 24th European Photovoltaic Solar Energy Conference*, 1–6.
68. T. Agui, H. Juso, A. Yoshida, K. Nakaido, K. Sasaki, and T. Takamoto, 2010. High efficiency InGaP/GaAs/InGaAs triple junction solar cells. *Renewable Energy, Proceedings*, Yokohama; O-Pv-5-4.
69. J. M. Olson, S. R. Kurtz, and A. E. Kibbler, 1988. *Proceedings of the 20th IEEE Photovoltaics Specialists Conference*, 777–780.
70. R. R. King, D. C. Law, K. M. Edmondson, C. M. Fetzer, G. S. Kinsey, H. Yoon, R. A. Sherif, and N. H. Karam, 2007. 40% efficient metamorphic GaInP/GaInAs/Ge multijunction solar cells, *Appl. Phys. Lett.* 90: 183516.
71. J. F. Geisz, D. J. Friedman, J. S. Ward, A. Duda, W. J. Olavarria, T. E. Moriarty, J. T. Kiehl, M. J. Romero, A. G. Norman, and K. M. Jones, 2008. 40.8% efficient inverted triple-junction solar cell with two independently metamorphic junctions, *Appl. Phys. Lett.* 93: 123505.
72. Solar Junction, www.sj-solar.com.
73. R. R. King, D. C. Law, K. M. Edmondson, C. M. Fetzer, G. S. Kinsey, H. Yoon, R. A. Sherif et al., 2007. Multijunction solar cells with over 40% efficiency and future directions in concentrator PV, *22nd European Photovoltaic Solar Energy Conference*, 11–15.
74. R. R. King, D. Bhusari, A. Boca, D. Larrabee, X.-Q. Liu, W. Hong, C. M. Fetzer, D. C. Law, and N. H. Karam, 2011. Band gap-voltage offset and energy production in next generation multijunction solar cells, *Prog. Photovolt.: Res. Appl.* 19: 797–812.
75. K. W. J. Barnham and G. Duggan, 1990. A new approach to high-efficiency multi-band-gap solar cells, *J. Appl. Phys.* 67: 3490.

76. A. H. Ip, S. M. Thon, S. Hoogland, O. Voznyy, D. Zhitomirsky, R. Debnath, L. Levina et al., 2012. Hybrid passivated colloidal quantum dot solids, *Nat. Nanotechnol.* 7: 577–582.
77. M. Yuan, D. Zhitomirsky, V. Adinolfi, O. Voznyy, K. W. Kemp, Z. Ning, X. Lan et al., 2013. Doping control via molecularly engineered surface ligand coordination. *Adv. Mater.* 25: 5586–5592.
78. M. S. Harari, N. Y. Gross, D. Steiner, A. Aharoni, U. Banin, O. Millo, and N. Tessler, 2008. Tuning energetic levels in nanocrystal quantum dots through surface manipulations, *Nano Lett.* 8: 678–684.
79. J. Jasieniak, M. Califano, and S. E. Watkins, 2011. Size-dependent valence and conduction band-edge energies of semiconductor nanocrystals, *ACS Nano* 5: 5888–5902.
80. C. H. M. Chuang, P. R. Brown, V. Bulović, and M. G. Bawendi, 2014. Improved performance and stability in quantum dot solar cells through band alignment engineering, *Nat. Mater.* 13: 796–801.
81. Colin Jeffrey, 2014. New record efficiency for quantum-dot photovoltaics, *Gizmag*, May 27, http://www.gizmag.com/quantum-dot-solar-cell-record-efficiency/32277.
82. R. Schaller and V. Klimov, 2004. High efficiency carrier multiplication in PbSe nanocrystals: Implications for solar energy conversion, *Phys. Rev. Lett.* 92: 186601.
83. R. J. Ellingson, M. C. Beard, J. C. Johnson, P. Yu, O. I. Micic, A. J. Nozik, A. Shabaev, and A. L. Efros, 2005. Highly efficient multiple exciton generation in colloidal PbSe and PbS quantum dots, *Nano Lett.* 5: 865.
84. H. Gerischer, M. Michel-Beyerle, E. Rebentrost, and H. Tributsch, 1968. Sensitization of charge-injection into semiconductors with large band gap, *Electrochim. Acta* 13: 1509–1515.
85. H. Tributsch, 1972. Reaction of excited chlorophyll molecules at electrodes and in photosynthesis. *Photochem. Photobiol.* 16: 261–269.
86. B. C. O'Regan and M. Gratzel, 1991. A low-cost, high-efficiency solar cell based on dye-sensitized colloidal TiO_2 films, *Nature* 353: 737–740.
87. A. Yella, H. W. Lee, H. N. Tsao, C. Yi, A. K. Chandiran, M. K. Nazeeruddin, E. W. G. Diau, C. Y. Yeh, S. M. Zakeeruddin, and M. Gratzel, 2011. Porphyrin-sensitized solar cells with cobalt (II/III)-based redox electrolyte exceed 12 percent efficiency, *Science* 334: 629–634.
88. S. Mathew, A. Yella, P. Gao, R. H. Baker, B. F. E. Curchod, N. A. Astani, I. Tavernelli, U. Rothlisberger, M. K. Nazeerudin, and M. Gratzel, 2014. Dye-sensitized solar cells with 13% efficiency achieved through the molecular engineering of porphyrin sensitizers, *Nat. Chem.* 6: 242.
89. A. K. Gosh and T. Feng, 1978. Merocyanine organic solar cells, *J. Appl. Phys.* 49: 5982.
90. C. W. Tang, 1986. Two-layer organic photovoltaic cell, *Appl. Phys. Lett.* 48: 183.
91. G. Yu, Y. Gao, J. C. Hummelen, F. Wudl, and A. J. Heeger, 1995. Polymer photovoltaic cells: Enhanced efficiencies via a network of internal donor–acceptor heterojunctions, *Science* 270: 1789–1791.
92. C. C. Chen, W. H. Chang, K. Yoshimura, K. Ohya, J. You, J. Gao, Z. Hong, and Y. Yang, 2014. An efficient triple junction polymer solar cell having a PCE exceeding 11%, *Adv. Mater.* 26: 5670.
93. Y. Liu, J. Zhao, Z. Li, C. Mu, W. Ma, H. Hu, K. Jiang, H. Lin, H. Ade, and H. Yan, 2014. Aggregation and morphology control enables multiple cases of high-efficiency polymer solar cells, *Nat. Commun.* 5:5293: 1–8.
94. G. Dennler and C. J. Brabec, 2008. *Organic Photovoltaics: Materials, Device Physics, and Manufacturing Technologies*, Wiley-VCH: Weinheim.
95. W. U. Hyunh, J. J. Dittmer, and A. P. Alivisatos, 2002. Hybrid nanorod-polymer solar cells, *Science* 295: 2425.
96. W. J. E. Beek, M. M. Wienk, and R. A. J. Janseen, 2004. Efficient hybrid solar cells from zinc oxide nanoparticles and a conjugated polymer, *Adv. Mater.* 16: 1009–1013.
97. P. A. van Hal, M. M. Wienk, J. M. Kroon, W. J. H. Verhees, L. H. Slooff, W. J. H. van Gennip, P. Jonkheijm, and R. A. J. Janssen, 2003. Photoinduced electron transfer and photovoltaic response of a MDMO-PPV:TiO_2 bulk-heterojunction, *Adv. Mater.* 15: 118–121.

98. W. J. E Beek, M. M. Wienk, M. Kemerink, X. Yang, and R. A. J. Janssen, 2005. Hybrid zinc oxide conjugated polymer bulk heterojunction solar cells, *J. Phys. Chem. B* 109: 9505–9516.
99. Z. Liang, K. L. Dzienis, J. Xu, and Q. Wang, 2006. Covalent layer by layer assembly of conjugated polymers and CdSe nanoparticles: Multilayer structure and photovoltaic properties. *Adv. Funct. Mater.* 16: 542–548.
100. B. Sun, E. Marx, and N. C. Greenham, 2003. Photovoltaic devices using blends of branched CdSe nanoparticles and conjugated polymers, *Nano Lett.* 3: 961–963.
101. H. L. Hu, S. C. Kung, L. M. Yang, M. E. Nicho, and R. M. Penner, 2009. Photovoltaic devices based on electrochemical-chemical deposited CdS and poly3-octylthiophene thin films, *Sol. Energy Mater. Sol. Cells* 93: 51–54.
102. K. Inpor, V. Meeyoo, and C. Thanachayanont, 2011. Enhancement of photovoltaic performance using hybrid CdS nano-rods and MEH-PPV active layer in ITO/TiO_2/MEH-PPV:CdS/Au devices, *Curr. Appl. Phys.* 11: S171–S174.
103. H. C. Liao, N. Chantarat, S. Y. Chen, and C. H. Peng, 2011. Annealing effect on photovoltaic performance of hybrid P3HT/in-situ grown CdS nanocrystal solar cells, *J. Electrochem. Soc.* 158: E67–E72.
104. M. J. Greaney, S. Das, D. H. Webber, S. E. Bradforth, and R. L. Brutchey, 2012. Improving open circuit potential in hybrid P3HT:CdSe bulk heterojunction solar cells via colloidal tert-butyl thiol ligand exchange, *ACS Nano* 6: 4222–4230.
105. S. Dayal, N. Kopidakis, D. C. Olson, D. S. Ginley, and G. Rumbles, 2010. Photovoltaic devices with a low band gap polymer and CdSe nanostructures exceeding 3% efficiency, *Nano Lett.* 10: 239–242.
106. Y. Y. Feng, D. Q. Yun, X. Q. Zhang, and W. Feng, 2010. Solution-processed bulk heterojunction photovoltaic devices based on poly(2-methoxy,5-octoxy)-1,4-phenylenevinylene-multiwalled carbon nanotubes/PbSe quantum dots bilayer, *Appl. Phys. Lett.* 96, doi: 10.1063/1.3337100.
107. L. Baeten, B. Conings, H. G. Boyen, J. D'Haen, A. Hardy, M. D'Olieslaeger, J. V. Manca, and M. K. Van Bael, 2011. Towards efficient hybrid solar cells based on fully polymer infiltrated ZnO nanorod arrays, *Adv. Mater.* 23: 2802–2805.
108. J. Boucle, S. Chyla, M. S. P. Shaffer, J. R. Durrant, D. D. C. Bradley, and J. Nelson, 2008. Hybrid bulk heterojunction solar cells based on blends of TiO_2 nanorods and P3HT, *Comptes Rendus Phys.* 9: 110–118.
109. K. F. Jeltsch, M. Schadel, J. B. Bonekamp, P. Niyamakom, F. Rauscher, H. W. A. Lademann, I. Dumsch, S. Allard, U. Scherf, and K. Meerholz, 2012. Efficiency enhanced hybrid solar cells using a blend of quantum dots and nanorods, *Adv. Funct. Mater.* 22: 397–404.
110. M. Shukla and N. Brahme, 2011. Improved efficiency of MEHPPV:PCBM solar cells by the use of ZnS nano-particles, *Poly. Bull.* 67: 709–718.
111. J. F. Lin, G. Y. Tu, C. C. Ho, C. Y. Chang, W. C. Yen, S. H. Hsu, Y. F. Chen, and W. F. Su, 2013. Molecular structure effect of pyridine-based surface ligand on the performance of P3HT:TiO_2 hybrid solar cell, *ACS Appl. Mater. Interface* 5: 1009–1016.
112. A. Stavrinadis, R. Beal, J. M. Smith, H. E. Assender, and A. A. R. Watt, 2008. Direct formation of PbS nanorods in a conjugated polymer, *Adv. Mater.* 20: 3105–3109.
113. N. I. Hammer, T. Emrick, and M. D. Barnes, 2007. Quantum dots coordinated with conjugated organic ligands: New nanomaterials with novel photophysics, *Nanoscale Res. Lett.* 2: 282–290.
114. J. Seo, W. J. Kim, S. J. Kim, K. S. Lee, A. N. Cartwright, and P. N. Prasad, 2009. Polymer nanocomposite photovoltaics utilizing CdSe nanocrystals capped with a thermally cleavable solubilizing ligand, *Appl. Phys. Lett.* 94, doi: 10.1063/1.3110969.
115. C. Fang, X. Y. Qi, Q. L. Fan, L. H. Wang, and W. Huang, 2007. A facile route to semiconductor nanocrystal-semiconducting polymer complex using amine-functionalized rod-coil triblock copolymer as multidentate ligand, *Nanotechnology* 18, doi: 10.1088/0957-4484/18/3/035704.
116. L. Zhao, X. Pang, R. Adhikary, J. W. Petrich, M. Jeffries-El, and Z. Lin, 2011. Organic–inorganic nanocomposites by placing conjugated polymers in intimate contact with quantum rods. *Adv. Mater.* 23: 2844–2849.

117. L. X. Reynolds, T. Lutz, S. Dowland, A. MacLachlan, S. King, and S. A. Haque, 2012. Charge photogeneration in hybrid solar cells: A comparison between quantum dots and *in situ* grown CdS, *Nanoscale* 4: 1561–1564.
118. Z. Liu, Y. Sun, J. Yuan, H. Wei, X. Huang, L. Han, W. Wang, H. Wang, and W. Ma, 2013. High-efficiency hybrid solar cells based on polymer/PbS_xSe_{1-x} nanocrystals benefiting from vertical phase segregation, *Adv. Mater.* 25: 5772–5778.
119. B. R. Saunders and M. L. Turner, 2008. Nanoparticle polymer photovoltaic cells, *Adv. Colloid Interface Sci.* 138: 1–23.
120. Y. F. Zhou, M. Eck, and M. Kruger, 2010. Bulk-heterojunction hybrid solar cells based on colloidal nanocrystals and conjugated polymers, *Energy Environ. Sci.* 3: 1851–1864.
121. M. Wright and A. Uddin, 2012. Organic–inorganic hybrid solar cells: A comparative review. *Sol. Energy Mater. Sol. Cells* 107: 87–111.
122. R. Liu, 2014. Hybrid organic/inorganic nanocomposites for photovoltaic cells, *Materials* 7: 2747–2771.
123. G. Hodes, 2013. Perovskite-based solar cells, *Science* 342: 317.
124. S. E. Shaheen, C. J. Brabec, N. S. Sariciftci, F. Padinger, T. Fromhert, and J. C. Hummelen, 2001. 2.5% efficient organic plastic solar cells, *Appl. Phys. Lett.* 78: 841.
125. J. You, L. Dou, K. Yoshimura, T. Kato, K. Ohya, T. Moriarty, K. Emery et al., 2013. A polymer tandem solar cell with 10.6% power conversion efficiency, *Nat. Comm.* 4: 1446.
126. A. Kojima, K. Teshima, Y. Shirai, and T. Miyasaka, 2009. Organometal halide perovskites as visible-light sensitizers for photovoltaic cells, *J. Am. Chem. Soc.* 131: 6050.
127. National Renewable Energy Laboratory, Best research-cell efficiencies, Graphic, http://www.nrel.gov/ncpv/images/efficiency_chart.jpg.
128. H. S. Kim, C. R. Lee, J. H. Lm, K. B. Lee, T. Moehl, A. Marchioro, S. J. Moon et al., 2012. Lead iodide perovskite sensitized all-solid-state submicron thin film mesoscopic solar cell with efficiency exceeding 9%, *Sci. Rep.* 2: 591.
129. J. Burschka, N. Pellet, S. J. Moon, R. H. Baker, P. Gao, M. K. Nazeeruddin, and M. Gratzel, 2013. Sequential deposition as a route to high-performance perovskite-sensitized solar cells, *Nature* 499: 316–319.
130. M. M. Lee, J. Teuscher, T. Miyasaka, T. N. Murakami, and H. J. Snaith, 2012. Efficient hybrid solar cells based on meso-superstructured organometal halide perovskites, *Science* 338: 643–647.
131. G. E. Eperon, V. M. Burlakov, P. Docampo, A. Goriely, and H. J. Snaith, 2014. Morphological control for high performance, solution-processed planar heterojunction perovskite solar cells, *Adv. Funct. Mater.* 24: 151–157.
132. M. Liu, M. B. Johnston, and H. J. Snaith, 2013. Efficient planar heterojunction perovskite solar cells by vapour deposition, *Nature* 501: 395–398.
133. J. A. Christians, R. C. M. Fung, and P. V. Kamat, 2014. An inorganic hole conductor for organo-lead halide perovskite solar cells: Improved hole conductivity with copper iodide, *J. Am. Chem. Soc.* 136: 758.
134. H. Zhou, Q. Chen, G. Li, S. Luo, T. Song, H. S. Duan, Z. Hong, J. You, Y. Liu, and Y. Yang, 2014. Interface engineering of highly efficient perovskite solar cells, *Science* 345: 542–546.
135. S. D. Stranks, G. E. Eperon, G. Grancini, C. Menelaou, M. J. P. Alcocer, T. Leijtens, L. M. Herz, A. Petrozza, and H. J. Snaith, 2013. Electron–hole diffusion lengths exceeding 1 micrometer in an organometal trihalide perovskite absorber, *Science* 342: 341–344.
136. J. H. Noh, S. H. Im, J. H. Heo, T. N. Mandal, and S. I. Seok, 2013. Chemical management for colorful, efficient, and stable inorganic–organic hybrid nanostructured solar cells, *Nano Lett.* 13: 1764.
137. G. E. Eperon, S. D. Stranks, C. Menelaou, M. B. Johnston, L. M. Herz, and H. J. Snaith, 2014. Formamidinium lead trihalide: A broadly tunable perovskite for efficient planar heterojunction solar cells, *Energy Environ. Sci.* 7: 982.
138. G. G. Kopp and J. Lean, 2011. A new, lower value of total solar irradiance: Evidence and climate significance. *Geophys. Res. Lett.* 38: L01706.

139. R. C. Willson and A. V. Mordvinov, 2003. Secular total solar irradiance trend during solar cycles 21–23, *Geophys. Res. Lett.* 30: 1199.
140. Halt & Hass, International test standards, http://www.halthass.co.nz/library-and-useful-resources/international-test-standards.
141. S. M. Sze, 1981. *Physics of Semiconductor Devices*, New York: John Wiley & Sons.
142. W. Shockley and H. J. Queiser, 1961. Detailed balance limit of efficiency of a *p-n* junction solar cell, *J. Appl. Phys.* 32: 510.
143. V. Fthenakis, 2012. How long does it take for photovoltaics to produce the energy used? *PE Magazine*, 16–17.
144. M. J. de Wild-Scholten, 2013. Energy payback time and carbon footprint of commercial photovoltaic systems, *Sol. Energy Mater. Sol. Cells* 119: 296–305.
145. A. D. Ramos, M. Held, R. Aldaco, M. Fischer, and A. Irabian, 2010. Carbon footprint assessment of photovoltaic modules manufacture scenario, *20th European Symposium on Computer Aided Process Engineering – ESCAPE20.*
146. Fraunhofer Institute for Solar Energy System ISE, *Photovoltaics Report*, October 2014.

2

Degradation in Different Solar Cell Technologies

2.1 Introduction

It has been observed that every solar cell or solar module gradually loses its performance over time. In some cases the deterioration in performance is very slow, and in other cases it is very fast and happens because of degradation in solar cells and modules. It is a very serious issue and a big obstacle for long-term energy production from these systems. Degradation alone is keeping some of the efficient and really cost-effective solar cell technologies away from commercialization. Degradation issues must be overcome before successful introduction of these solar cell technologies into the market. There are lots of factors responsible for degradation of a solar cell and module. The degradation happens irrespective of its use but it is faster during its operation. In some cases the electrode materials decompose and diffuse into the active materials, whereas in other cases the active materials themselves decompose and lose their optical and electrical properties. Sunlight, temperature, humidity, and oxygen are some of the important parameters that accelerate degradation in solar cells. For reliable and sustainable solar cell technology, the prevention of degradation is a must and for this purpose the detection and understanding of degradation mechanisms in a solar cell is of great importance. This chapter covers the degradation mechanisms in different solar cell technologies and modules.

2.2 How Different Solar Cells Degrade

Degradation in solar cells is a complicated issue as it depends not only on different kinds of stress applied to the cell like light, temperature, atmospheric conditions, bias, humidity, oxygen, and their combinations but it also depends on the structure and fabrication procedures. Different kinds of solar cells degrade differently. Even the identical solar cells prepared in different ways have different degradation mechanisms. For maximum stability and minimum uncertainty in lifetimes of different solar cells and modules, better understanding of their degradation mechanisms is of utmost importance. The reliability of a solar module is inextricably related to cell degradation therefore understanding of cell-level degradation is primary for understanding of the module degradation.

2.2.1 Degradation in Silicon (Si) Wafer Solar Cells

Though crystalline silicon (c-Si) is a very stable material, prolonged illumination of solar radiation on the solar cells leads to some losses in the photovoltaic performance. The degradation effect is directly reflected in the illuminated current density–voltage (*J–V*) characteristics of the solar cells. The degradation caused by solar radiation is known as light-induced degradation (LID) [1–3]. A prolonged exposure to light causes formation of the boron–oxygen (B–O) defect complex, which reduces the minority carrier lifetime and increases the charge recombination in bulk Si [4–6]. Formation of this complex takes place at room temperature and it is driven by photogenerated or externally injected charge carriers. In c-Si the observable LID happens in two distinct fast and slow steps. Fast process happens within a few seconds, whereas the slow process happens on the scale of several hours [7]. Initially the two processes lead to two distinct states, but after a long time both processes transform into the state of slow process, therefore the initial fast process is usually ignored. A model was proposed by Palmer et al. that described the kinetics of slow process, which involves diffusion of positively charged O atoms toward negatively charged B impurity atoms [8]. When the O atoms reach the B atoms they automatically form the B–O complex. It is a thermally activated reaction and more complexes are generated at higher temperatures. However, this complex becomes unstable at very high temperatures and can be decomposed by thermal annealing. Decomposition of the B–O complex would cause a full recovery in minority carrier lifetime [5].

In some cases, the LID is fully recoverable in c-Si solar cells via thermal annealing when the solar cells are annealed at around 200°C for about 30 min [1,7,9]. However, there are also some light-induced effects that are not recoverable. The degradation happening at different interfaces particularly at the SiN/Si interface is not recoverable. This interface has lots of defects states that are affected by the light exposure [10]. A study under a high-energy UV source shows that the main reason behind degradation in the cell performance is the degradation taking place in the SiN coating and SiN/Si interface [11]. Degradation in the SiN layer happens due to generation of positive charge by high-energy UV light. Corrosion and delamination of electrodes, migration of electrode materials through the *p-n* junction and degradation of antireflection coating (ARC), and discoloration in ethyl vinyl acetate (EVA) are some other reasons behind deterioration in the performance of c-Si solar cells [12,13]. Some of the examples of degradation in ARC, corrosion of busbars, delamination of electrodes, and EVA discoloration in c-Si solar cells are shown in Figure 2.1. Moisture ingress is found to be the main reason behind electrode/contact degradation.

2.2.2 Degradation in Hydrogenated Amorphous Silicon (a-Si:H) Solar Cells

Hydrogenated amorphous silicon (a-Si:H) thin-film solar cells are very cost effective compared to conventional Si wafer solar cells. But they possess comparatively lower power conversion efficiency (PCE) and faster degradation that make these solar cells less attractive for large-scale energy production. Because of the amorphous nature of light-absorbing materials, the solar cells exhibit the Staebler–Wronski effect, which reduces the cells' PCE by 10%–15% as soon as we start using them [14,15]. a-Si possesses lots of defects and dangling bonds, and to saturate the dangling bonds and fix the problem of defects in a-Si, H_2 is used along with other precursor gases during deposition and a-Si:H is obtained. Still there are lots of defects remaining in a-Si:H. Also exposure to high-energy UV and blue light breaks the Si–H bonds, which results back in the generation of dangling bonds that deteriorate the cell performance. Electron paramagnetic resonance

FIGURE 2.1
(a) Degradation in the ARC, (b) corrosion of busbars, (c) delamination of top electrode, and (d) EVA discoloration in c-Si solar cells. (Modified from E. Kaplani, 2012. *J. Eng. Sci. Technol. Rev.* 5: 18–23.)

(EPR) studies show that the defects in a-Si:H could be native or light induced [16–18]. Both kinds of defects are due to threefold-coordinated Si atoms possessing dangling bonds. Dangling bonds are actually unsaturated bonds that are occupied by a single electron only and thus are paramagnetic in nature [16–18] and play a crucial role in degradation of a-Si solar cells. Prolonged light exposure causes increment in deep-level defects that induce additional electronic states in the band gap of a-Si. The increment in defects is caused by the recombination of photoinduced charge carriers and it reduces the lifetime of the charge carriers that leads to enhanced recombination and reduced PCE of the solar cells. The microstructure of native defects is not very clear and it makes their elimination difficult, however, they could be minimized to some extent by improving the fabrication techniques. The light-induced defects are metastable and could be removed by thermal annealing, but the native defects generated during film processing are stable with respect to annealing [16,19].

A number of models have been proposed to explain the Staebler–Wronski effect, but still the microscopic process is unclear [20,21]. Most models are based on the understanding of light-induced breaking of Si–Si bonds or formation of metastable H complexes. a-Si:H is quite inhomogeneous and possesses a lot of microvoids and vacancies [22–24]. These microvoids control the solar cell performance and play an important role in light-induced degradation [25]. According to the model proposed by D. E. Carlson, the photogenerated holes in a-Si:H are trapped near the microvoids that allow the H atoms to migrate on the internal surface of microvoids [26]. This migration of H atoms causes the breaking of weak dimeric Si–Si bonds and formation of dangling bonds. The model proposed by

D. E. Carlson for light-induced formation of dangling bonds and defects in a-Si:H is shown schematically in Figure 2.2. Initially the internal surface of microvoids is saturated with H atoms and almost no dangling bond is observed. The light-induced holes (h^+) cause H migration and break the Si–Si atomic bonds to generate defects in terms of dangling bonds. Fehr et al. performed an EPR study to understand the defects in a-Si:H solar cells in more detail [27]. With the help of electron spin echo (ESE) relaxation measurements in the light soaked and annealed states of a-Si:H, they found that there were two types of defects with different ESE relaxation. One type of defect was concentrated in clusters at internal surfaces of microvoids giving fast ESE decay, whereas the other type of defects were isolated and uniformly distributed in the bulk giving slow ESE decay that is dominated by ^{1}H nuclear spin dynamics. The observed difference in ESE decays has been proposed to indirect flip-flops of adjacent electron spin pairs. The electron spins could be divided into excited and nonexcited spins (see Figure 2.2). The nonexcited spins occur due to random fluctuations by spin lattice relaxation or by mutual spin flip-flops. The nonexcited spins cause a loss in phase coherence of excited spins. The defect density in the light-soaked state was more than that in the annealed state due to light-induced increment in both kinds of defects. Around 21% of light-induced metastable defects are of the first type, whereas around 79% are of second type.

To reduce the light-induced degradation in a-Si:H solar cells, the incident light should have no high-energy photons, and this could be achieved by using UV filters, but this would reduce the number of photons absorbed by the cell and would result in lower efficiency. Alternately, the light-induced degradation in a-SI:H could be reduced by reducing the a-Si:H film thickness so that the photogenerated charge carriers need not to move long distances before the collection at respective electrodes. But still the reduced thickness suffers from inefficient light absorption, and optical confinement techniques are employed to improve light absorption. The best solution is to fabricate the tandem solar cell where both cells are made of a-Si:H. The film thickness of the top cell is kept thin so that it works like a high-energy filter and absorbs all the high-energy photons, whereas the low-energy photons are absorbed by the bottom cell. The fast drift of charge carriers before recombination in the top cell reduces degradation. This is the commonly used method in commercial a-Si:H solar cells to avoid degradation. However, the degradation problem could also be reduced by reducing the native defects in a-Si film. The number of native defects in a-Si can be reduced by improving the film quality. A high-quality film would have

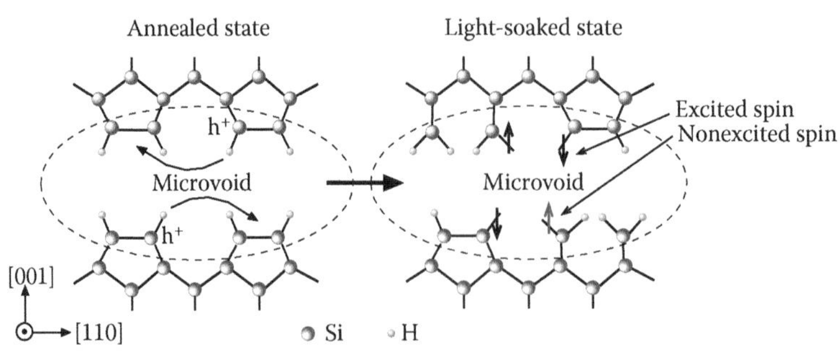

FIGURE 2.2
Formation of light-induced defects in a-Si:H, where (a) represents the initial state (corresponding to annealed state) and (b) represents the light-soaked state. This structure corresponds to ($\bar{1}$11) plane and the orientations corresponding to c-Si are shown in left corner. (Adapted from D. E. Carlson, 1986. *Appl. Phys. A* 41: 305; M. Fehr et al., 2014. *Phys. Rev. Lett.* 112: 66403.)

minimum voids and dangling bonds, and Si would require little H_2 to bound at the surface. Wang et al. proposed the improved stability of a-Si solar cells using chemical annealing of the intrinsic layer with helium [28]. For this purpose, the devices were prepared in a layer-by-layer fashion and the helium ions were bombarded on each layer. Annealing with helium ions resulted in reduced surface defects and less voids in the a-Si film. Chemically annealed samples exhibited less degradation compared to unannealed samples. Stradins et al. from NREL, United States, introduced a technique to reduce the dangling bonds in a-Si [29]. They annealed the a-Si:H films at very high temperatures up to 400°C for 30 min and found that the number of dangling bonds is reduced during light-induced degradation. Initially the annealed film had larger density of dangling bonds compared to the untreated sample, but after some time of light exposure the treated film exhibited less degradation and less dangling bonds. For very high stability of a-Si:H solar cells, new methods are being developed for improved film quality with reduced dangling bonds.

2.2.3 Degradation in Copper Indium Gallium Diselenide (CIGS) Solar Cells

The degradation in copper indium gallium diselenide (CIGS) solar cells is not completely understood and different types of structures degrade differently. However, the moisture and elevated temperatures have been found to have a negative effect on the performance of CIGS solar cells. A combination of moisture with high temperature is commonly known as damp heat (DH). In some cases, the light exposure has shown degradation in CIGS solar cells, whereas in some cases light exposure has shown no effect or improvement in performance [30–34]. For commercial viability of a solar cell, a DH test is a must. DH testing is performed at 85°C with 85% relative humidity (RH). According to the IEC standard, the cell or module should not degrade to more than 10% of its initial PCE in 1000 h of DH testing [30]. DH testing is used as an accelerating lifetime test for terrestrial solar modules [35,36]. However, the solar cells exposed to high temperature in dry atmosphere degrade slowly. Along with degradation in light-absorbing layer, both the front and the back contacts can encounter degradation. There are many reports in the literature where degradation and stability of CIGS solar cells have been studied [37–40]. During the D.H. test of CIGS solar cells, Powalla et al. observed corrosion in molybdenum contact and degradation in the ZnO layer, which were the dominating factors causing degradation in these cells [37–38,41,42]. ZnO is moisture sensitive and degrades in the humid environment. Degradation in contact materials increases the cell's series resistance that results in deterioration in cell performance. Lee et al. performed degradation studies on 475 × 375 mm encapsulated CIGS solar modules and observed degradation in the ZnO:Al layer and discoloration from the outside edge to the center of module during D.H. testing [43]. Degradation in ZnO:Al resulted in a considerable increment in its resistivity, which led to degradation in the fill factor (*FF*) and PCE of the modules. A detailed characterization revealed that the discoloration of modules and degradation in ZnO:Al happened due to chemical reactions caused by oxygen adsorption and moisture diffusion in the modules. XPS (x-ray photoelectron spectroscopy) and FTIR (Fourier transform infrared spectroscopy) studies suggested the formation of $Zn(OH)_2$ and carbolic acids by a combination of EVA, ZnO:Al, and water molecules.

In a later study Lee et al. performed a detailed failure analysis of encapsulated and nonencapsulated CIGS solar cells and modules [30]. The samples were prepared on soda-lime glass substrate in Mo/CIGS/CdS/ZnO:Al structure. The modules were encapsulated by glass, EVA, and a back sheet. For failure analysis, the modules were subjected to DH testing (85°C, 85% RH) and high temperature (HT) (85°C) conditions for 1000 h. Some of

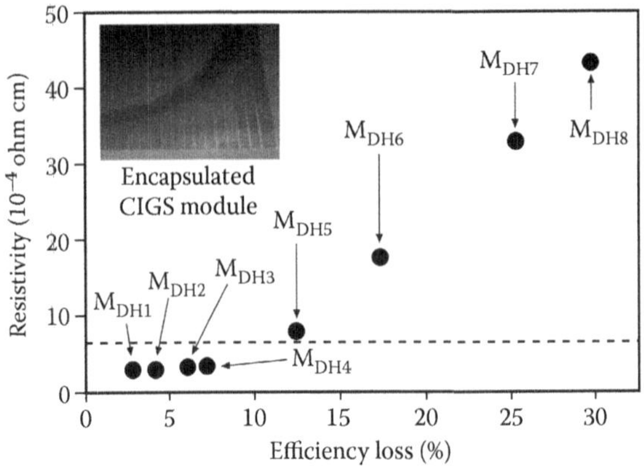

FIGURE 2.3
Variation in the sheet resistance of ZnO:Al layers as a function of efficiency loss after DH testing of the encapsulated CIGS solar modules. The inset shows discoloration in the module M_{DH8} after DH testing. (Modified from D. W. Lee et al., 2015. *Prog. Photovolt.: Res. Appl.* 23: 829–837.)

the modules exposed to DH conditions were discolored, however, none of the modules exposed to HT conditions exhibited discoloration. The resistivity of ZnO:Al layers of eight encapsulated CIGS modules (M_{DH1}–M_{DH8}) was measured before and after the DH testing, and the results are shown in Figure 2.3 as a function of PCE loss. These measurements were performed after DH testing, where some of the modules were discolored (M_{DH5}–M_{DH8}) and some were not (M_{DH1}–M_{DH4}). The discoloration of modules was associated with the increment in resistivity of the ZnO:Al layer, which was attributed to moisture-induced degradation in ZnO:Al. Studies were also performed on resistivity variation in CIGS and Mo (molybdenum) films. The resistivity of these films was observed to increase after both the DH and HT tests and the trend in increment was the same, which suggested that the degradation in CIGS and Mo films was associated with temperature but not with humidity. The resistivity of the ZnO:Al layer was also measured at different positions in the most degraded modules after DH and HT tests. In discolored modules the resistivity changed significantly with the position, however, in nondiscolored modules there was no significant change in resistivity with position.

In the nonencapsulated solar cells, the ZnO:Al layer was not discolored after DH testing and the resistivity did not change with position. These results suggested that the discoloration of modules happened due to encapsulation failure. Studies on the variation in resistivity of ZnO:Al, CIGS, and Mo layers conferred that the moisture-induced degradation is associated with discoloration and increased resistivity in the ZnO:Al layer, whereas temperature-induced degradation is associated with increment in the resistivity of CIGS and Mo layers. HT caused adsorption of O_2 in the CIGS and Mo layers that led to increased resistivity. The increment in resistivity of individual layers caused increment in overall series resistance of the cell that led to deterioration in its *FF* and PCE.

2.2.4 Degradation in Cadmium Telluride (CdTe) Solar Cells

The lifetime testing of cadmium telluride (CdTe)/cadmium sulfide (CdS) solar cells shows deterioration in their performance with significant losses in V_{oc} and *FF* due to formation of blocking interfaces at the electrodes, increased charge carrier recombination at the junction, and increment in overall bulk resistivity of the cell [44–48]. Back contact has been observed to play an important role in determination of the cell's performance. Most people used Cu-added graphite paste for this purpose, but it had serious stability issues. Though Cu addition provides a low resistive contact via *p*+ type doping of CdTe interface and make it pseudo-ohmic for efficient current collection [49], initial investigations showed that diffusion of Cu into CdTe/CdS films from the top electrode was the main reason behind degradation in the cell's performance. The grain boundaries in polycrystalline CdTe layer assist fast diffusion of Cu to the CdTe/CdS junction [50,51] that causes poisoning to the junction, increased space charge recombination, and depletion of the back contact. On the other hand Cu loss from the favorable contact results in roll over in the *I–V* characteristics. Cu also forms Cu_2Te at the CdTe interface and affects the cell performance in various ways. Low Cu dose gave better stability but poor cell performance, whereas higher Cu dose caused faster degradation in both the V_{oc} and *FF* [52,53]. High Cu dose forms shunt paths that generate large recombination centers under illumination [54,55]. Cu also dopes CdS *p*-type and deteriorates the junction properties [56]. Choice of suitable ohmic contact with *p*-CdTe still remained a critical issue in terms of efficiency and stability. Formation of a thin insulating layer of oxide at CdTe back surface due to O_2/H_2O diffusion develops a barrier for hole transportation. Along with Cu, the diffusion and accumulation of Cl and other materials to the CdTe/CdS junction were also expected to degrade the cell performance. Cu-free contacts solar cells were also proposed to overcome the degradation problem, but these solar cells also exhibited degradation under stress [57]. A systematic study on stability of CdTe solar cells at elevated temperatures under various biased conditions in dark and under illumination suggested that degradation in CdTe solar cells is bias dependent and device specific [47]. Solar cells with thick CdTe and no Cu in the back electrode exhibited stable solar cells. Dobson et al. discussed some of the issues affecting stability of CdTe/CdS solar cells [58]. The role of Cu and non-Cu related issues in degradation of CdTe/CdS solar cells have also been discussed.

A concern was rising in the minds of the researchers of whether Cu was really detrimental to the stability of CdTe/CdS solar cells. Dobson et al. performed some *I–V* measurements, secondary ion mass spectroscopy (SIMS), x-ray photoelectron and Auger electron spectroscopy (XPS and AES), photoluminescence (PL), and surface photovoltage spectroscopy (SPS) studies to determine the role of Cu in degradation of CdTe/CdS solar cells and found that Cu by itself is not detrimental to the cell performance [59]. To formulate the cell, CdS was deposited on TCO-coated glass by closed space sublimation and then CdTe was deposited on CdS film. The CdTe/CdS structure was treated with $CdCl_2$ to improve the crystallinity. Then Cu-HgTe doped graphite paste was applied on the CdTe layer and annealed in dry N_2 at 250°C for 25 min. The back contact was completed with the use of Ag paste. Though there was a combined doping of Cu and Cl in CdS and at the CdTe/CdS interface, this did not cause degradation in cell performance as long as O_2 and H_2O were rigorously absent. Though both the Cu and Cl were observed to diffuse to CdS, the characteristics under illumination did not show any rollover or degradation in the performance of the solar cell unless the testing was done in air. Therefore, Cu/Cl diffusion alone was not the cause of degradation, but the air oxidation played an important role. Other studies also showed that the light-induced degradation and the degradation caused by H_2O/O_2 molecules are very important in CdTe/CdS solar cells [58]. Though the H_2O/O_2 ingress can be prevented with proper encapsulation that may lead to enhanced stability, the encapsulated solar cells may still degrade due to excess Cu doping causing generation of more recombination centers, top electrode depletion because of Cu loss, and because of CdTe–CdS intermixing caused by high temperature stress.

2.2.5 Degradation in III–V Multijunction Solar Cells

Due to their high efficiency, high thermal stability, and high resistivity toward the damages caused by celestial radiations, multijunction solar cells are used for space applications. There is not much field data available on the long-term stability of III–V multijunction solar cells, but they are highly stable having a lifetime of more than 25 years in space. Short-term field data on the reliability of III–V multijunction solar cells can be found in the reports [60–62]. After a long time these solar cells also exhibit some loss in their performance due to irradiation of high-energy space particles, UV light, thermal cycling, and thermal expansion. The degradation in multijunction solar cells is significantly much lower than that in c-Si solar cells. However, the packaging and design of the solar cell has a crucial effect on cell stability. If the thermal expansion of substrate and deposited layers are matched, there is stress relief at the interfaces, thermal resistance to heat sink is low, mechanical stresses are isolated, and if the cell and connections are moisture protected, the solar cells would exhibit very high stability. The interconnecting junctions play an important role in long-term stability of multijunction solar cells. At high illumination (around 1000 suns) the solar cells would operate at high current density (~10 A/cm^2) and high temperatures. For longer stability the interconnecting tunnel junctions should be stable enough to bear such high currents. As the solar cells are expected to be exposed to several hundred suns, an IEC62108 standard has been developed for indoor testing and qualification of concentrated multijunction solar cells for space applications. The solar cells are subjected to many accelerated tests for determination of their lifetime in space conditions.

2.2.6 Degradation in Dye-Sensitized Solar Cells (DSSCs)

Dye-sensitized solar cells (DSSCs) are quite promising for cost-effective energy production, but their short-span lifetime is keeping them from proper utilization. These cells exhibit deterioration in their performance over time. Though the degradation mechanisms in DSSCs are not completely understood, there are a number of publications that have reported degradation studies in DSSCs, suggesting that the performance depends on their construction and constituent materials. As mentioned in the previous chapter, DSSCs could be liquid electrolyte-based or based on a solid-state medium. DSSCs having a solid-state medium for charge transportation have shown comparatively higher stability but they suffer with low efficiencies, whereas those based on liquid electrolyte have shown maximum efficiency but poor stability [63,64]. In general the degradation in DSSCs could be physical as well as chemical. Physical degradation is related to leakage or drying of the electrolyte medium, whereas chemical degradation is related to decomposition of dye and electrolyte components. Degradation in photoelectrodes and corrosion of counter electrodes are some other important reasons behind the deterioration in cell performance over the time. The leakage of liquid electrolyte could be minimized with proper sealing, but mostly the chemical degradations due to photoelectrochemical reactions and thermal stimulations are irreversible. Usually the electrolyte medium possesses the I^-/I_3^- redox couple, where I_3^- ions react with the dye and counter electrode causing inferior properties and reduced energy conversion. Cell degradation gets accelerated under illumination, humidity, and elevated temperatures [65]. High temperature increases the exchange reaction rate. During outdoor operation the temperature of solar cells may reach around 70°C, and there are only a few molecular dyes that can bear such a high temperature. Elevated temperatures are difficult to avoid in normal operating conditions, therefore it is most critical for long-term stability of DSSCs. Also the photooxidation in organic dyes due to side-chain reactions involving cation radical and triplet states is quite a common problem causing degradation in cell performance [66,67].

Kohle et al. performed degradation studies on DSSCs based on nanocrystalline TiO_2 sensitized with *cis*-(2,2′-bipyridyl-4,4′-dicarboxylic acid)$_2$(SCN)$_2$Ru(II) dye and 0.5 M KI/50 mM I_2 as the redox couple in glutaronitrile electrolyte. They used a 395 UV cutoff filter and observed no deterioration in the cell performance over a time period of 7000 h under continuous illumination at 35°C [67]. This stability corresponded to 6 years of outdoor working lifetime in Central Europe. (Please note that for calculation of actual life of the long lasting solar cells, the appropriate accelerating lifetime tests are performed and the useful extrapolations give the anticipated lifetime in outdoor conditions.) The cells exhibited degradation when they were kept at elevated temperatures. At elevated temperatures the side reaction takes place even in dark and degrades the cell's performance [68]. Xue et al. investigate the thermal stability and degradation mechanisms in DSSCs based on bis(tetrabutylammonium) *cis*-bis(thiocyanato)bis(2,2′-bipyridine-4-carboxylic acid, 4′-carboxylate)ruthenium(II) (N719) dye [69]. The solar cells were subjected to degradation studies in two temperature conditions: in one condition the temperature was kept at 25°C, whereas in the other condition the sample was kept in a thermal cycle of −20°C for 12 h and then subsequently at 25°C for 24 h. The cells were tested up to about 1000 h where both maintained about 80% of their initial PCE, though the solar cell subjected to the thermal cycle exhibited slightly slower degradation in short-circuit current density (J_{sc}). The degradation in cell performance was mainly because of degradation in J_{sc}, however, the V_{oc} and *FF* exhibited little improvement. Reduction in J_{sc} was because of degradation in N719 dye. Though the TiO_2 photoelectrode exhibited increased cracks that would decrease charge collection efficiency, due to reduction in I_3^- concentration with time, the back electron transfer effect was suppressed and the overall charge collection efficiency was not affected much. The thermal cycle would cause expansion and contraction in the solar cell, which would decrease collection efficiency, but due to reduced degradation of N719 at low temperature the overall degradation in J_{sc} was reduced.

Kern et al. investigated the stability of DSSCs and found that chemical compositions of the cell and electrolyte solvents play a very important role in deciding the cell stability [70]. They performed investigations on the solar cells based on *cis*-(2,2′-bipyridyl-4,4′-dicarboxylic acid)$_2$(SCN)$_2$Ru(II) dye with the electrolytes possessing 0.5 M LiI/50 mM I_2 and 4-*tert*-butylpyridine in different solvents. The investigations showed that propionitrile was the best solvent for longer stability and other solvents like acetonitrile and methoxyacetonitrile degraded faster. In some cases water was deliberately added in the electrolyte to see the effect on stability of the cell, and it was found that to some extent the content of water could also be tolerated in the presence of 4-*tert*-butylpyridine. Later Hinsch et al. performed long-term stability studies of DSSCs under visible light soaking, UV illumination, and thermal stress and found that the degradation mechanisms in DSSCs are electrochemical, photochemical, and purely chemical in nature [71]. The basic components of the cells were the same as those used by Kern et al. [70]. The visible light soaking alone was not a dominating factor, but thermal stress was very critical behind degradation. Device engineering led to good stability with only a 15% decrease in maximum power after 3400 h of combined thermal and light soaking stress (45°C, 1 sun). This stability corresponded to 3 years of outdoor operation without UV filter. The light soaking at elevated temperatures showed different stability figures with different solvents. Thermal stress results in change in the chemical composition of electrolytes.

Wang et al. used a quasi-solid-state gel electrolyte possessing 3-methoxypropionitrile mixed with a fluorine polymer and achieved remarkable stability under temperature and light soaking conditions [72]. The degradation was below 5% over a period of 1000 h at 85°C and 55°C under 1 sun. Agrell et al. [73] performed degradation studies on N719-dye-based DSSCs under illumination, increased temperature, humidity, and air, and found that the thiocyanate ion ligand is lost in air at temperatures above 135°C. The loss of thiocyanate ion got accelerated under illumination of visible light. Due to water contents in the electrolyte, the thiocyanate ion is exchanged with H_2O/OH^- that accelerated under illumination of visible light. Some other studies showed that the thermal stress decreases the electron lifetime in TiO_2 film of photoelectrode [74,75]. In some cases the repeatedly and varying temperature has shown regenerative effects [76]. Electrochemical impedance spectroscopy revealed that the electron transport resistance of TiO_2 film increases with both the time and temperature that causes deterioration in cell performance over the time [77].

Reduction in electron lifetime in TiO_2 or changes in TiO_2 crystal structure causes trapping and recombination of electrons with I_3^- in electrolyte that results in loss of photocurrent. Degradation in dye causes reduction in its electron injection efficiency and change in its absorption spectrum. Electrolyte decomposition causes adsorption of new products on TiO_2 and desorption of dye molecules. The Pt catalyst at the counter electrode diffuses into the electrolyte to the TiO_2 surface and enhances the internal recombination. These factors cause a reduced photocurrent in the solar cell. The degradation in liquid electrolyte, diffusion of Pt to TiO_2, formation of side products on the TiO_2 surface, and thermally induced changes at the TiO_2/electrode interface are the main reasons behind deterioration in the open-circuit voltage (V_{oc}) of the cell. However, more investigations are still required to properly understand the degradation mechanisms in DSSCs. The aforementioned degradation phenomena are just guidelines for degradation analysis and should not be taken as inevitable events happening in DSSCs, however, controlling the water content in the cell, proper encapsulation, and filtering of UV light can make DSSCs more durable and robust.

2.2.7 Degradation in Organic Solar Cells (OSCs)

Although organic solar cells (OSCs) are quite fascinating and highly cost effective, they suffer from rapid degradation in air. Unfortunately, they are least stable among the solar cell technologies known so far. Degradation in OSCs is a complicated issue and is not known completely yet, but it is mainly caused by adsorption and diffusion of ambient O_2/H_2O molecules that react with cell components to make them lose their electrical/optical properties. Even the electrode materials diffuse into the active layers and react with organic molecules. Degradation happens throughout the solar cells from the top to bottom electrode in dark as well as under illumination. The degradation gets accelerated in the presence of UV light. Oxidation of molecules in the presence of UV light causes photobleaching and decomposition of organic molecules. Decomposition of organic molecules results in formation of new molecules with inferior properties, causing degradation in photovoltaic performance. Protection from the ambient with proper encapsulation in inert atmosphere imparts them longer stability and application of less sensitive molecules with more robust architectures may lead to better stability in OSCs. Detailed information about degradation mechanisms in OSCs can be found in Chapter 5.

2.3 Degradation in Solar Modules

The degradation in individual solar cells would obviously cause deterioration in the module performance. Apart from degradation in solar cells, there could be many other factors responsible for the failure of a solar module. Therefore, the stability of a solar module can be divided into two categories: first, the inherent stability of solar cells and, second, the overall stability of the module. Degradation mechanisms of individual solar cells or their inherent stability were discussed earlier, therefore, here I discuss the other factors that could cause failure to a solar module.

2.3.1 Short-Circuit and Open-Circuit Failure

Short-circuit and open-circuit failures are the most common failures in solar modules. In conventional solar modules the individual cells are wire connected and short-circuiting might occur due to interconnection of the anode and cathode of individual cells. Sometime diffusion of electrode materials through pinholes or damaged cell areas also cause short-circuiting in the cells. Short-circuiting via diffusion of electrode materials is more often observed in thin-film solar cells, where thickness of the active layer is very small. Open-circuiting might occur via disconnection of wire contacts due to thermal stress or external pressure by the wind. In case of short-circuiting, the cells or wire connections burn out and the module stops working. Similarly, in the case of open-circuiting the module stops working. Such problems might also occur due to failure of an individual cell. The problem of short-circuit and open-circuit failure could be easily fixed via replacement of the damaged cell or fixing the burned wires and disconnected contacts. However, to avoid such failures the module itself should be very robust and the connections should be very strong. Also, the modules should be mounted on a sturdy platform so that it could bear wind pressure without any damage.

2.3.2 Hot-Spot Failure

In a solar module the cells are connected in series and/or parallel. A bad cell or shading of a cell in the module causes lots of electrical energy to dissipate in that cell. The energy dissipated in the cell converts into heat and increases its temperature. Generation of tremendous heat in a particular cell corresponds to generation of a hot spot in the module that leads to its failure [78]. Generation of a hot spot and failure of the module can be understood from Figure 2.4a, which schematically shows a module made of a series connection of 16 cells where 1 cell is shaded by some object. The shading would cause that cell to generate no or much less current. This develops a mismatch in the short-circuit current (I_{sc}) of the series connected solar cells. The mismatch in I_{sc} of the cells could also be from their inherent properties. The shaded cell or the cell that generates low current limits the current throughout the series. The extra current generated by the unshaded or good cells makes them forward biased. Under module operation the shaded cell gets reversed biased by the high voltage generated by unshaded cells that cause the shaded cell to heat up. A large power is dissipated in the shaded cell that causes local heating and generates a hot-spot, which leads to destructive effects like cell degradation, cracking of the cell, cracking of the module glass, and delamination of the cell contacts.

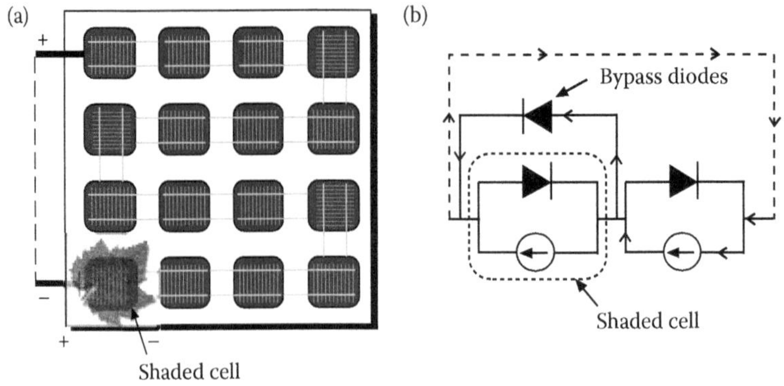

FIGURE 2.4
(a) Schematic representation of hot-spot generation in a PV module. (b) Protection from hot-spot failure using a bypass diode.

The hot-spot effect can be circumvented by the use of a bypass diode, connected in parallel but with opposite polarity to the shaded solar cell. When the shaded cell is reversed biased by the unshaded cells, the bypass diode is in forward bias and conducts the current. The bypass diode allows the current from unshaded cells to flow in the external circuit. The bypass diode reduces the voltage drop across the shaded cell to a single diode drop and prevents hot-spot heating. The operation of a bypass diode is shown in Figure 2.4b. When the bypassed solar cell gets unshaded, the bypass diode becomes reverse biased and behaves as an open circuit. Whenever the reverse voltage is more than the knee voltage of a cell, the bypass diode turns on and allows the current to flow. Therefore, if each cell in the solar module is bypassed, it will reduce the possibility of formation of hot-spots anywhere in the module, but in general one bypass diode per cell is very expensive so cells are bypassed in groups. Though the problem of a hot-spot in the module could be overcome by a bypass diode, when, for the purpose of high-power generation, several solar modules are connected in series and parallel together in an array, the same hot-spot would now become a problem for the entire array.

The problem caused by a hot-spot in a PV array can be understood from Figure 2.5, where two parallel-connected modules are connected in series with other two parallel-connected modules. The corresponding bypass diodes of the modules are also shown in the figure. Partial shading of a module might reduce its output voltage, output current, or both. Such reduction in output current or voltage would create a mismatch between parallel-connected modules leading to a mismatch between series-connected strings. This mismatch would cause the parallel-connected bypass diodes to conduct a high current of the array, and the diodes would burn out if they are not rated to bear such a large current. Irrespective of the shaded module this high current would pass through the diodes having lower resistance and they would burn out. Burning of bypass diodes would allow damage to the PV array. To avoid such mismatch damages along with the bypass diodes, an additional diode called the blocking diode is also used with each parallel-connected string. When a battery is connected with the module for charging during daytime, the blocking diode additionally prevents the module from back flow of current from the battery during night. The blocking diode also inhibits the current flow from one parallel string to other lower currents carrying parallel string and reduces the mismatch losses caused by parallel connection of the modules. Sometimes the diode temperature also gets increased due

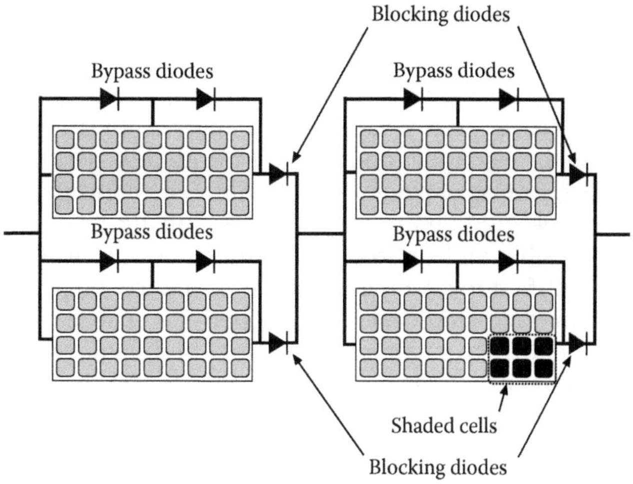

FIGURE 2.5
Schematic representation of protection of a solar array from hot-spot failure using bypass and blocking diodes.

to heating by the solar radiation and it leads to bypass diode failure and ultimately to the array damage. The problem could be overcome by keeping the diode and module temperatures at lower values.

2.3.3 Encapsulant Failure

Encapsulation of a solar module is very crucial for its long-term stability. The purpose of encapsulation is to prevent the entrance of the entities that could degrade the solar cells. In most solar modules moisture causes rapid degradation in solar cells and corrosion of the contacts. An encapsulant should be highly transparent to the light absorbed by the cell and a high barrier to ingress of any vapor or gas. Unfortunately the encapsulant itself starts degrading and losing its properties over time. Sometimes delamination of encapsulant is also observed. Degradation in encapsulant happened due to elevated temperature and long-term UV exposure. To ensure maximum stability, the UV absorbers and some stabilizers are used in the encapsulating materials. The concentration of stabilizers depletes due to radiation, and when the concentration is below a critical value the encapsulating material degrades rapidly. Though EVA is used as an encapsulant for many solar modules, it is not ideal as it transports the water vapors and also degrades with time. Coloration of EVA encapsulant is another problem that reduces the amount of light absorbed by the cells [12]. Kaplani [12] performed a study on the degradation of c-Si solar modules under natural field conditions for over 22 years and with external agents inducing acceleration in degradation. The natural degradation was mild and uniform over the module, whereas the localized-induced degradation was stronger and nonuniform. For natural degradation the coloration of EVA imparted a golden brown color, which was observed on the entire surface region of all the solar cells. Degradation in EVA and other aging effects in c-Si solar modules become more prominent when the modules are placed in hot and dry conditions with high ambient temperature and large UV flux [79,80]. EVA is not a suitable encapsulant for many solar cell technologies and alternately polyvinyl butyral (PVB) is used as an encapsulant [81,82]. The issues related to failure of OSCs models are discussed in Chapter 6.

2.3.4 Cracking of Solar Cells and Solar Modules

Sometimes cracks might also occur in the solar cells or cover glass due to stress like high temperature or high pressure. Cracking of solar cells might cause an open-circuit failure, whereas cracking of the cover glass would result in ingress of water vapors into the module that would corrode the electrodes and degrade the module performance.

Shading by an object, soiling of front surface, failure of an individual cell, and change in operating point of the module are other factors causing failure of a solar module. These factors are reversible and the reduction in power output of the module due to these facts could be rectified to give the initial performance. Edge sealing and framing are also very important and should be done properly to ensure maximum stability of the solar modules. Shielding from air, moisture, and high temperature is a must for all solar cell technologies to ensure the maximum stability. The shielding should be good enough to prevent exchange of any external entity between the cell/module and the ambient.

References

1. H. Fischer and W. Pschunder, 1973. Investigation of photon and thermal induced changes in silicon solar cells, *Conference Record of the 10th IEEE Photovoltaic Specialists Conference*, 404.
2. V. G. Weizer, H. W. Brandhorst, J. D. Broder, R. E. Hart, and J. H. Lamneck, 1979. Photon-degradation effects in terrestrial silicon solar cells, *J. Appl. Phys.* 50: 4443.
3. L. I. Murin, E. A. Tolkavheva, V. P. Markevich, A. R. Peaker, B. Hamilton, E. Monakhov, B. G. Svensson et al., 2011. The oxygen dimmer in Si: Its relationship to the light induced degradation of Si solar cells, *Appl. Phys. Lett.* 98: 182101.
4. L. C. Kimerling, M. T. Asom, J. L. Benton, P. J. Drevinsky, and C. E. Caefer, 1989. Interstitial defect reaction in silicon, *Mater. Sci. Forum* 38–41: 141–150.
5. J. Schmidt and K. Bothe, 2004. Structure and transformation of metastable boron and oxygen related defect center in crystalline silicon, *Phys. Rev. B* 69: 24107.
6. A. Herguth, G. Schubert, M. Kaes, and G. Hahn, 2008. Investigations on the long time behavior of the metastable boron–oxygen complex in crystalline silicon, *Prog. Photovolt.: Res. Appl.* 16: 135.
7. B. Lim, F. Rougieux, D. Macdonald, K. Bothe, and J. Schmidt, 2010. Generation and annihilation of boron oxygen related recombination centers in compensated *p*- and *n*-type silicon, *J. Appl. Phys.* 108: 103722.
8. D. W. Palmer, K. Bothe, and J. Schmidt, 2007. Kinetics of the electronically stimulated formation of a boron–oxygen complex in crystalline silicon, *Phys. Rev. B* 76: 035210.
9. J. Schmidt, K. Bothe, D. Macdonald, J. Adey, R. Jones, and D. W. Palmer, 2006. Electronically stimulated degradation of silicon solar cells, *J. Mater. Res.* 21: 5–12.
10. B. Sopori, 2005. Dielectric films for Si solar cell applications, *J. Electron. Mater.* 34: 564.
11. H. Kim, P. Choi, K. Kim, H. Kuh, D. Beak, J. Lee, J. Yi, and B. Choi, 2014. Performance degradation of *c*-Si solar cells under UV exposure, *J. Nanosci. Nanotechnol.* 14: 3561.
12. E. Kaplani, 2012. Degradation effects in sc-Si PV modules subjected to natural and induced ageing after several years of field operation, *J. Eng. Sci. Technol. Rev.* 5: 18–23.
13. M. A. Quintana, D. L. King, T. J. McMahon, and C. R. Osterwald, 2002. Commonly observed degradation in field-aged photovoltaic modules, *Conference Record of the Twenty-Ninth IEEE Photovoltaic Specialists Conference 2002*: 1436–1439, doi: 10.1109/PVSC.2002.1190879.
14. D. L. Staebler and C. R. Wronski, 1977. Reversible conductivity changes in discharge-produced amorphous Si, *Appl. Phys. Lett.* 31: 292.

15. D. L. Staebler and C. R. Wronski, 1980. Optically induced conductivity changes in discharge-produced hydrogenated amorphous silicon, *J. Appl. Phys.* 51: 3262.
16. M. Stutzmann, W. B. Jackson, and C. C. Tsai, 1985. Light-induced metastable defects in hydrogenated amorphous silicon: A systematic study, *Phys. Rev. B* 32: 23.
17. M. Stutzmann and D. K. Biegelsen, 1989. Microscopic nature of coordination defects in amorphous silicon, *Phys. Rev. B* 40: 9834.
18. M. Fehr, A. Schnegg, B. Rech, K. Lips, O. Astakhov, F. Finger, G. Pfanner et al., 2011. Combined multifrequency EPR and DFT study of dangling bonds in a-Si:H, *Phys. Rev. B* 84: 245203.
19. H. Dersch, J. Stuke, and J. Beichler, 1981. Light-induced dangling bonds in hydrogenated amorphous silicon, *Appl. Phys. Lett.* 38: 456.
20. H. Fritzsche, 2001. Development in understanding and controlling the Staebler–Wronski effect in a-Si:H, *Annu. Rev. Mater. Res.* 31: 47–79.
21. M. J. Powell, S. C. Deane, and R. B. Wehrspohn, 2002. Microscopic mechanisms for creation and removal of metastable dangling bonds in hydrogenated amorphous silicon, *Phys. Rev. B* 66: 155212.
22. J. Melskens, A. H. M. Smets, M. Schouten, S. W. H. Eijt, H. Schut, and M. Zeman, 2013. New insights in the nanostructure and defect states of hydrogenated amorphous silicon obtained by annealing, *IEEE J. Photovoltaics* 3: 65–71.
23. W. Beyer, W. Hilgers, P. Prunici, and D. Lennartz, 2012. Voids in hydrogenated amorphous silicon materials, *J. Non-Cryst. Solids* 358: 2023–2026.
24. A. H. Mahan, D. L. Williamson, B. P. Nelson, and R. S. Crandall, 1989. Characterization of microvoids in device-quality hydrogenated amorphous silicon by small-angle x-ray scattering and infrared measurements, *Phys. Rev. B* 40: 12024.
25. S. Guha, J. Yang, S. J. Jones, C. Yan, and D. L. Williamson, 1992. Effect of microvoids on initial and light-degraded efficiencies of hydrogenated amorphous silicon alloy solar cells, *Appl. Phys. Lett.* 61: 1444.
26. D. E. Carlson, 1986. Hydrogenated micro voids and light-induced degradation of amorphous-silicon solar cells, *Appl. Phys. A* 41: 305.
27. M. Fehr, A. Schnegg, B. Rech, O. Astakhov, F. Finger, R. Bittl, C. Teutloff, and K. Lips, 2014. Metastable defect formation at micro voids identified as a source of light-induced degradation in a-Si:H, *Phys. Rev. Lett.* 112: 66403.
28. N. Wang and V. L. Dalal, 2006. Improving stability of amorphous silicon using chemical annealing with helium, *J. Non-Crystal. Solids* 352: 1937–1940.
29. D. C. Bobela, H. M. Branz, and P. Stradins, 2011. Anneal treatment to reduce the creation rate of light-induced metastable defects in device-quality hydrogenated amorphous silicon, *Appl. Phys. Lett.* 98: 201908.
30. D. W. Lee, W. J. Cho, J. K. Song, O. Y. Kwon, W. H. Lee, C. H. Park, K. E. Park, H. Lee, and Y. N. Kim, 2015. Failure analysis of Cu(In,Ga)Se2 photovoltaic modules: Degradation mechanism of Cu(In,Ga)Se2 solar cells under harsh environmental conditions, *Prog. Photovolt.: Res. Appl.* 23: 829–837.
31. A. G. Aberle, 2009. Thin-film solar cells, *Thin Solid Films* 517: 4706–4710.
32. R. P. Kenny, A. I. Chatzipanagi, and T. Sample, 2012. Preconditioning of thin-film PV module technologies for calibration, *Prog. Photovolt.: Res. Appl.* 22: 166–172, doi: 10.1002/pip.2234.
33. C. Radue, E. E. van Dyk, and E. Q. Macabebe, 2009. Analysis of performance and device parameters of CIGS PV modules deployed outside, *Thin Solid Films* 517: 2383–2385.
34. R. P. Kenny, M. Nikolaeva-Dimitrova, and E. D. Dunlop, 2006. Performance measurements of CIS modules: Outdoor and pulsed simulator comparison for power and energy rating, *Proceedings of the 4th WCPEC*: 2058–2061.
35. International Electrotechnical Commission, 1996. IEC 61646 Thin-film terrestrial photovoltaic (PV) modules—Design qualification and type approval. Technical report, Geneva, Switzerland.
36. International Electrotechnical Commission, 2005. IEC 61215 Crystalline Silicon Terrestrial Photovoltaic (PV) Modules—Design Qualification and Type Approval. Technical report, Geneva, Switzerland.

37. M. Bär, J. Reichardt, I. Sieber, A. Grimm, I. Kötschau, I. Lauermann, S. Sokoll, T. P. Niesen, T. P. Lux-Steiner, and C. H. Fischer, 2006. ZnO layers deposited by the ion layer gas reaction on Cu(In,Ga)(S,Se)2 thin film solar cell absorbers-Impact of "damp-heat" conditions on the layer properties, *Prog. Photovolt.: Res. Appl.* 15: 187–198.
38. C. Heske, U. Groh, L. Weinhardt, O. Fuchs, B. Holder, E. Umbach, C. Bostedt et al., 2002. Damp-heat induced sulfate formation in Cu(In,Ga)(S,Se)2-based thin film solar cells, *Appl. Phys. Lett.* 81: 4550–4552.
39. M. Schmidt, D. Braunger, R. Schäffler, H. W. Schock, and U. Rau, 2000. Influence of damp heat on the electrical properties of Cu(In,Ga)Se2 solar cells, *Thin Solid Films* 361–362: 283–287.
40. D. J. Coyle, H. A. Blaydes, J. E. Pickett, R. S. Northey, and J. O. Gardner, 2009. Degradation kinetics of CIGS solar cells, *Proceedings of 34th IEEE photovoltaic Specialists Conference (PVSC)* 1943–1947.
41. M. Powalla and B. Dimmler, 2001. Process development of high performance CIGS modules for mass production, *Thin Solid Films*, 387: 251–256.
42. D. J. Coyle, I. M Blaydes, M. Wimbor, and J. Wennergerg, 2002. The change of the electronic properties of CIGS devices induced by the "damp heat" treatment, *Thin Solid Films* 403–404: 320–324.
43. D. W. Lee, O. Y. Kwon, J. K. Song, C. H. Park, K. E. Park, S. M. Nam, and Y. N. Kim, 2012. Effects of ZnO:Al films on CIGS PV modules degraded under accelerated damp heat, *Sol. Energy Mater. Sol. Cells*, 105: 15–20.
44. P. V. Meyers, S. Asher, and M. M. Al-Jassim, 1996. A search for degradation mechanisms of CdTe/CdS solar cells, in *Proceedings of the Materials Research Society Symposium*, Cambridge University Press, pp. 317–324.
45. S. S. Hegedus and B. E. McCandless, 2005. CdTe contacts for CdTe/CdS solar cells: Effect of Cu thickness, surface preparation and recontacting on device performance and stability, *Sol. Energy Mater. Sol. Cells* 88: 75–95.
46. D. S. Albin, 2008. Accelerated stress testing and diagnostic analysis of degradation in CdTe solar cells, *Proceedings SPIE* 7048: 70480N.
47. J. F. Hiltner and J.R. Sites, 1999. Stability of CdTe solar cells at elevated temperatures: Bias, temperature, and Cu dependence, *AIP Conf. Proc.* 462: 170–175.
48. I. Visoly-Fisher, K. D. Dobson, J. Nair, E. Bezalel, G. Hodes, and D. Cahen, 2002. Factors affecting the stability of CdTe/CdS solar cells, deduced from stress tests at elevated temperature, http://www.nrel.gov/pv/thin_film/docs/cahen_factors_afm.pdf.
49. T. L. Chu, S. S. Chu, C. Ferekides, C. Q. Wu, J. Britt, and C. Wang, 1991. 13.4% efficient thin-film CdS/CdTe solar cells, *J. Appl. Phys.* 70: 7608.
50. H. C. Chou, A. Rohatgi, N. M. Jokerst, E. W. Thomas, and S. Kamra, 1996. Copper migration in CdTe heterojunction solar cells, *J. Electron. Mater.* 25: 1093–1098.
51. S. E. Asher, F. S. Hasoon, T. A. Gessert, M. R. Young, P. Sheldon, J. Hiltner, and J. Sites, 2000. Determination of Cu in CdTe/CdS devices before and after accelerated stress testing, *28th IEEE Photovoltaic Specialist Conf.* 479–482, doi: 10.1109/PVSC.2000.915876 2000.
52. S. S. Hegedus, B. E. McCandless, and R. W. Birkmire, 2000. Analysis of stress-induced degradation in CdS/CdTe solar cells, *28th IEEE Photovoltaic Specialist Conf.* 535–538.
53. D. J. Coyle, 2012. The effect of copper on accelerated life test performance of CdTe solar cells, *NREL PV Module Reliability Workshop*, February 28, Golden, Colorado.
54. J. Lee, T. H. Myers, N. C. Giles, B. C. Dean, and C. J. Johnson, 1994. Optical quenching of bound excitons in CdTe and $Cd_{1-x}Zn_xTe$ alloys: A technique to measure copper concentration, *J. Appl. Phys.* 76: 537.
55. H. C. Chou, A. Rohatgi, E. W. Thomas, S. Kamra, and A. K. Bhat, 1995. Effects of Cu on CdTe/CdS heterojunction solar cells with Au/Cu contacts, *J. Electrochem. Soc.* 142: 254–259.
56. P. J. Sebastian and M. Ocampo, 1995. Modification of structural and opto-electronic properties of CdS thin films by Cu doping, *J. Appl. Phys.* 77: 4548.
57. B. Ghosh, S. Purakayastha, P. K. Datta, R. W. Miles, M. J. Carter, and R. Hill, 1995. Formation of a stable ohmic contact to CdTe thin films through the diffusion of P from Ni-P, *Semicond. Sci. Technol.* 10: 71.

58. K. D. Dobson, I. V. Fisher, G. Hodes, and D. Cahen, 2000. Stability of CdTe/CdS thin-film solar cells, *Sol. Energy Mater. Sol. Cells* 62: 295.
59. K. D. Dobson, I. V. Fisher, G. Hodes, and D. Cahen, 2001. Stabilizing CdTe/CdS solar cells with Cu-containing contacts to p-CdTe, *Adv. Mater.* 13: 1495.
60. P. J. Verlinden, A. Lewandowski, C. Bingham, G. S. Kinsey, R. A. Sherif, and J. B. Lasich, 2006. *4th IEEE World Conference on Photovoltaic Energy Conversion*, 592–597.
61. A. Hakenjos, J. Wullner, and H. Lerchenmuller, 2007. Field performance of Flatcon high concentration photovoltaic systems, *22nd European Photovoltaic Solar Energy Conference*, 156–159.
62. A. W. Bett, B. Burger, F. Dimroth, G. Siefer, and H. Lerchenmuller, 2006. High-concentration PV using III–V solar cells, *IEEE 4th World Conference on Photovoltaic Energy Conversion*, 615–620.
63. S. Mathew, A. Yella, P. Gao, R. H. Baker, B. F. E. Curchod, N. A. Astani, I. Tavernelli, U. Rothlisberger, M. K. Nazeerudin, and M. Gratzel, 2014. Dye-sensitized solar cells with 13% efficiency achieved through the molecular engineering of porphyrin sensitizers, *Nat. Chem.* 6: 242.
64. A. Yella, H. W. Lee, H. N. Tsao, C. Yi, A. K. Chandiran, M. K. Nazeeruddin, E. W. G. Diau, C. Y. Yeh, S. M. Zakeeruddin, and M. Gratzel, 2011. Porphyrin-sensitized solar cells with cobalt (II/III)-based redox electrolyte exceed 12 percent efficiency, *Science* 334: 629–634.
65. M. I. Asghar, K. Miettunen, J. Halme, P. Vahermaa, M. Toivola, K. Aitola, and P. Lund, 2010. Review of stability for advanced dye solar cells, *Energy Environ. Sci.* 3: 418.
66. R. Griinwdld and H. Tributsch, 1997. Mechanisms of instability in Ru-based dye sensitization solar cells, *J. Phys. Chem. B* 101: 2564–2575.
67. O. Kohle, M. Gratzel, A. Meyer, and T. B. Meyer, 1997. The photovoltaic stability of bis(isothiocyanato)ruthenium(11)-bis-2,2′-bipyridine-4,4-dicarboxylic acid and related sensitizers, *Adv. Mater.* 9: 904.
68. P. M. Sommeling, M. Späth, H. J. P. Smit, N. J. Bakker, and J. M. Kroon, 2004. Long-term stability testing of dye-sensitized solar cells, *J. Photochem. Photobiol. A: Chem.* 164: 137–144.
69. G. Xue, Y. Guo, T. Yu, J. Guan, X. Yu, J. Zhang, J. Liu, and Z. Zou, 2012. Degradation mechanisms investigation for long-term thermal stability of dye-sensitized solar cells, *Int. J. Electrochem. Sci.* 7: 1496–1511.
70. R. Kern, N. Van Der Burg, G. Chmiel, J. Ferber, G. Hasenhindl, A. Hinsch, R. Kinderman, et al., 2000. Long term stability of dye-sensitized solar cells for large area power applications, *Opto-Electron. Rev.* 8: 284–288.
71. A. Hinsch, J. M. Kroon, R. Kern, I. Uhlendorf, J. Holzbock, A. Meyer, and J. Ferber, 2001. Long-term stability of dye-sensitized solar cells, *Prog. Photovolt.: Res. Appl.* 9: 425.
72. P. Wang, S. M. Zakeeruddin, J. E. Moser, M. K. Nazeeruddin, T. Sekiguchi, and M. Gratzel, 2003. A stable quasi-solid-state dye-sensitized solar cell with an amphiphilic ruthenium sensitizer and polymer gel electrolyte, *Nat. Mater.* 2: 402.
73. H. G. Agrell, J. Lindgren, and A. Hagfeldt, 2003. Degradation mechanisms in a dye-sensitized solar cell studied by UV–VIS and IR spectroscopy, *Sol. Energy* 75: 169–180.
74. Q. Wang, J. E. Moser, and M. Gratzel, 2005. Electrochemical impedance spectroscopic analysis of dye-sensitized solar cells, *J. Phys. Chem. B* 109: 14945–14953.
75. R. Kern, R. Sastrawan, J. Ferber, R. Stangl, and J. Luther, 2002. Modeling and interpretation of electrical impedance spectra of dye solar cells operated under open-circuit conditions, *Electrochim. Acta* 47: 4213–4225.
76. M. Toivola, L. Peltokorpi, J. Halme, and P. Lund, 2007. Regenerative effects by temperature variations in dye-sensitized solar cells, *Sol. Energy Mater. Sol. Cells* 91: 1733–1742.
77. M. Toivola, J. Halme, L. Peltokorpi, and P. Lund, 2009. Investigation of temperature and aging effects in nanostructured dye solar cells studied by electrochemical impedance spectroscopy, *Int. J. Photoenergy* 786429, doi:10.1155/2009/786429.
78. M. Simon and E. L. Meyer, 2010. Detection and analysis of hot-spot formation in solar cells, *Sol. Energy Mater. Sol. Cells* 94: 106–113.
79. A. W. Czanderna and F. J. Pern, 1996. Encapsulation of PV modules using ethylene vinyl acetate copolymer as a pottant: A critical review, *Sol. Energy Mater. Sol. Cells* 43: 101–181.

80. J. Schlothauer, S. Jungwirth, M. Koehl, and B. Roder, 2012. Degradation of the encapsulant polymer in outdoor weathered photovoltaic modules: Spatially resolved inspection of EVA ageing by fluorescence and correlation to electroluminescence, *Sol. Energy Mater. Sol. Cells* 102: 75–85.
81. B. Koll, 2007. Alternative encapsulation for double-glass PV modules, Glass Performance Days 2007, http://www.glassfiles.com.
82. H. Schmidhuber and K. Krannich, 2001. Why using EVA for module encapsulation if there is a much better choice? *Proceedings of 17th European Photovoltaic Solar Energy Conference*, 662–663.

3

Organic Solar Cells

3.1 Introduction

Though a brief introduction of organic solar cells (OSCs) has already been given in Chapter 1, here I will discuss more about the materials, processing, limitations, and recent developments in OSCs. As mentioned earlier, OSCs are based on organic light absorbers, which are special types of hydrocarbons. These materials possess energy band gaps ranging from about 1.2 eV to about 3.5 eV and are known as organic semiconductors. These materials possess alternate single and double bonds among the C atoms, because of which they are also known as conjugated semiconductors. In pure form these materials are insulators, but they possess many properties similar to semiconductors like band gaps and temperature-dependent electrical conductivity, and therefore are called semiconductors. Though organic semiconductors have been known since the 1950s, their potential was realized in the early 1970s. Realization of very high conductivity in doped conjugated organic semiconductors generated a great interest in these materials for their applications in various electronic devices such as light-emitting diodes, solar cells, photodiodes, transistors, lasers, and sensors.

The thought of first realization of high conductivity in conjugated organic semiconductors recalls the interesting story that led to this accidental discovery. A graduate student working under the supervision of Professor Hideki Shirakawa from University of Tsukuba, Japan, synthesized a silvery film of *trans*-polyacetylene by accidentally adding a thousand times more catalyst to the reaction vessel than the amount required to produce polyacetylene. Polyacetylene is a very simple linear conjugated polymer that possesses two chemical forms, namely, *trans* and *cis* (see Figure 3.1). The produced film looked like silver film but stretched like a thin plastic sheet. Around the same time Professor Alan G. MacDiarmid, a chemist, and Professor Alan J. Heeger, a physicist from the University of Pennsylvania, were experimenting with a metallic-looking film of the inorganic polymer sulfur nitride (SN). Amazingly the two men, Shirakawa and MacDiarmid met at a seminar and discussed their findings. This event led to Shirakawa being invited to the University of Pennsylvania where he and MacDiarmid modified polyacetylene by oxidation with iodine vapor. Heeger was then asked to look for changes in the optical properties during the oxidation process. Conductivity measurements of iodine-doped *trans*-polyacetylene were done and the incredible increase of a billion times the original conductivity was discovered [1]. This discovery gave birth to a new class of materials called conducting polymers. This marked the beginning of an era for a new class of conducting materials, which are typically inexpensive and easy to process. For the discovery of conducting polymers, these three scientists were awarded the Nobel Prize in Chemistry in 2000.

Conducting polymers are often considered quasi-one-dimensional metals due to the fact that the strong intrachain interaction (strong covalent bonding along the chain) and

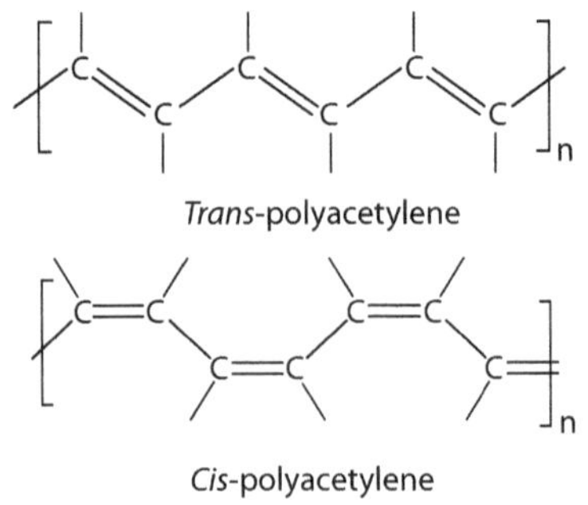

FIGURE 3.1
Chemical structure of *trans*- and *cis*-polyacetylene.

the weak van der Waals-type interchain coupling interactions lead to delocalization of π-electrons along the polymer chain. The conductivity of various materials, including doped conducting polymer, is shown in Figure 3.2. The doped conducting polymers have shown conductivity nearly as good as that of Cu. High conductivity in organic semiconductors stimulated lots of research interest. These materials have been tested and proven to be potential materials for application in various electronic devices and have emerged as very strong competitors to inorganic semiconductors. Substituted polythiophenes and substituted poly(*p*-phenylene-vinylene) have been used to construct various electronic devices. Realization of efficient photovoltaic and light-emitting devices from undoped conjugated semiconductors in the 1980s also revived interest in undoped semiconductors [2,3]. Then successful fabrication of thin-film transistors could also be demonstrated using undoped organics semiconductors [4–6]. Electronic devices based on organic semiconductors could

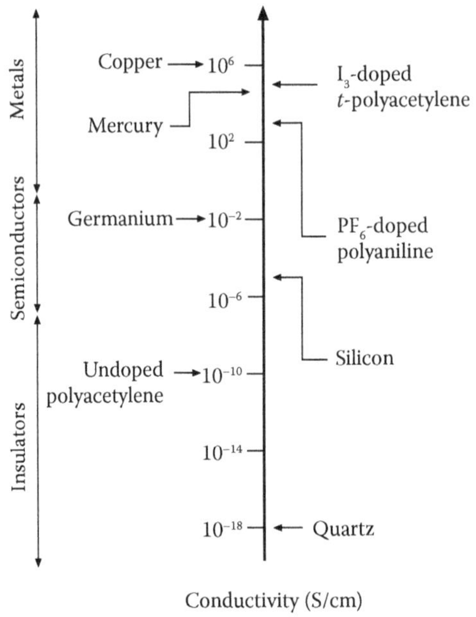

FIGURE 3.2
A comparison of the conductivity of some of doped conjugated polymers with that of other materials.

be easily processed at room temperature using various inexpensive techniques that led to their low-cost production, a key deriving factor for choosing these materials for various applications. Thin, lightweight, ease in large-area processing, and mechanical flexibility are some other attractive features of the devices based on these materials.

3.2 Organic Semiconductors

Organic semiconductors are a special class of unsaturated hydrocarbons that can easily be classified into two categories named conjugated small molecules and conjugated polymers. In both systems the C atoms are sp^2 hybridized. Small molecules are the materials with molecular weights up to several hundred daltons and do not contain repeating units like polymers, which have a molecular weight of several 100,000 daltons. Small molecular materials have shown poor solubility in common organic solvents therefore, their thin films are usually prepared via thermal evaporation in high vacuum. On the other hand, polymers exhibit decomposition and degradation upon heating, therefore their thin films are not prepared by thermal evaporation. Polymer thin films can easily be prepared with several other methods, such as spin coating, screen printing, spray coating, or inkjet printing from their solutions, allowing large area, ultrathin, flexible, and very low-cost devices. Though there have been some efforts to develop soluble small molecular and dendimer materials, the concept of efficient devices based on complete solution-processed small molecules has yet to be proven. In the last two decades, the efforts in both academic and research organizations led to rapid development in organic light-emitting diodes (OLEDs). OLED displays are now used on a large scale in commercial products like TVs, cameras, mobiles, MP3 players, razors, and many toys. Solar cells based on organic semiconductors are now of special interest, and elaborate research is being carried out in this field, as they can play a major role in generating clean and cheap energy.

3.2.1 Origin of Semiconducting Behavior

Based on the number and type of the valence electrons involved in the chemical bonding between consecutive carbon atoms, the organic compounds can be classified into saturated and unsaturated hydrocarbons. A neutral carbon atom has six electrons with a ground state electronic configuration $1s^2$, $2s^2$, $2p^2$, where s and p represent the atomic subshells. s and p subshells have one and three orbitals, respectively, and each orbital can have two electrons of opposite spins. In a C atom, out of three p_x, p_y, and p_z orbitals, p_x and p_y orbitals are occupied by electrons whereas p_z remains vacant. When the carbon atoms form covalent bonds with each other, one 2s electron is promoted to the vacant $2p_z$ orbital, and the 2s and 2p orbitals are combined together to form the hybrid orbitals of equal energy. Depending on the carbon–carbon bonding, the hybridization can be classified into sp^3, sp^2, and sp hybridizations. Figure 3.3 shows the formation of sp^3 and sp^2 hybridization in C atoms. The hybrid orbitals consisting of one s orbital and three p orbitals are known as sp^3 hybrid orbitals. The sp^3 hybrid orbitals allow a strong degree of overlap in bond formation with another atom and this produces high bond strength and stability in the molecules. The arrangement of bonds resulting from overlap with sp^3 hybridized orbitals on adjacent atoms gives rise to the tetrahedral structure that is found in the lattice of diamond and alkane molecules such as methane, and ethane. In these structures all the available electrons are tied up in strong

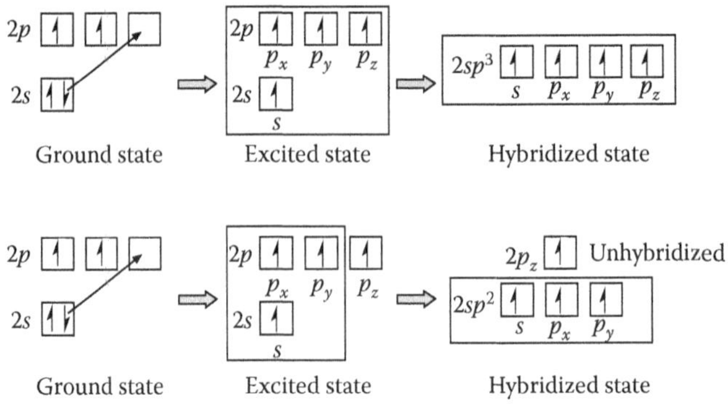

FIGURE 3.3
Schematic representation of sp^3 (upper scheme) and sp^2 (lower scheme) hybridization in carbon atoms.

covalent bonds, named σ-bonds. Carbon compounds containing sp^3 hybridized orbitals and σ-bonds are termed as saturated hydrocarbons. Saturated hydrocarbons are perfect insulators since all four valence electrons of the carbon atom are used up in covalent bonds. The saturated hydrocarbons have high band gaps.

The organic semiconductors are unsaturated hydrocarbons, which are chemically conjugated aromatic hydrocarbons. Conjugation in organic semiconductors refers to the regular alternation of single and double bonds between adjacent C atoms. The alternating single and double bonds are formed by the sp^2 hybrid orbitals. The sp^2 hybridization contains one s orbital combined with two p_x and p_y orbitals to form three sp^2 hybrid orbitals of equal energy and the p_z orbital is left unhybridized (see Figure 3.3). To minimize the electron repulsion, the hybrid orbitals make an angle of 120° with each other in the same molecular plane and the unhybridized lone p_z orbital lies with its axis perpendicular to the plane of the molecule (see Figure 3.4a). The family of conjugated organic compounds can be regarded as derivatives of benzene (C_6H_6) and the conducting behavior of organic semiconductors can be understood by understanding the properties of benzene molecules. In benzene, one of the hybridized orbitals in C is bound with the H atom and the other two are bound with the neighboring C atoms (see Figure 3.4b). All the sp^2 hybridized orbitals (or σ orbitals) form covalent σ bonds and the lone p_z orbitals (or π orbitals) overlap to form the π bonds.

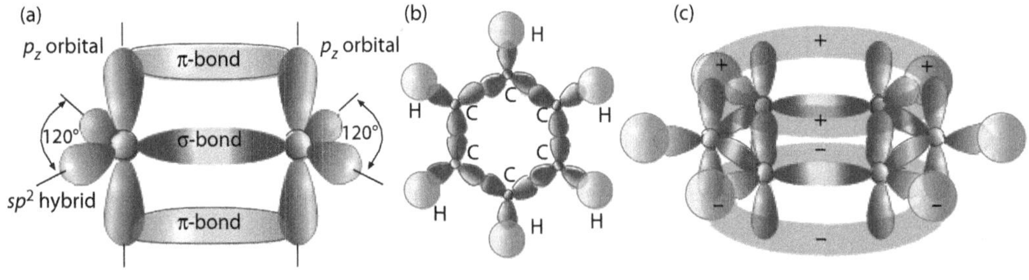

FIGURE 3.4
Schematic representation of (a) σ bonding between sp^2 hybridized and π bonding between unhybridized p_z orbitals in C atoms, (b) overlap of the C and H atom orbitals in benzene molecule, and (c) charge cloud layers above and below the plane of the benzene molecule.

The electron charge density is symmetrically distributed in the π orbitals to from charge cloud layers one above and the other below the plane of the molecule (see Figure 3.4c). The electrons that occupy σ orbitals are called σ electrons and are localized, whereas the electrons that occupy π orbitals are known as π electrons and do not participate in any bond formation and are delocalized over all C atoms. These delocalized electrons are responsible for the electric current flow in the conjugated hydrocarbons. They are mobile electrons and can move freely inside the molecule. One electron moves from atom 1 to atom 2 but another electron moves from atom 2 to atom 1, constituting equal currents in the opposite directions and the net current remains zero. With the help of π electrons the electrical influences can easily propagate from one part of the molecule to another. The π-bonds are thus considered the basic source of transport bands in the conjugated systems or organic semiconductors. The overlap of two lone p_z orbitals form two π molecular orbitals named as π and π* molecular orbitals, and the electrons in the p_z orbital are stabilized by occupying the lower energy π orbital. In the ground state the occupied orbitals are also known as bonding orbitals, whereas unoccupied orbitals are known as antibonding orbitals. The antibonding orbitals are represented by an asterisk (*) in the corresponding bonding orbital. The bonding orbitals exist lower in energy than the antibonding orbitals. σ bonds are strong and constitute the backbone of the molecules, whereas π bonds are significantly weak.

Figure 3.5 shows the formation and distribution of bonding and antibonding orbitals in a di-carbon atomic unsaturated hydrocarbon. The saturated hydrocarbons possess σ bonding and antibonding orbitals only, and the energy difference between orbitals is beyond the visible energy. Therefore, the saturated hydrocarbons are practically insulators. The unsaturated hydrocarbons possess π bonding and antibonding orbitals, and the energy difference between these orbitals is much smaller and the electronic transition can take place with absorption of visible or near-visible light like in inorganic semiconductors. In the case of the benzene molecule, there are 6C atoms in the ring and the unhybridized electrons are delocalized over all of them. In the benzene molecule there are 6 p_z and 18 sp^2 hybridized orbitals. The 6 p_z orbitals form 3 occupied π molecular orbitals and 3 unoccupied π* molecular orbitals. On the other hand 18 sp^2 hybridized orbitals form 9 occupied σ molecular orbitals and 9 unoccupied σ* molecular orbitals. All the 12 bonding orbitals are fully occupied by outermost 24 valance electrons. For the molecules having a large number of delocalized C atoms, there will be many occupied and unoccupied molecular orbitals, out of which π and π* molecular orbitals take part in optical and electrical activities. In case of molecular single crystals, the molecules are arranged in a fashion with close coupling of orbitals that form some sort of conduction and valence bands. The antibonding orbitals located higher in energy (π* and σ*) form a conduction band, whereas the lower energy

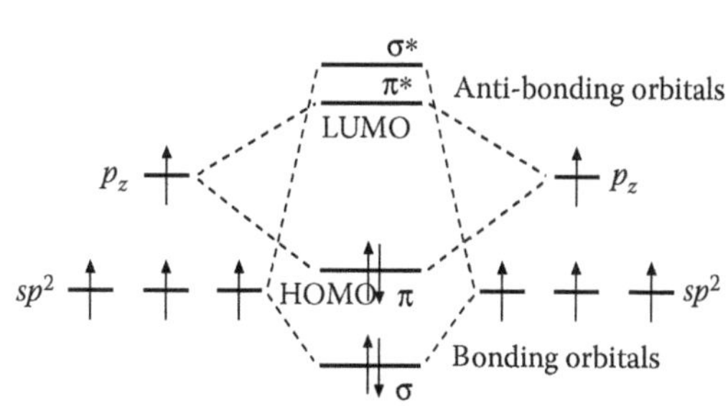

FIGURE 3.5
Schematic representation of formation of bonding and antibonding orbitals in an unsaturated hydrocarbon.

bonding orbitals (π and σ) form the valance band. π* molecular orbitals are lower in energy in the conduction band and therefore are designated as lower unoccupied molecular orbitals, whereas π molecular orbitals are higher in energy in the valance band and are known as higher occupied molecular orbitals. The molecular crystals possess remarkable transport properties like band transport with high mobility. In polycrystalline and amorphous materials the charge transport happens via hopping and the mobilities are low. In conjugated polymers there is a long chain of C atoms and there are numerous higher occupied (π) and lower unoccupied (π*) molecular orbitals. Figure 3.6 shows the scheme of formation of bonding and antibonding orbitals in a benzene and poly(*p*-phenylene-vinylene) (PPV) molecule. The electrons in polymers are delocalized along the chain through unoccupied and occupied molecular orbitals. It is like a one-dimensional electronic system and the charge transport within the chain takes place through higher occupied and lower unoccupied molecular orbitals. The interchain charge transport happens via the thermal-assisted hopping phenomenon.

The two separate bands of occupied and unoccupied molecular orbitals are characterized by two quite important energy levels, namely, electron affinity (E_a) and ionization potential (I_p). E_a is the energy released during an electronic transfer from the vacuum level to the lowest energy level in the conduction band. On the other hand, I_p is the energy required to release an electron from the highest energy level in the valance band. Therefore, E_a corresponds to the energy of lowest state of the conduction band (π* orbital) or the lowest unoccupied molecular orbital (LUMO). Similarly, I_p represents the energy of upper state of the valence band (π orbital) and corresponds to the highest occupied molecular orbital (HOMO). The separation between LUMO and HOMO of the semiconductor is known as band gap (E_g). The band gaps of organic semiconductors lie within the range of ~1.2–3.5 eV, which covers the spectral range from the infrared to ultraviolet region. In organic semiconductors, the charge transport takes place in π bonding and antibonding orbitals or in other terms in HOMO and LUMO levels. The HOMO and LUMO levels of individual molecules are Gaussianly distributed in bulk of the semiconductor.

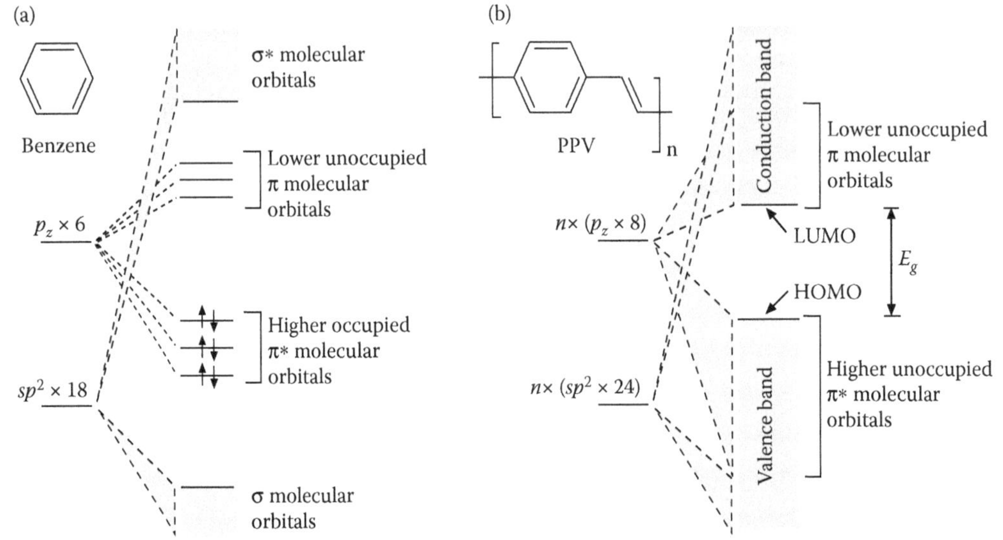

FIGURE 3.6
Schematic representation of formation of conduction and valence bands in (a) benzene (small molecule) and (b) PPV (polymer) molecules.

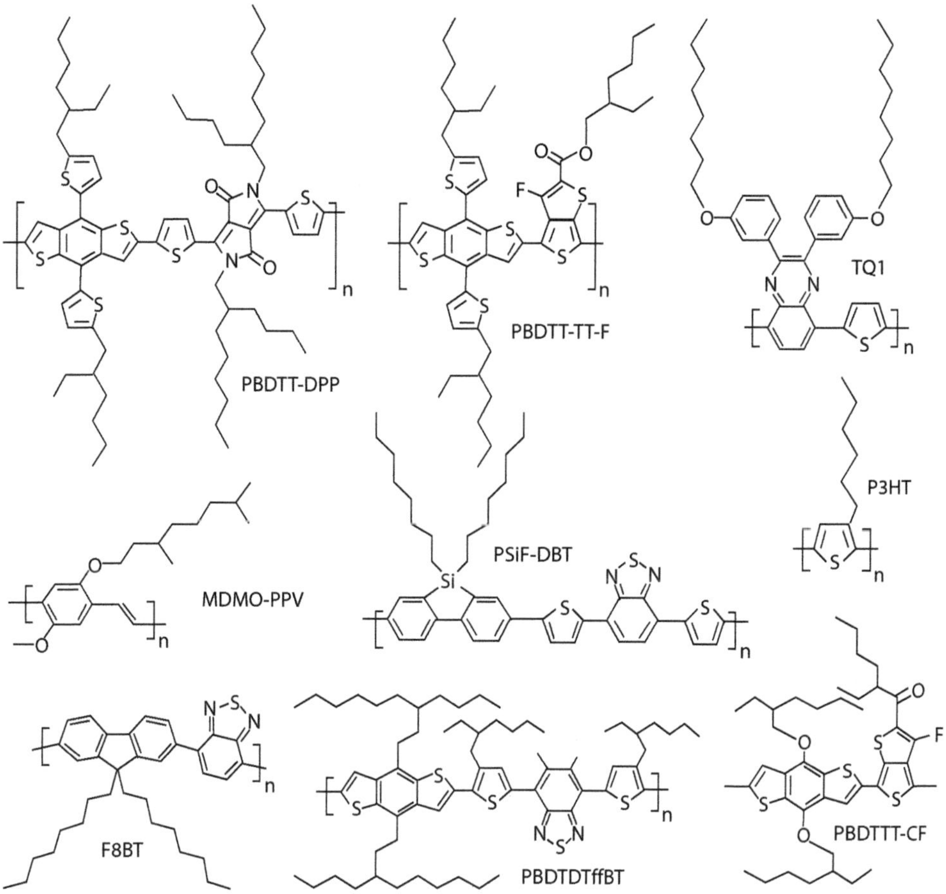

FIGURE 3.7
Chemical structure of some of the important conjugated polymers.

Each and every conjugated organic semiconductor has a unique chemical structure, which determines its optical and electrical behavior. The chemical structures of some of the most important conjugated polymers and small molecular materials specifically used for OSCs are illustrated in Figures 3.7 and 3.8, respectively. The chains of conjugated polymers are not straight but rather make twists and coiled structures resulting in an amorphous nature of the polymers. This disordered morphology limits the delocalization length of the π-cloud of electrons to a definite length known as a conjugation length. The conjugation length segments have random distributions leading to different energies of the π-electrons. This is clearly manifested in the featureless, broad absorption, and emission spectra of conjugated polymers. Assuming a simple one-dimensional particle-in-a-box picture, the longer segments will have a low π–π^* energy gap whereas the gap of the shortest segments will be much higher. The electronic properties of these materials depend on the conjugation length and the presence of electron withdrawing and donating groups. There are lots of possibilities to tune the optoelectronic properties of these materials and the chemists can easily synthesize new organic molecules with desired optoelectronic properties. The emission spectrum is highly Stokes shifted because excitons on high-energy segments will undergo rapid energy transfer to lower energy segments so that nearly all the emission comes from

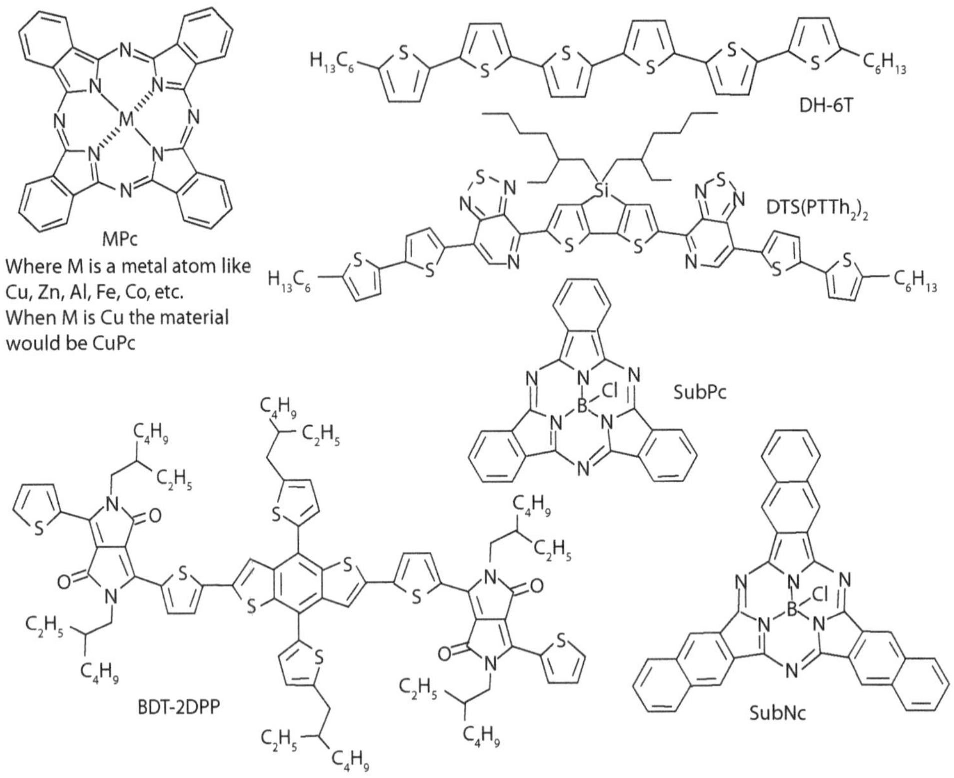

FIGURE 3.8
Chemical structure of some of the important small molecular materials.

low energy, long conjugation length segments. Chain distributions, therefore, determine the morphology, optical, and electrical behavior of a conjugated polymer. The distribution of chain length can be controlled to a certain extent through well-maneuvered synthesis steps. The functionality of conducting conjugated polymers can easily be tailored through careful molecular design and synthesis giving rise to new, specific optical and electrical properties to the polymers. Although the organic semiconductors have shown various electronic applications, in this book I shall focus on the understanding of organic-semiconductor-based solar cells and their future prospects only.

3.2.2 Electrical and Optical Properties of Organic Semiconductors

Electrical conductivity of a material is given by the product of total charge in the material per unit volume and the mobility of charge carriers. If the charge carrier density in a material is zero or the charge carriers are immobilized, the material is considered to be an insulator. The intrinsic hole density in an undoped semiconductor is calculated from $p = N_v \exp(-E_g/2kT)$, where N_v is effective density of states in the valance band and E_g is the band gap of the material. Considering a band gap of ~2 eV and charge carrier mobility of ~1 cm^2/Vs, the conductivity of an undoped molecular organic solid comes out to be less than 10^{-12} S/cm (see Figure 3.2), suggesting the materials are practically insulators. However, for conventional semiconductors where band gap is about 1.1 eV and charge carrier mobility is around 1000 cm^2/Vs, the conductivity of about 10^{-6} S/cm is predicted.

The organic semiconductors are inherently insulators, but they are call semiconductors because they can conduct electrical current if the charge carriers could be injected or created in them. Also their electrical conductivity increases exponentially with temperature, and some other properties like band gap are similar to those of the inorganic semiconductors. But the organic semiconductors differ from their inorganic counterparts in many ways. Organic semiconductors are mechanically very soft and possess low melting points. In inorganic semiconductors like Si, Ge, and GaAs the neighboring atoms in the lattice are strongly coupled together via covalent or ionic bonding in a long-range order that causes a strong electronic wave-function overlap over the atoms. Electronic wave-function overlap of the atoms forms the allowed conduction and valance bands. Due to electronic wave-function overlap, the charge carriers are delocalized within the energy bands and can transport from one atom to another very easily via strong exchange interaction of overlapped atomic orbitals. Such a transport mechanism is known as the band transport mechanism. In quantum mechanical terms the transport of charge carriers in inorganic semiconductors is described by Bloch functions, wave vector-space, and dispersion functions. On the other hand, the organic semiconductors are disordered and possess lots of structural and chemical defects and the intermolecular interactions are quite weak. The intramolecular interactions are mainly covalent, whereas the intermolecular interactions are weak van der Waals and London forces. The overlap of the molecular orbitals and the intermolecular electron exchange are very small and the energy levels in the conduction bands are discrete. In amorphous films, the surrounding polarization of the molecules varies spatially in a random manner that causes a random distribution in absolute values of molecular energies. The transport bands are quite narrow and the HOMO and LUMO levels can easily be disrupted. The energy levels in the bands are not continuous and the allowed band concept has limited validity. Weak electronic exchange among the molecules directly affects the optical and electrical properties of the semiconductors. The charge transport sites are proposed to be Gaussianly distributed in energy. The shape of the density of states (DOS) is Gaussian. Figure 3.9 shows the distribution of DOS in a typical inorganic and organic semiconductor. The charge carriers usually transport by hopping from one molecule to another. Hopping is a phonon-assisted tunneling mechanism and depends on the spatial and energetic difference between the hopping sites. The hopping transport phenomenon makes the charge carrier mobilities in these materials very

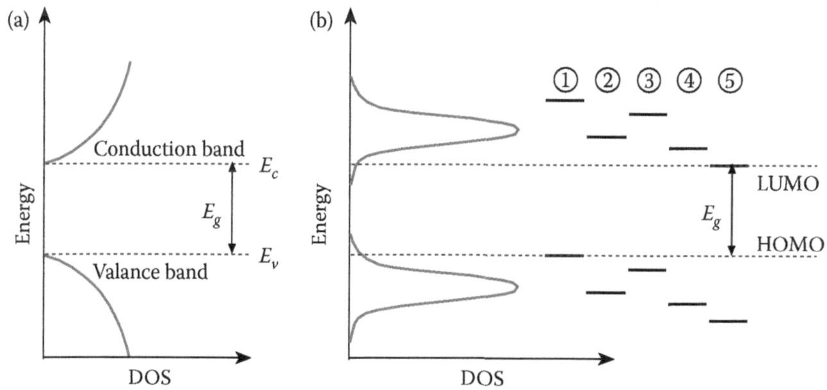

FIGURE 3.9
Schematic representation of distribution of DOS in (a) inorganic and (b) organic semiconductors. The numbers in circles correspond to individual molecules in the semiconductor.

low and strongly dependent on the temperature and electric field. However, for crystalline molecular solids the interaction between adjacent molecules can be relatively strong and instead of hoping the band transport can be observed. The charge transport phenomenon in organic semiconductors will be discussed in more detail in the next chapter. The dielectric constant of organic semiconductors is about 3 and that for conventional semiconductors is about 13. Due to low dielectric constants, the lifetime of the charge carriers in organic semiconductors is also very low, which leads to their small diffusion lengths.

Due to opposite charge on the electrons and holes, they are bound together by electrostatic coulombic attraction and this bound electron–hole pair is known as exciton. Exciton is an electrically neutral quasi-particle and is formed in semiconductors and insulators. It may be created by injection or generation (due to photo or thermal excitation) of charge carriers. In an exciton, the electron in conduction band (bound with a hole in the valance band) also feels some repulsion force from the electrons surrounding the hole in the valance band, therefore the exciton energy is slightly less than that of the unbound electron and hole. The lifetime of an exciton is limited by the wave-function overlap of the electron and hole in the exciton. Depending on the properties of medium material the excitons can be Wannier–Mott or Frenkel type. For the materials of high dielectric constants the electric field screening reduces the attraction force between electrons and holes, and the binding energy of excitons is very low on the order of 0.01 eV. Such excitons are named Wannier–Mott excitons after Gregory Wannier and Nevill Francis Mott. The radius of Wannier–Mott excitons is larger than the lattice spacing. Wannier–Mott excitons are formed in dense medium like inorganic semiconducting crystals. The attraction force between electrons and holes are screened by neighboring atoms that results in large radius. On the other hand, in the materials of low dielectric constants, the coulombic attraction between electron and hole is quite strong and results in a small exciton radius. The exciton radius is of the same size of the unit cell. The binding energy of such excitons is on the order of 0.1–1 eV and these excitons are known as Frenkel excitons named after Yakov Frenkel. These excitons are highly localized usually on a single molecule and are formed in molecular solids like organic semiconductors. The weak intermolecular interaction in organic semiconductors confines the electronic wave functions to the potential wells caused by their own positive charges. This causes the generated electrons and holes to be on the same molecule with large exciton binding energy. For better understanding the Wannier–Mott and Frenkel excitons are shown schematically in Figure 3.10. The excitons are also considered as mobile excited states.

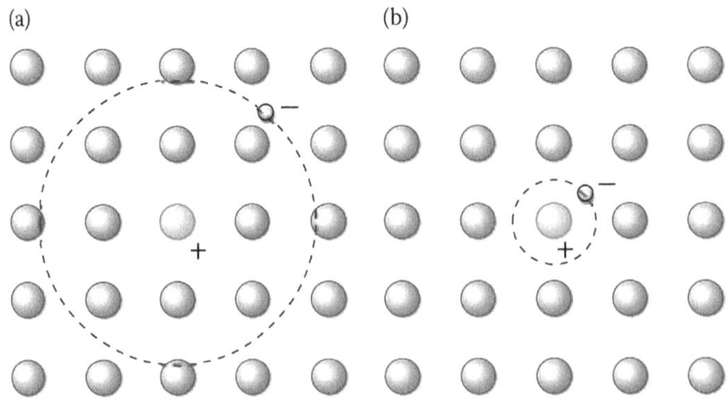

FIGURE 3.10
Schematic representation of (a) Wannier–Mott and (b) Frenkel excitons.

The binding energy of excitons can given by $E_B = (q_1q_2/4\pi\varepsilon\varepsilon_0 r)$, where q_1 and q_2 are the two opposite charges (which are same in magnitude), ε is the dielectric constant of the semiconductor, ε_0 is the permittivity of free space, and r is the distance between two charges. A rough calculation for an exciton of diameter ~1 nm in a medium of dielectric constant 3 (organic semiconductors) gives the binding energy of ~0.5 eV. On the other hand, the binding energy of excitons in inorganic semiconductors comes out to be ~0.01 eV. Therefore, the excitons in inorganic semiconductors break into free electrons and holes at room temperature (thermal energy ~0.026 eV), but those in organic semiconductors are tightly bound and do not break. The excitons in single organic semiconductors could be broken by increasing the temperature (which imparts thermal instability) or increasing the exciton radius. In OSCs, for efficient breaking of excitons, the electron donor–acceptor concept was introduced where two materials with different electron donating and accepting properties are used [2]. The energy offset at the donor–acceptor interface provides sufficient force to break the excitons. This phenomenon will be discussed in more detail later.

It is worth mentioning that a hydrocarbon possesses several bonding and antibonding orbitals, and depending upon the type of hydrocarbon, light absorption may result in different kinds of electronic transitions like $\sigma \rightarrow \sigma^*$, $n \rightarrow \sigma^*$, $\pi \rightarrow \pi^*$, and $n \rightarrow \pi^*$. The $\sigma \rightarrow \sigma^*$ transition is a high-energy process and occurs in the saturated hydrocarbons where all the valence shell electrons are involved in the formation of σ bonds. The $n \rightarrow \sigma^*$ transition takes place in saturated compounds containing one hetero atom with an unshared pair of electrons (n electrons) (like saturated halides, alcohols, ethers, amines, etc.). Such transitions require less energy compared to the $\sigma \rightarrow \sigma^*$ transition. The $\pi \rightarrow \pi^*$ transition occurs in unsaturated hydrocarbons (alkenes and alkynes). The excitation of π electrons requires low energy and hence this transition occurs at a longer wavelengths. The $n \rightarrow \pi^*$ transition takes place when an electron of an unshared electron pair on a hetero atom gets excited to the π^* antibonding orbital. Saturated aldehydes show both type of transitions $n \rightarrow \pi^*$ and $\pi \rightarrow \pi^*$. Due to weak intermolecular interactions, organic semiconductors have well-defined singlet and triplet spin states as in individual molecules. Usually the ground state is a singlet state (S0) and the photon absorption causes electron excitation to the first excited singlet state (S1). The electronic transitions in organic semiconductors that occur during both the absorption and emission processes correspond to $\pi \rightarrow \pi^*$ transitions. The lifetime of the S1 state is of a few nanoseconds and the excited electron returns back to the ground S0 state with radiative or nonradiative transition. There is also little probability that interstate crossing takes place, and the electrons in the triplet state (T1) return to ground state S0 with radiative or nonradiative transitions. The lifetime of triplet states is relatively longer in milliseconds.

As the LUMO and HOMO levels in organic semiconductors are not continuous, the absorption spectrum of organic semiconductors is expected to possess sharp peaks corresponding to the promotion of electrons from discrete ground electronic levels to discrete excited levels. But actually sharp peaks are seldom observed and instead broad absorption bands are observed. It can be understood from the distribution of DOS in these materials that is Gaussian in conduction and valence bands. The excitations of electrons are accompanied by the constant vibratory and rotatory motion of the molecules. A molecule in a particular electronic state possesses quantized vibrational and rotational states. Each electronic state possesses vibrational levels and each vibrational level possesses rotational levels. During excitation the electrons move from lowest vibrational and rotational level of the ground state to some vibrational and rotational level of the excited state. In this way there will be a large number of possible transitions among electronic, vibrational, and rotational states. Hence, a large number of wavelengths that are close enough are absorbed

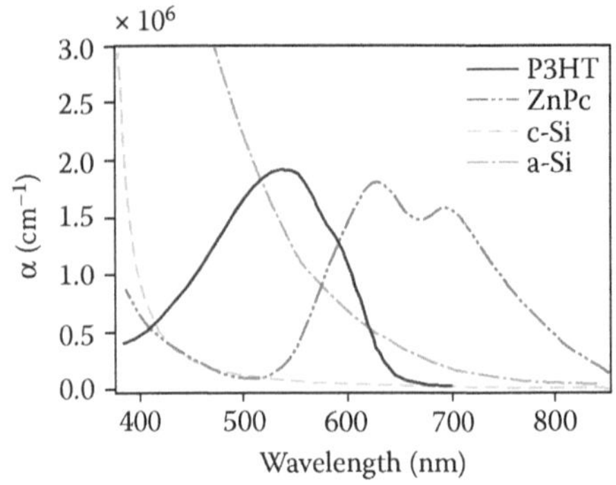

FIGURE 3.11
Absorption coefficients of some of the organic and inorganic semiconductors.

and an absorption band is formed. In complex molecules, which contain a large number of atoms, the multiplicity of vibrational sublevels and their closeness results in discrete bands that join together to form a broad absorption band. Usually emission spectra of organic semiconductors is red shifted compared to their absorption spectra. This red shift in emission spectrum is explained by the Franck–Condon principle, according to which the molecule in the excited electronic state quickly relaxes to the lowest vibrational level of an excited level by releasing some energy as heat. Then, the molecule relaxes to the ground state giving off photon emission. This transition is vertical and the molecule does not hit its zero vibrational state, therefore the molecule loses some energy to come to its ground state. Due to energy losses during relaxation of the molecule within the same electronic level, the emission is always red shifted (lower in energy) compared to absorption. This shift in the emission spectrum is known as Stokes shift.

Organic semiconductors have a very high absorption coefficient compared to inorganic semiconductors. The absorption coefficient is the measure of distance that is required to absorb the incident light photons. The relation between light intensity $I(\lambda, x)$ at any distance x in the medium and absorption coefficient of the medium is given by $I(\lambda, x) = I_0(\lambda) \exp(-\alpha(\lambda)x)$, where $I_0(\lambda)$ is the incident light intensity and $\alpha(\lambda)$ is the absorption coefficient. Therefore, the absorption coefficient of a material is the inverse of the distance traveled by incident light in that material by which the light intensity reduces to $1/e$ times of its original value. The absorption coefficient of important organic semiconductors exceeds 10^5 cm^{-1} in the visible spectrum that gives the required film thickness of less than 100 nm for sufficient absorption of incident light. In the case of inorganic semiconductors, the absorption coefficients are about 2 orders of magnitude less, therefore about 100 times thicker films are required for efficient light absorption. Figure 3.11 shows the absorption coefficients of some important organic and inorganic semiconductors.

3.3 Architecture of Organic Solar Cells (OSCs)

OSCs are actually bipolar devices that possess two electrodes with active organic materials sandwiched between them. The two electrodes are of different work functions and the difference in the work functions develops a built-in electric field, directed from a low work function (cathode) to a high work function (anode) electrode. For incident light to reach the light-absorbing organic semiconductors, at least one of the electrodes should be optically transparent and for this purpose the transparent conducting oxide (TCO) materials like indium tin oxide (ITO) and fluorine doped tin oxide (FTO) are the best options, and ITO is the most frequently used TCO for OSCs. ITO has not only high optical transparency but also high electrical conductivity. As far as basic geometry is concern, the OSCs are always prepared in superstrate configuration where substrates are either glass or plastic sheets. ITO substrates enable the OSCs to be prepared in both the normal as well as reverse geometries. In normal geometry ITO is used as a hole-collecting electrode (anode), whereas in inverted geometry ITO collects electrons (cathode) under illumination.

3.3.1 Normal Geometry of OSCs

In normal cell geometry the ITO-coated substrates are coated with some suitable surface modifier or buffer layer to smooth the ITO interface and to increase the effective work function of the anode. This buffer layer should possess good hole-transport properties and a work function higher than that of ITO. Poly(ethylene-dioxythiophene):poly-styrene sulfonate (PEDOT:PSS), MoO_3, V_2O_5, and NiO are some of the materials used for this purpose [7–15]. PEDOT:PSS is actually a PSS-doped PEDOT polymer that possesses high conductivity and high work function. An increase in anode work function enhances the built-in electric field improving the performance of the cell. Thin films of PEDOT:PSS can be prepared by spin coating, dip coating, or inkjet printing of its water-based suspension. Thin films of other buffer materials like MoO_3 and V_2O_5 are prepared either by spin coating from their solutions or by thermal evaporation or e-beam evaporation in vacuum [8–15]. Some groups have also used MoO_3-PEDOT:PSS composite films as the hole-transport buffer layers and achieved considerable improvement in cell performance and its stability [7,16]. On the hole-transport buffer layer, the light-absorbing active layers of organic semiconductors are deposited by some suitable means. The methods of deposition of different active layers are discussed later in detail. The active layer is finally coated with electron collecting top electrode (cathode) by e-beam or thermal evaporation in vacuum. Usually the cathode materials are low work-function metals like Ca, Mg, Al, Ag, and their combinations. OSCs can also possess electron buffer layers like LiF and $CsCO_3$ before the cathode for better performance [16,17]. The top electrodes are opaque to light, therefore light is illuminated from the ITO side unless the solar cells are transparent where both the top and bottom electrodes are optically transparent and light could be illuminated from either side [18]. Light absorption in the active layer generates electron–hole pairs and the built-in electric field directs the photogenerated charge carriers to transport and collect at respective electrodes. Figure 3.12a shows the schematic structure of a typical OSC in normal geometry. Though there are numerous publications in the literature on high-efficiency OSCs with normal geometry, for an example the schematic structure of an OSC prepared by Liang et al. in normal geometry that exhibited about 7.4% PCE [19] is shown in Figure 3.12b.

3.3.2 Inverted Geometry of OSCs

Inverted OSCs are also prepared on ITO-coated substrates, but the directions of collection of respective photogenerated charge carriers are opposite to that in the normal geometry. Here ITO is used as the electron-collecting cathode, whereas the top electrode collects the holes. For efficient electron collection, ITO is coated with low work function and a high band gap transition metal oxide like ZnO, TiO_x, and TiO_2, or polymers like

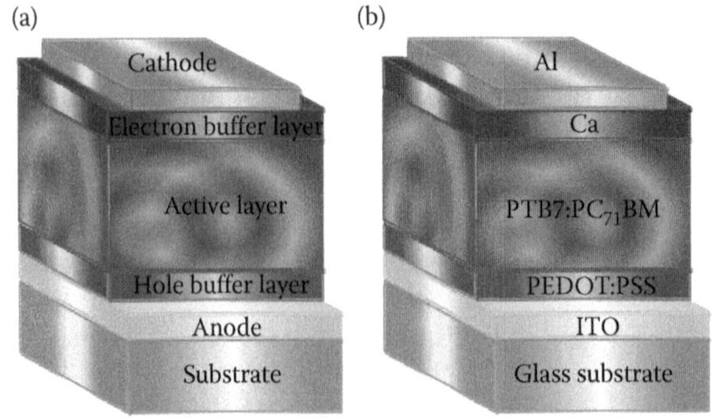

FIGURE 3.12
(a) Schematic structure of a typical OSC with normal geometry. (b) Schematic structure of the normal OSCs. (Data from Y. Liang et al., 2010. *Adv. Mater.* 22: E135–E138.)

poly[(9,9-bis(3′-(*N*,*N*-dimethylamino)propyl)-2,7-fluorene)-alt-2,7-(9,9-dioctylfluorene)] (PFN) [20–24]. Then, like in normal geometry, the active layer is coated by some suitable means that is followed by deposition of a thin film of high work function hole-transport buffer layer like PEDOT:PSS, MoO_3, and V_2O_5. Finally, a thin layer of Al, Ag, or Au is deposited by e-beam or thermal evaporation [20–24]. The difference in work function of the two electrodes develops a built-in electric field that directs the photogenerated electrons toward ITO and holes toward the top electrode. Figure 3.13a shows the schematic structure of a typical inverted OSC. Like normal geometry, the inverted geometry has shown equally efficient but highly stable solar cells [20,22,24]. Yang et al. used ZnO and a polymer PFN-Br to modify the ITO electrode and got an efficiency of 8.4% with a PBDT-DTNT:$PC_{71}BM$ active layer and MoO_3/Ag anode [24]. Similarly, He et al. used PFN polymer to modify the ITO surface and achieved 9.2% PCE in inverted cell geometry with a PTB7:$PC_{71}BM$ active layer and MoO3/Ag anode [20]. The structure of an inverted OSC prepared by He et al. [20] is schematically shown in Figure 3.13b. Many structures have been employed in both the normal and inverted geometries of OSCs and are discussed in the following sections.

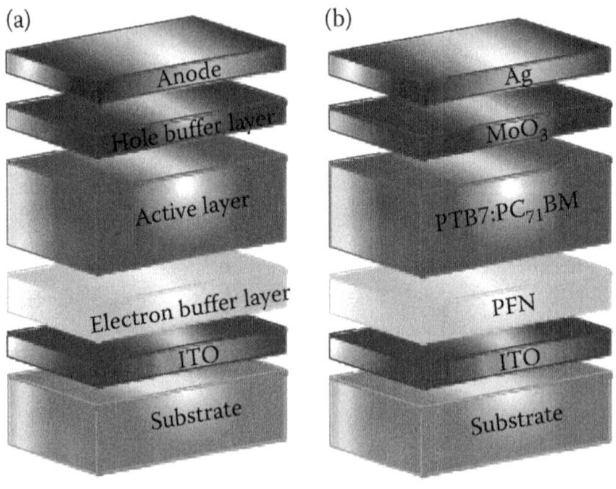

FIGURE 3.13
(a) Schematic structure of a typical OSC with inverted geometry. (b) Schematic structure of the inverted OSCs. (Data from Z. He et al., 2012. *Nat. Photon.* 6: 591–595.)

The first OSC was based on single organic material and due to low charge carrier mobilities and small diffusion lengths of charge carrier these solar cells suffered from great recombination losses and exhibited very low PCE of ~0.5% [25]. In a single organic layer solar cell only one light-absorbing organic layer is sandwiched between the two electrodes. Each absorbed photon generates one exciton, which requires some force to break into a free electron and hole. For dissociation the excitons have to diffuse to the cathode interface where excitons may break by transferring the electrons to cathode. The holes are left in the HOMO of the organic layer, which transport to the anode via drift and diffusion processes. But due to short exciton diffusion lengths in organic semiconductors, most of the excitons recombine before they reach the cathode. Still most of those surviving and reaching the cathode interface quench through interface states. Some of the surviving excitons dissociate into electrons and holes; however, due to very low hole mobility only a few holes get collected at the anode giving a little photocurrent. Therefore, due to lots of recombination losses in single organic layer solar cells, the PCE remained extremely low. In 1986, Tang introduced the electron donor and acceptor concept, which caused enhanced exciton dissociation and enhanced PCE [2]. The excitons diffused to the donor–acceptor interface and dissociated ultrafast within ~10^{-15} s into free electrons and holes. The donor and acceptor materials were chosen in a way that the acceptor had higher electron affinity compared to the donor, and the donor had lower ionization potential compared to the acceptor. The energy offset between electron affinities of donor and acceptor materials provided the necessary force to break the excitons. Electrons transferred to the LUMO of the acceptor, whereas holes left in the HOMO of the donor. Electrons transported through acceptor channel and holes transported through donor channel to collect at respective electrodes. This concept proved the importance of donor and acceptor materials in OSCs that could either be employed in planar heterojunction or bulk-heterojunction structures.

3.3.3 Bilayer/Planar Heterojunction Structure

A bilayer or planar heterojunction structure is the one where donor and acceptor materials are deposited one on another to form a heterojunction interface. The first planar heterojunction OSC was prepared by Tang, where he employed small molecular copper phthalocyanine (CuPc) and a perylene derivative 3,4,9,10-perylene tetracarboxylic bis-benzimidazole (PTCBI or Im-PTC) as the donor and acceptor materials, respectively [2]. This was also the first demonstration of donor–acceptor OSCs that exhibited ~1.0% PCE. Figure 3.14 schematically shows the structure of the bilayer solar cell prepared by Tang and the corresponding energy level diagram with working mechanism. This demonstration of efficient exciton dissociation at the donor–acceptor interface revived the interest in OSCs and developed the hope of fabrication of efficient OSCs. Initial efforts to make efficient planar heterojunction solar cells using conjugated polymers were not successful, because thin films of conjugated polymers could be prepared by solution processing only and the processing of one layer over another damaged the layer underneath due to solubility in common solvents. Therefore, the planar heterojunction structure was generally employed for small molecular materials, which were deposited via thermal evaporation in vacuum. However, new methods were later developed to deposit one polymer layer over another to make polymer planar heterojunction solar cells [26–29].

The materials with better optical and electrical properties were expected to give better PCE, therefore, Forrest et al. used CuPc as the donor and perylene tetracarboxylic

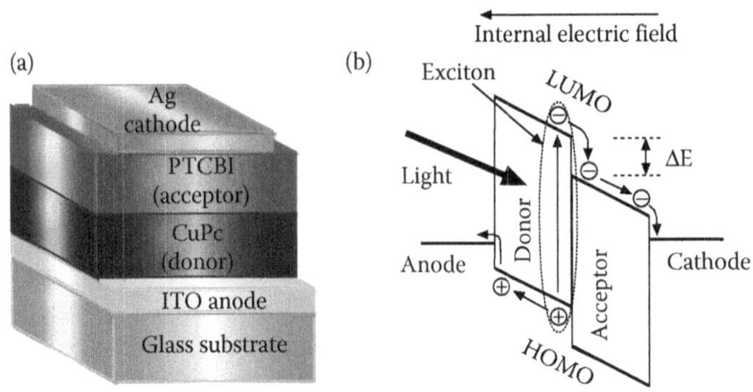

FIGURE 3.14
(a) Schematic structure of the bilayer OSC. (Data from C. W. Tang, 1986. *Appl. Phys. Lett.* 48: 183.) (b) The corresponding schematic energy level diagram with the representation of exciton generation and dissociation.

dianhydride as the acceptor and achieved an efficiency of 1.8% in the planer structure [30]. Later fullerene (C_{60}) was identified as a potential material for acceptor applications due to its larger exciton diffusion length compared to perylene derivatives, energy levels compatible to most donor materials, and better electronic properties [31]. Although C_{60} did not show strong absorption in the visible range of solar spectrum, it was still better than other acceptors. Later C_{60} was used with many other small molecular donors such as zinc phthalocyanine (ZnPc) [32], boron subphthalocyanine (Sub-Pc) [33,34], tin phthalocyanine (SnPc) [35], metal free phthalocyanine (H_2Pc) [36], pentacene [37], bis(2,2-dicyanovinyl)-quinquethiophene [38] to give high-efficiency solar cells.

The planar heterojunction solar cells still suffered from lots of recombination problems that kept their efficiency very low. The optical absorption length of organic semiconductors is ~100 nm, but due to small diffusion length of excitons (L_D ~ 5–10 nm) most of the excitons get recombined and only those generated in the close vicinity of the donor–acceptor interface have efficient dissociation. Also, the excitons generated in the acceptor get quenched at the metal–organic interface by the induced defect states [39]. To prevent exciton quenching at the metal organic interface, Peumans et al. introduced an exciton blocking layer (EBL) of bathocuproine (BCP) between the active organic layer and the metal cathode, which improved the PCE from 1.0% to 2.4% [40]. EBL benefits the solar cells in many ways; it prevents the organic layer from damage by cathode deposition, eliminates exciton quenching, blocks and confines the excitons in active layer, reduces the organic/cathode interface resistance, and works as an optical spacer to give enhanced absorption and improved PCE [39,41–43]. Later, many new ideas were implemented to ensure enhanced efficiency in planar heterojunction devices by minimizing the optical and electrical losses. For example, Chan et al. doped the donor and/or acceptor layers with 5,6,11,12-tetraphenylnaphthacene (rubrene) and a tremendous improvement in the efficiency was observed [44]. The device prepared in ITO/CuPc (10 nm)/CuPc:rubrene (10 nm)/C_{60}:rubrene (10 nm)/C_{60} (30 nm)/BPhen (5 nm)/Al configuration exhibited an efficiency of 5.58% under 1-sun illumination. This improvement in efficiency was because of improved spectral coverage of the cell as rubrene possesses the complementary absorption in the range where CuPc and C_{60} did not absorb.

Small diffusion lengths of excitons in organic semiconductors limit the thickness of the active layer to a few hundred angstroms, which is further limited by the low charge carrier mobilities. On the other hand, very thin films suffer from inefficient light absorption. Small interfacial area between donor and acceptor materials in the planar heterojunction structure caused reduced dissociation of excitons, therefore for enhanced exciton dissociation, a larger interfacial area was needed. The solution to this dilemma was found to be the mixing of the donor and acceptor materials into a single layer where numerous heterojunctions are formed throughout the bulk and ensure maximum dissociation of excitons [45,46].

3.3.4 Bulk-Heterojunction (BHJ) Structure

Intermixing of donor–acceptor materials generates numerous donor and acceptor interfaces with numerous interpenetrating donor–acceptor networks. Such a structure with interpenetrating donor–acceptor networks is known as a bulk-heterojunction (BHJ) structure. The BHJ concept was first used for polymer solar cells in 1995, where the donor was a polymer and the acceptor was a derivative of fullerene [45,46]. Heeger et al. used poly[2-methoxy-5-(2-ethylhexyloxy)1,4-phenylene-vinylene] (MEH-PPV) as the donor and phenyl [6,6] C_{61} butyric acid methyl ester (PCBM) as the acceptor and got a PCE of ~1.5% [45]. This efficiency was almost 50% more than that reported by Tang in planar heterojunction configuration [2]. Because of low entropy of mixing, the blend has phase separation between donor–acceptor materials and if the length scale of the separated phases is of the order of the diffusion length of the excitons, almost all the excitons generated in the blend get dissociated at the donor–acceptor interface. After exciton dissociation the electrons transport in the acceptor phase, whereas holes transport in the donor phase to collect at the cathode and anode, respectively. There are lots of loss/recombination processes that take place from exciton generation to collection of the charge carriers at corresponding electrodes that will be discussed later in detail.

Compared to planar heterojunction structure the BHJ structure essentially extends the photoactive layer thickness and decreases the charge carrier recombination probability, leading to high short-circuit current density (J_{sc}) [47]. Figure 3.15 shows the schematic structure of the BHJ solar cell prepared by Heeger et al. [45] and the corresponding

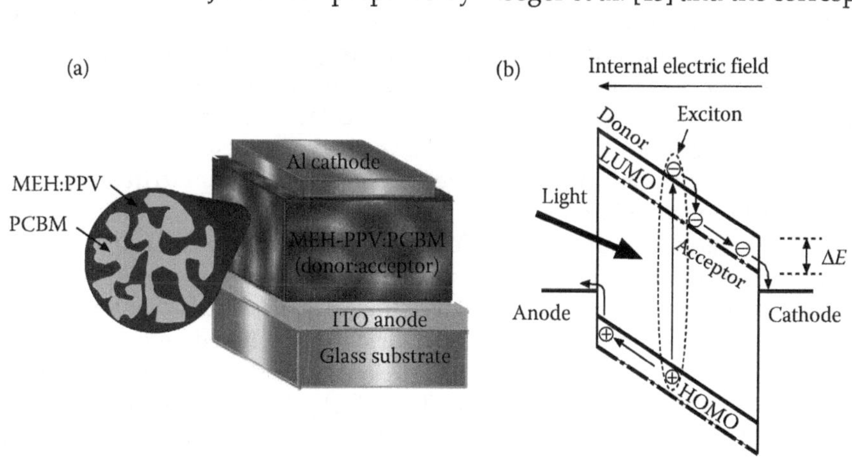

FIGURE 3.15
(a) Schematic structure of the BHJ solar cell. (Data from G. Yu and A. J. Heeger, 1995. *J. Appl. Phys.* 78: 4510–4515; G. Yu 1995. *Science* 270: 1789–1791.) (b) The corresponding schematic energy level diagram with the representation of exciton generation and dissociation.

schematic energy level diagram. It was expected that the materials with broad absorption spectrum and higher charge carrier mobilities would give better performance, but interestingly the nanoscale morphology of the active layer and the separated phases were also found to be very crucial in determination of BHJ cell performance. In 2001, Shaheen et al. used poly(2-methoxy-5-(3,7,dimethyloctyloxy)-1,4-phenylenevinylene) (MDMO-PPV) and PCBM as the donor and acceptor materials, respectively, and prepared the blend solutions in two different solvents like toluene and chlorobenzene. The BHJ solar cells, processed with chlorobenzene exhibited a record jump in efficiency to 2.5% [48]. Since MDMO-PPV had better optical and electrical properties compared to MEH-PPV, the improvement in efficiency was quite expected, but more importantly the BHJ solar cells prepared with toluene as a solvent exhibited rather inferior photovoltaic performance. This variation in the cell performance with two different solvents was attributed to different nanoscale morphologies of the blend layers and the one with chlorobenzene was better. Since then, several donor–acceptor materials with better optical and electrical properties have been employed with improved nanostructured morphologies to give high-performance OSCs. Nanoscale morphology of the active layer depends upon a number of parameters that are discussed later in this chapter. BHJ structure could easily be employed in small molecular solar cells as well as by coevaporation of the donor and acceptor materials [49,50]. Today, the record efficiency in OSCs is owned by BHJ solar cells incorporating advanced materials [21].

3.3.5 Hybrid Planar-Mixed Heterojunction Structure

A hybrid planar-mixed heterojunction structure incorporates a BHJ layer sandwiched between the homogeneous donor and acceptor layers. This kind of structure results in efficient photon harvesting and better transport of photogenerated charge carriers to the respective electrodes. The thickness of each homogeneous layer is kept approximately equal to the diffusion length of excitons. In this way, the exciton dissociation occurs throughout the mixed layer and at the interfaces of the mixed layer and homogenous layers as well. It is noteworthy that for efficient photon harvesting the thickness of the mixed layer should be less than the charge collection length. The charge collection length is nothing but the sum of the hole and electron diffusion lengths, which depends on the mobilities of charge carriers. As the mean distance between two similar neighboring molecules in a mixed layer is more than that in a homogeneous layer, the mobility of charge carriers from one molecule to another in the mixed layer becomes less than that in the homogenous layer. This lower mobility in the mixed layer causes a lower drift length that would lead to incomplete collection of photogenerated charge carriers. The collection length also depends upon the applied voltage and decreases rapidly for a thicker mixed layer. It is therefore important to have high charge carrier mobilities in the mixed layers so that the collection length is more than the mixed layer thickness and full advantage of the mixed layer could be availed. Xue et al. fabricated a hybrid planar-mixed heterojunction solar cell where the mixed photoactive layer of CuPc and C_{60} was sandwiched between the homogeneous layers of CuPc and C_{60}. The cell exhibited a PCE of ~5.0% [47]. The mixed layer was composed of a 1:1 ratio of CuPc and C_{60}, and the thicknesses of the donor, mixed, acceptor layers were 15, 10, and 35 nm, respectively. This device exhibited ~40% higher efficiency compared to planar heterojunction and mixed heterojunction solar cells [42,50]. Though similar structures were also reported much earlier, the advantage of a hybrid planar-mixed heterojunction structure could not be encased because of the large thickness of the mixed layers, which exceeded the charge collection lengths [51,52].

3.3.6 *p-i-n* Structure

OSCs can also be prepared in *p-i-n* structure. Due to low diffusion lengths of excitons in most organic semiconductors, the active region is kept very narrow, and to avoid shorts and recombination some additional transport layers are employed that make the solar cells reasonably thick [40]. These additional layers do not contribute to photocurrent but limit the electric current by introducing additional series resistance. If these layers could be made highly conducting, the collection of photogenerated charge carriers and cell performance would improve. For this purpose the noncontributing transport layers are replaced by doped wide band-gap materials [32]. Therefore, the active organic layers are sandwiched between two doped wide band-gap *p*-type and *n*-type materials. Such a solar cell is known as *p-i-n* solar cells where the active organic layer is termed the intrinsic layer. Only the intrinsic layer, which could either be bilayer or BHJ, absorbs the incident light and the doped layers help in charge transportation. There have been various reports on *p*-type and *n*-type dopings of organic semiconductors [32,53–55]. F4TCNQ and rhodamine B are the important materials used, respectively, for *p*-type and *n*-type doping of various organic materials.

Gebeyehu et al. [56] fabricated the organic *p-i-n* solar cells where donor–acceptor blends of small molecular materials were sandwiched between two *p*-type and *n*-type doped wide band-gap charge transport layers. ZnPc and C_{60} were used as the donor and acceptor materials, respectively. For the *p*-type transport layer, *N,N,N′,N′*,-tetrakis (4-methoxyphenyl)-benzidine (MeO-TPD) was doped with 2% of F4TCNQ, whereas the *n*-type transport layer was obtained by 2% doping of rhodamine B in C_{60}. The device with the *n*-type C_{60} layer exhibited an efficiency of 1.9%. This efficiency was almost 2 times of that reported for similar cells without any doped charge transport layer [57]. Later the *p-i-n* concept was also employed in tandem structures and a PCE of 3.8% was achieved that was almost double of that achieved with a single *p-i-n* solar cell [32]. In *p-i-n* OSCs, the *p*-type and *n*-type layers are only transport layers but play various roles in determination of cell performance. The doped layers do not absorb light but help in efficient absorption of incident light by redistributing the optical field inside the cell. The active layer can be varied to enough thickness without limitations due to shorts. For efficient devices, the interfaces of doped layers and active organic layers should be free from any kind of barrier for charge transfer; therefore alignment in the energy levels is very important. There should not be transfer of the minority carriers to the transport layers and only the right type of carriers should leave the active layer to collect at the electrodes. The main reason for this approach not being used frequently is the difficulty in production of stable and reproducible doped organic layers. A nice review on organic *p-i-n* solar cells can be found in Maennig et al. [58].

3.3.7 Tandem Structure

To overcome the limited absorption of incident light in single OSC, the solar cells are prepared in tandem structure where two or more cells are stacked one over another so that the fraction of light not absorbed by one cell is absorbed by other. Tandem structure is nothing but a series connection of two or more solar cells. A tandem structure can have higher optical density over a wider range of solar spectrum than that of a single cell without increasing the internal resistance. This concept is well proven in most efficient III–V multijunction solar cells [59]. In OSCs this concept was used by Hiramoto et al. in 1990, where two thin small molecular cells were stacked in series by a noncontinuous ultrathin layer of Au [60]. The tandem cell exhibited PCE more than double that of the individual cells. Later this concept was used vastly for both small molecular and polymeric systems [26–29,61–67]. Initially the tandem structure was somewhat difficult for polymer solar cells as one polymer film damaged the other film underneath due to its solubility in common solvents. But now, this problem has been fixed with suitable bridging interlayers [26–29,66]. The interlayer connects the two cells and provides the junction for recombination of electrons from one cell and holes from other. It acts as cathode for one cell and anode for the other cell. It also enhances the optical field in the cell and controls the cell performance.

Usually in inorganic tandem or multijunction solar cells the front cell is made of a higher band gap semiconductor and the back cell possesses lower band gap semiconductors so that lower wavelength photons (higher absorption coefficient) are absorbed in the front and higher wavelength photons (lower absorption coefficient) are absorbed in the back cell. To balance the photocurrent in two cells, the front cell is made thinner than the back cell. But in the case of OSCs, it is not only a matter of sequence but also the matter of optimization [32,68,69]. There have been a number of reports in the literature on efficient tandem OSCs, and for an example Figure 3.16a shows the schematic structure of then most efficient polymeric tandem OSC prepared by Kim et al. [68]. The tandem cell possessed two subcells each made of polymeric BHJ network. The two cells possessed materials of different absorption characteristics to harvest a wider range of solar spectrum. The front cell possessed a composite of lower band-gap poly[2,6-(4,4-bis-(2-ethylhexyl)-4*H*-cyclopenta[2,1-b;3,4-b′]dithiophene)-alt-4,7-(2,1,3-benzothiadiazole)] (PCPDTBT) and $PC_{61}BM$, whereas the back cell was made of the composite of higher band gap P3HT and $PC_{71}BM$. A thin transparent layer of wide band gap TiO_x connected the two cells. The connecting TiO_x layer provided a strong foundation for the fabrication of another cell on it. The reason behind taking the lower band gap PCPDTBT cell as the front cell was that optimized thickness for PCPDTBT:$PC_{60}BM$ is less than that for P3HT:$PC_{71}BM$ [68], and for a balanced current the front cell is required to be thin compared to the back cell. PEDOT:PSS was used as the hole-transport layer for the two cells. Figure 3.16b schematically shows the corresponding energy level diagram and the mechanism of photocurrent generation in the cell. The photogenerated holes in the front cell transport through the PEDOT:PSS layer and collect at the ITO anode, whereas the photogenerated electrons transport through the TiO_x layer and reach the TiO_x–PEDOT:PSS interface. Similarly the holes from the back cell transport and collect at the TiO_x–PEDOT:PSS interface, whereas

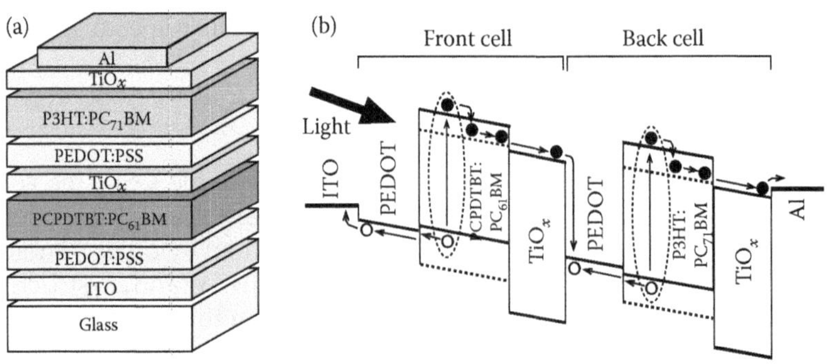

FIGURE 3.16
(a) Schematic structure of the tandem solar cell. (Data from J. Y. Kim et al., 2007. *Science* 317: 222–225.) (b) The corresponding energy level diagram and representation of photocurrent generation.

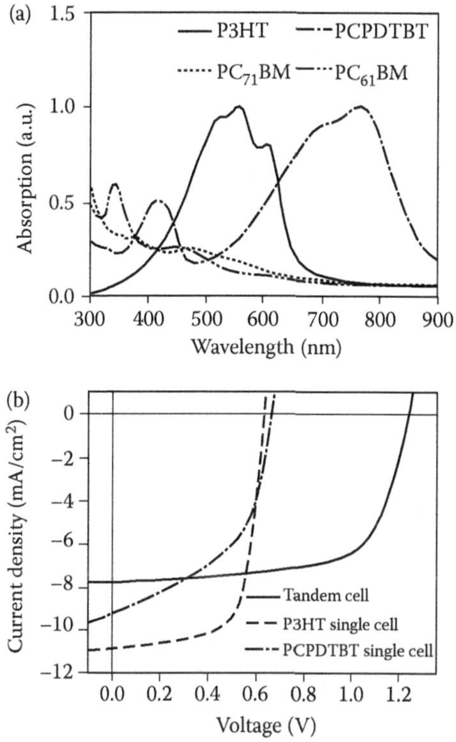

FIGURE 3.17
(a) Absorption spectra of thin films of P3HT, PCPDTBT, $PC_{71}BM$, and $PC_{61}BM$. (b) *J–V* characteristics of single and tandem solar cells made of P3HT:$PC_{71}BM$ and PCPDTBT:$PC_{61}BM$ composites (Modified from J. Y. Kim et al., 2007. *Science* 317: 222–225.)

electrons transport through the TiO_x layer and collect at the Al cathode. The TiO_x layer serves several functions here, for example, it is hydrophilic and provides a strong base for the next PEDOT:PSS coating on it, it breaks the symmetry in the first cell and generates V_{oc}, it works as an electron-transport layer, it works as a hole blocking layer for the back cell, and it also acts as an optical spacer that redistributes the light intensity in the cell for maximum efficiency. The electrons from the front cell combine with the holes from the back cell at the TiO_x–PEDOT:PSS interface. The tandem cell exhibited much better PCE compared to single BHJ solar cells made of PCPDTBT:$PC_{61}BM$ and P3HT:$PC_{71}BM$ composites. Figure 3.17a shows the absorption spectra of the two polymers along with that of $PC_{61}BM$ and $PC_{71}BM$. The absorption spectra of the two polymers are complementary to each other. Due to light absorption in different spectral regions by the two cells, the photon harvesting improved. The *J–V* characteristics of single and tandem solar cells using PCPDTBT:$PC_{61}BM$ and P3HT:$PC_{71}BM$ composites are shown in Figure 3.17b. The single PCPDTBT:$PC_{61}BM$ solar cell yielded $J_{sc} = 9.2$ mA/cm², $V_{oc} = 0.66$ V, $FF = 0.50$, and $\eta = 3.0\%$, whereas the P3HT:$PC_{71}BM$ single cell yielded $J_{sc} = 10.8$ mA/cm², $V_{oc} = 0.63$ V, $FF = 0.69$, and $\eta = 4.7\%$. In the case of the optimized tandem cell, the performance parameters were observed to be $J_{sc} = 7.8$ mA/cm², $V_{oc} = 1.24$ V, $FF = 0.67$, and $\eta = 6.5\%$.

As the subcells are connected in series, the V_{oc} of a tandem cell is given by summation of the V_{oc}s of subcells. But according to Kirchoff's law, the total current of the cell is limited by the subcell generating the lowest current, and the total current remains less than or equal to the current generated by that subcell. Because of imbalanced generation of charge carriers in the front and back subcells, the excess charge carriers do not contribute to the photocurrent and accumulate in the cell, which reduces its V_{oc}. Therefore, to exploit the maximum advantage of tandem structure, the current from the bottom cell should match with that of the top cell. This approach is based on the optimization of thicknesses of active layers with respect to their charge transport properties.

3.4 Materials and Processing

For separation of photogenerated charge carriers in the active layer, the electrons in the donor should be donated to the acceptor and the holes should be left in the donor itself. Even if the excitons are generated in the acceptor, the electrons should remain on the acceptor and the holes should be transferred to the donor. The electron donating and accepting properties of the materials are decided by their electron affinities (E_a) and ionization potentials (I_p). The materials having low I_p can easily donate the electrons and therefore can serve as electron donors, whereas those having high E_a can easily take electrons and can serve as electron acceptors. However, their compatibility with each other is decided by their processing methods and relative energy levels. Where LUMO offset should be at least equal to the exciton binding energy, the energy offset between HOMOs should be high enough to prevent hole transfer from donor to acceptor. The electron acceptors should possess high electron mobilities, whereas the donors should have high hole mobilities. Apart from the active layer, the transport layers are also equally important and should have high charge carrier mobilities with energy levels that introduce no barriers for charge extraction. During device fabrication, the deposition of one material should not affect in any way the underlying film.

Thermal evaporation in vacuum and solution processing are the most frequently used methods to fabricate OSCs. Usually the small molecular materials are not soluble in common organic solvents, therefore, thin films are prepared via thermal evaporation in vacuum. Under high vacuum these materials get sublimed before they melt or reach their decomposition temperature. Deposition of thin films via thermal evaporation does not affect the quality of the underlying film and is quite an easy way to deposit one layer on another with precise controlling of film thickness. The rate of film growth can also be precisely controlled. Thermal evaporation gives us the liberty to deposit as many layers of different material as we want, in any sequence. Using thermal evaporation, the device design and its engineering became very easy. But unfortunately, it is not a cost-effective method for large-scale production and also has substrates size limitation that is limited by the size of evaporation chamber. On the other hand, thin films of conjugated polymers cannot be prepared by thermal evaporation as they suffer from chain breaking and decomposition upon heating. Therefore, thin films of polymers are prepared by solution processing. However, there have been some polymers that do not dissolve in any organic solvent. Such polymers are either polymerized in situ or modified via attachment of side groups in the polymer backbone. Side group attachment makes the polymers soluble in most organic solvents. C_{60} is not soluble in organic solvent, therefore its thin films are prepared by thermal evaporation. However, its derivatives are soluble in most organic solvents and their films are prepared by solution processing. That is why C_{60} is mostly used in small molecular solar cells, whereas its derivatives are used in solution-processable polymer solar cells. Before we learn something about the materials, let us understand the processing techniques first.

3.4.1 Thermal Evaporation Technique

The thermal evaporation technique is a kind of physical vapor deposition. It is a very simple technique where evaporant material is placed in a source, which could be electrically heated to the sublimation/melting point of the depositing material. The source could be a ceramic crucible wrapped with a heating W wire, a Ta, Mo, or graphite boat, or a W wire coil. The source is always in an evaporation chamber, which is evacuated to high vacuum before the evaporation is started. Each material has some finite vapor pressure, which increases with an increase in its temperature. Depending upon their properties, some solids evaporate before they melt (sublimation) and some evaporate through the liquid phase. Usually the organic semiconductors sublime on heating, however, the metals melt first and then evaporate. If no collision with other molecules takes place, the vapor molecules move in straight trajectories from the source to the depositing surface. The evaporated molecules get condensed on the surrounding walls of the chamber and the substrate placed above the source. Condensation of the evaporated molecules on the relatively cold surface forms a thin film. A high vacuum is essentially required for the growth of high-quality films. In high vacuum, the vapor of the depositing material does not react or get scattered by other gas molecules. High vacuum increases the mean free path of the vapor molecules and reduces the possibility of impurities in the film that could be incorporated from the residual gas in the chamber.

Thermal evaporation can be better understood if the source is a Knudsen cell. It possessed a source surrounded with an isothermal enclosure having a small opening. The enclosure maintains an equilibrium vapor pressure inside the enclosure. The evaporant material is placed in the source and heated to evaporation. The opening of the enclosure is very small compared to the surface area of the evaporant. The evaporation through an orifice is known as effusion and such a cell is known as a Knudsen cell, effusion cell, or K-cell. Effusion is a sort of directional evaporation that follows the Lambertian angular distribution. Considering A_e is the area of the orifice, $P^*(T)$ is the vapor pressure of the evaporant at a temperature T, and P is the ambient pressure acting on it, the evaporation rate from the Knudsen cell will be proportional to $(P^*(T) - P)$ and given by

$$\frac{dN_e}{dt} = A_e \frac{1}{\sqrt{2\pi m k_B T}} (P^*(T) - P), \tag{3.1}$$

where N_e is the of number evaporated atoms, k_B is the Boltzmann constant, and m is the molecular or atomic mass of the evaporant. Now the mass evaporation rate through the orifice will be given by

$$\Gamma = m \frac{d}{dA_e}\left(\frac{dN_e}{dt}\right) = \sqrt{\frac{m}{2\pi k_B T}} (P^*(T) - P). \tag{3.2}$$

The total mass of evaporated material (M_e) can be calculated from

$$M_e = \int_0^t \int_0^{A_e} \Gamma \, dA_e \, dt. \tag{3.3}$$

Considering the Lambertian angular distribution of evaporated material, the quantity of the material emitted into the solid angle $d\omega = dA_e/r^2$ at an angle ϕ relative to the surface normal can be given by

$$dM_e = M_e \cos\phi \frac{d\omega}{\pi}. \tag{3.4}$$

The spatial distribution of vapor current density $\Phi(\phi)$ will now be given by

$$\Phi(\phi) = M_e \cos\phi \frac{1}{\pi}. \tag{3.5}$$

Let us consider that the substrate is at a distance r from the source and tilted at an angle θ with respect to the source, the rate of deposition at an arbitrary point of the substrate will be given by

$$R(\phi) = \Phi(\phi)\cos\theta \frac{1}{r^2}, \tag{3.6}$$

or

$$R(\phi) = M_e \cos\phi \cos\theta \frac{1}{\pi r^2}. \tag{3.7}$$

This equation is known as the Hertz–Knudsen law and enables us to determine the rate distribution for any geometry. Let us consider three different cases of the substrate type and positions shown in Figure 3.18.

Case 1: Infinite plane substrate

The case of infinite plane substrate is shown schematically in Figure 3.18a. The substrate S is at a normal distance D from the source Q. For an arbitrary point B at an angle ϕ and a distance x from the normal on the substrate we have $\phi = \theta$, $r^2 = x^2 + D^2$, which gives the deposition rate distribution

$$R(\phi) = M_e \cos^4\phi \frac{1}{\pi D^2}. \tag{3.8}$$

Correspondingly the rate distribution on the infinite plane substrate is shown in Figure 3.18b, which shows that the deposition rate (or film thickness if the evaporation rate is stable) is not uniform on the substrate area.

Case 2: Hemisphere substrate with the source at the center

The case of hemisphere substrate S with source Q at the center of the hemisphere is shown schematically in Figure 3.18c. If D is the radius of hemisphere for a point B at an angle ϕ and a distance r from the source, we have $\theta = 0$ and $r = D$, which gives the deposition rate distribution

$$R(\phi) = M_e \cos\phi \frac{1}{\pi D^2}. \tag{3.9}$$

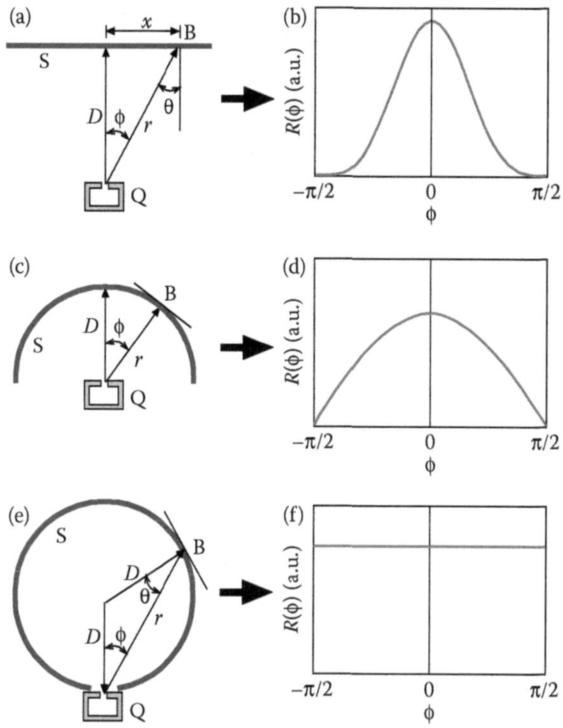

FIGURE 3.18
Schematic representation of geometries for thin film deposition on (a) infinite plane substrate, (c) hemispherical substrate, and (e) spherical substrate. (b), (d), and (f) show the rate distribution (or film thickness if evaporation rate is stable) for the respective cases.

Correspondingly the rate distribution on the hemisphere substrate is shown in Figure 3.18d, which shows that the deposition rate (or film thickness) is not uniform on the substrate area.

Case 3: Spherical substrate with the source at one pole

The case of spherical substrate S with source Q at the pole of the substrate is shown schematically in Figure 3.18e. If D is the radius of sphere for a point B at an angle ϕ from the normal to the source at a distance r, we have $\phi = \theta$ and $r = 2D\cos\phi$, which gives the deposition rate distribution

$$R(\phi) = M_e \frac{1}{4\pi D^2}. \qquad (3.10)$$

The deposition rate distribution is independent of the distribution angle or constant in all directions at any angle. The corresponding rate distribution is shown in Figure 3.18f. This geometry gives uniform film thickness on the spherical substrate and is used in industries where highly uniform films are required.

Usually the substrates are not of the size or shape mentioned earlier, but they are smaller in size and planar/straight in shape. The schematic diagram of a typical thermal evaporation chamber is shown in Figure 3.19. The chamber is evacuated with suitable vacuum

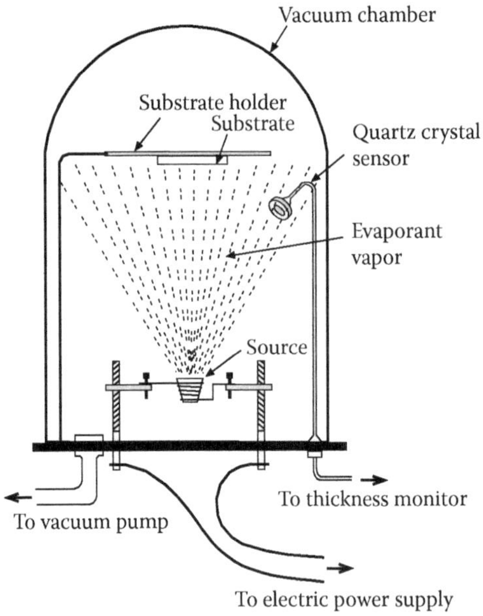

FIGURE 3.19
Schematic diagram of a typical thermal evaporation chamber.

pumps. Usually the thermal evaporation of different materials is done in the vacuum better than 10^{-5} Torr. If the source is made of conducting material like Mo, Ta, or W, the source is filled with the material to be evaporated and is heated directly by passing the electric current through it. If the source is made of ceramic crucible, it is wrapped with a W wire and the electric current is passed through the wire. This is called indirect heating of the source. Heating of the source causes the filled material to evaporate, and the rate of evaporation is controlled by controlling the temperature of the source that is controlled by controlling the electric current through it. The rate of deposition and the thickness of deposited film on the substrate are measured using a quartz crystal thickness monitor. The working of a quartz crystal thickness monitor is based on the piezoelectric effect. The thickness monitor possesses a quartz crystal sensor that is placed near to the substrate. The sensor is applied a free-running RF voltage and it starts vibrating at its natural frequency (~5–6 MHz). The vibrating frequency of the crystal depends on its mass and temperature. Deposition of some mass on the crystal reduces its frequency. The change in crystal frequency is easily detected by an oscillator circuit connected to the crystal. Knowing the *Z*-factor of the depositing material, the change in frequency gives the deposited mass per unit area and knowing the density of depositing material, the deposited mass is easily converted into the thickness of deposited film. The sensor reads out the rate of deposition and the deposited film thickness on it, and with the help of some calculations the corresponding values on the substrate are found.

For calculation of the rate of deposition and film thickness on the substrate, the tooling factor (*TF*) is important. *TF* is actually a correction factor for the thickness difference on the sensor and substrate. The *TF* compensates the necessary differences in placement of the substrate and the crystal. *TF* is calculated by measuring the deposited film thickness using a profilometer. For this purpose, an arbitrary value of the tooling factor, say, 1.0 is fed in the quartz crystal monitor and a film of thickness, say, T_i, as indicated by the quartz crystal monitor, is deposited. The sample is taken to a profilometer and the actual thickness of the film is measured. Let us say it comes out to be T_m. In that case the *TF* would be T_m/T_i. This value of *TF* is fed in the quartz crystal monitor and the aforementioned experiment is repeated again to get the new value of *TF*. The experiment is repeated until the difference between T_m and T_i becomes negligible. However, the quartz crystal thickness monitor does not give the accurate value of the film thickness on the substrate, but it is reasonably acceptable when the difference between T_m and T_i is below 2%. In general, the tooling factor can be calculated using

$$TF_{actual} = TF_{approx}\frac{T_m}{T_i}. \tag{3.11}$$

Note that the *TF* is a function of the relative positions of the substrate and sensor, and for their different positions it would be different. When the deposition is to be done on many substrates, for uniform depositions the substrates are placed on a rotatable substrate holder at the same distances from the center and the deposition is done under rotation of the substrate holder. In this condition, the *TF* would be different compared to that for a stationary substrate. Even for the rotating substrates, the *TF* is calculated following the method discussed earlier. *TF* is found to be independent of the rotation speed of the substrate holder.

For thin-film deposition via thermal evaporation the substrates are usually kept at room temperature, however, sometimes to achieve better film quality the substrates are heated to high temperatures with substrate heating mechanisms. The rate of evaporation and substrate temperature play an important role in controlling the film morphology. High evaporation rates give rough films, whereas slow evaporations result in smother films. This process encounters great material waste as the material also evaporates at the angles not having substrates and it gets deposited on the walls of the vacuum chamber. Nonuniformity in the deposited film thickness for large-area substrates is the main drawback of this technique. Also, it proves to be quite costly and limits the maximum size of the substrate to that of the evaporation chamber. Thermal evaporation technique is not suitable for production of large-area devices.

3.4.2 Solution Processing

For thin-film fabrication via solution processing, the materials are required to be soluble in some suitable solvent. Solution processing has proved to be a truly cost-effective method for fabrication of thin films. There are many ways to prepare thin films from solutions, but spin coating and inkjet printing are the most frequently used techniques. Sometimes dip coating and screen printing techniques are also used for this purpose. Inkjet printing is the most fascinating and important technique that can produce very large-area devices at ease with roll-to-roll (R2R) processing. Inkjet printing is a cost-effective way for production without any material wastage and it is the first choice for industrial production.

3.4.2.1 Spin Coating

Spin coating is a simple and low-cost process for fabrication of thin films. It is being used for several decades to produce smooth and uniform films on planar substrates. The spin coating process is depicted stepwise in Figure 3.20. The coating is carried out in a spin coater that possesses a rotatable vacuum chuck, where substrates are placed and held via

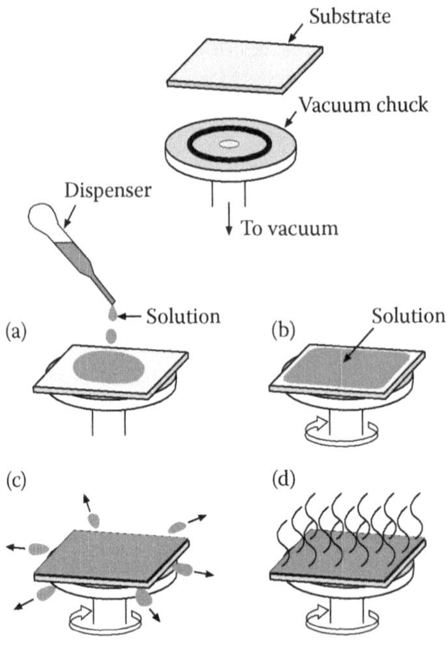

FIGURE 3.20
Schematic representation of the spin coating process: (a) dispensing of the solution, (b) spin up, (c) spin off, and (d) evaporation of the solvent.

vacuum. Holding of the substrate ensures its fixed position at the time of rotation. A solution of the depositing material is prepared in a suitable solvent with desired concentration and dispensed on the substrate held on the chuck (Figure 3.20a). The solution is dispensed in excess of the amount that is required for the film formation. The chuck is now rotated at a certain speed for the specific time (Figure 3.20b). The rotation of substrate exerts a centrifugal force on the solution and spreads it out all over the area of the substrate (Figure 3.20c). There is one resisting viscous force as well that acts on the solution radially inward. The excess amount of the solution is thrown off the substrate in the form of droplets. At high rotational speed, the centrifugal force is very strong and causes thin uniform coverage of the substrate. Fast rotations cause the solvent to evaporate rapidly and the spreaded film starts solidifying (Figure 3.20d). As the film becomes thinner, the rate of removal of excess solution slows down because the thinner the film, the greater the resistance to flow. After rotating for some time, a uniform, thin, solid film of dissolved material is achieved on the substrate. However, some residual solvent may still be trapped in the solid film that could be removed by heating for some time at a certain temperature. The film thickness is measured using a profilometer. According to Meyerhofer [70], the film thickness (d) shows $d \propto f^{-2/3}\eta^{1/3}e^{1/3}$, where f is the rotation speed, η is the viscosity of the solution, and e is the solvent evaporation rate that is proportional to $f^{1/2}$. Therefore, the spin coating process can be classified into four major steps: (1) dispensing of solution, (2) spin up, (3) spin off, and (4) evaporation of the solvent. The spin off stage is key for high-quality films.

Variation in processing conditions of spin coating leads to variation in film morphology. Speed of rotation, rotation time, and substrate temperature greatly affect the film morphology. The quality of films is also related to the solution concentration and solvents used. All of these parameters also affect the film thickness. A relatively thick film will be formed if the concentration of solution is increased and/or the rotation speed is reduced. For good-quality films, all of these parameters need optimization. For a given temperature and given solvent, the rotation time, rotation speed, and solution concentrations need to be optimized to achieve high quality film of desired thickness. The spin coating process is an important technique for preparation of high-quality thin films of organic materials that are not thermally evaporable. The spin coating process has given the most efficient OSCs. But unfortunately, this process encounters lots of material loss. The problem of materials loss is completely solved in the printing process that makes it the most cost-effective technique for OSCs.

3.4.2.2 Printing and Other Coating Processes

For printing and other coating processes, see Chapter 7.

3.5 Parameters That Control the Performance of OSCs

There are lots of parameters that control the performance of OSCs via lots of optical and electrical loss mechanisms. Optical losses include the processes, which lead to reduced number of photons absorbed by the cell, whereas electrical losses include the processes, which lead to a reduced number of charge carriers extracted out of the cell.

3.5.1 Optical Losses

Optical losses are quite decisive and affect the cell performance by reducing its J_{sc}. These losses cause loss in photons that could have absorbed and contributed to the photocurrent. Some of the important optical losses are discussed next.

3.5.1.1 Reflection Losses

Reflection of the incident light is an important optical phenomenon that always takes place at the surface and different interfaces in the cell. Reflection of incident light causes some of the incident photons to go back in the surrounding medium without absorption. A typical OSC has several organic/inorganic layers, which have different refractive indices and the reflection is accompanied by refraction at several interfaces. Figure 3.21 schematically shows the reflection and refraction at different interfaces in a typical OSC. Under illumination, some of the photons incident on the cell are reflected back into the same medium from the top surface of the substrate and the rest are refracted into the cell. At the interface of two mediums, the reflected angle is equal to the incident angle (θ), whereas the refracted angle ϕ is decided by the incident angle and the relative refractive indices of the two mediums as per the Snell's law $\sin(\varphi) = (n_1/n_2)\sin(\theta)$, where n_1 and n_2 are the refractive indices of two mediums. Before the incident light reaches the absorbing medium, it encounters several reflections and refractions at different interfaces. If the thickness of the active layer is less than the absorption length, not all the photons will be absorbed and the remaining photons will get reflected back from the back electrode. Some of these photons will get reabsorbed in the active layer and those still not absorbed will encounter further reflections and refractions as shown in Figure 3.21. These multiple reflections cause some of the photons to move out of the device from the front and edges that leads to reduced photocurrent, which

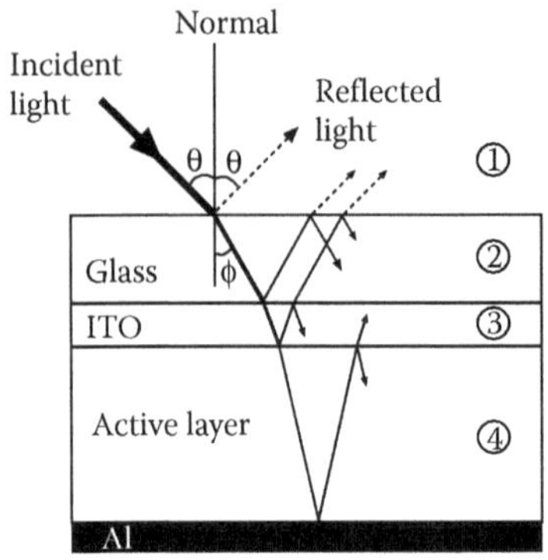

FIGURE 3.21
Schematic representation of reflection and refraction at interfaces of different mediums in an OSC.

would be higher otherwise. For transparent solar cells, the back electrode is not reflecting and the nonabsorbed photons get transmitted, though this causes relatively more loss in photocurrent. To ensure maximum photocurrent generation in the cell, the reflection losses should be reduced to zero by some means like an antireflection coating (ARC) on the top of the solar cell. ARC materials are high band-gap materials and do not absorb the light being absorbed by the cell. ARC not only reduces the refection losses from the top but also ensures maximum absorption by trapping the light inside the cell.

3.5.1.2 Inefficient Light Absorption

Most of the organic semiconductors absorb in the visible range of the solar spectrum, and ultraviolet (UV) and infrared (IR) radiations are left unused. For an example, Figure 3.22 compares the solar spectrum with the IPCE (incident photon to converted electron) spectra of some of the best solar cells based on P3HT:$PC_{61}BM$ and PTB7:$PC_{71}BM$ solar cells. The IPCE spectrum of the most efficient tandem solar cell based on P3HT and PCPDTBT is also shown in the figure for comparison. For the P3HT:$PC_{61}BM$ solar cell, a gap in photon harvesting beyond 650 nm can be clearly seen. Similarly the photons beyond 800 nm are not absorbed by PTB7:$PC_{71}BM$ solar cell. Even the P3HT/PCPDTBT tandem solar cell absorbs the light photons up to ~900 nm only and those beyond it are left unused. Harvesting of light photons only from specific regions of the solar spectrum is a main reason behind low efficiency in OSCs. To overcome this problem, new molecules with lower band gaps are being synthesized so that more energy could be harvested by broadening of the absorption spectrum. Considering all the photons having energy more than the band gap of the semiconductor are absorbed, the fraction of the incident optical power that is absorbed by the semiconductor or solar cell can easily be calculated from

$$P_{abs} = \frac{\int_0^{\lambda_g} \Phi(\lambda)(hc/\lambda)\,d\lambda}{\int_0^{\infty} \Phi(\lambda)(hc/\lambda)\,d\lambda}, \tag{3.12}$$

where $\Phi(\lambda)$ is the photon flux density and λ_g is the wavelength corresponding to the band gap of the semiconductor.

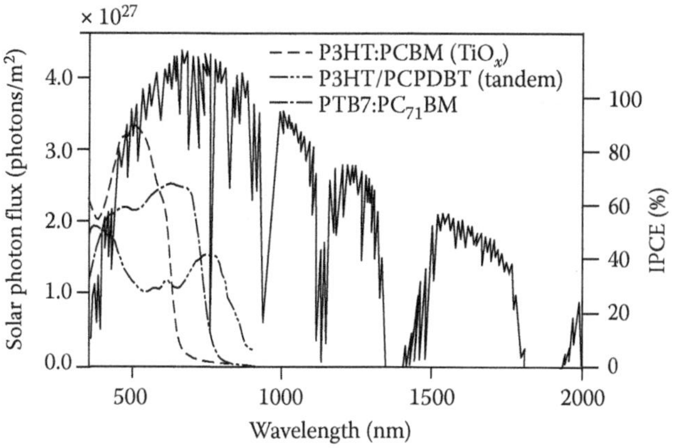

FIGURE 3.22
Photon flux from the sun (AM1.5G) (solid curve) and IPCE spectrum of the solar cells based on different active layer materials. The dashed curve is for P3HT:PCBM solar cell with an optical spacer. (Data from J. Y. Kim et al., 2006. *Adv. Mater.* 18: 572–576.) The dash-dotted curve is for IPCE spectrum of PTB7:$PC_{71}BM$ solar cell prepared by chlorobenzene mixed with 3% of 1,8-diiodooctane as the solvent. (Data from Y. Liang et al., 2010. *Adv. Mater.* 22: E135–E138.) The dash-double dotted curve corresponds to the IPCE spectrum of the P3HT/PCPDTBT tandem solar cell. (Data from J. Y. Kim et al., 2007. *Science* 317: 222–225.)

3.5.1.3 Thermalization Losses

In principle, only the photons having energy more than the band gap of the absorber are absorbed and contribute to electron–hole generation. The photons with lower energy are not involved in energy conversion. The photons having more energy than the band gap of the absorber excite electrons from the HOMO and/or from other deeper energy levels of the valance band to the LUMO and/or higher energy levels of the conduction band. Electrons have the tendency to occupy the lowest energy level, LUMO, and holes have the tendency to occupy the highest energy level, HOMO. The electrons and holes having more energy than the minimum energy of the corresponding energy bands in the semiconductor are known as hot carriers. The hot carriers relax to their corresponding minimum energy levels by releasing extra energy in the form of heat in the semiconductor. It is basically a kind of process of conversion of high energy into low energy. This process is known as thermalization and can be better understood from Figure 3.23, which schematically shows three different processes. Process 1 corresponds to nonabsorption of photons having less energy than the band gap of the semiconductor. These photons are transmitted through the semiconductor and no electron–hole pair is generated. Process 2 represents the absorption of photons having an energy equal to the band gap of the semiconductor. This photon absorption generates an electron–hole pair in the semiconductor. Process 3 represents the absorption of photons having more energy than the band gap and loss of excess energy through thermalization. Out of the absorbed power only a fraction is used in generating electron–hole pairs and the rest is lost in the thermalization process. The useful fraction of the absorbed energy can be calculated from

$$P_{use} = \frac{E_g \int_0^{\lambda_g} \Phi(\lambda)d\lambda}{\int_0^{\lambda_g} \Phi(\lambda)(hc/\lambda)d\lambda}, \tag{3.13}$$

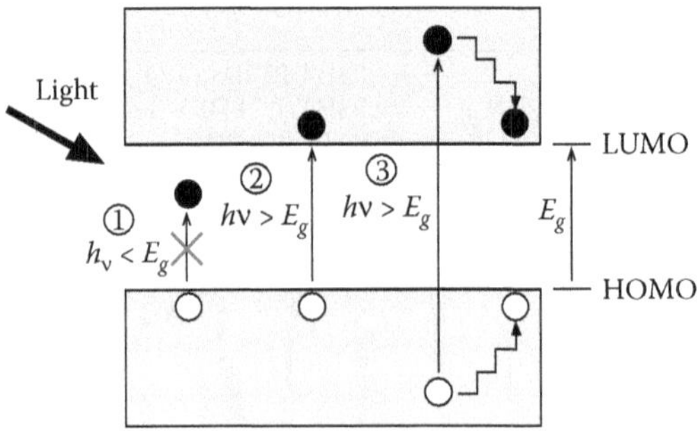

FIGURE 3.23
Schematic representation of three transition processes during light absorption: (1) $h\nu < E_g$ corresponds to no absorption, (2) $h\nu = E_g$ corresponds to absorption, and (3) $h\nu > E_g$ corresponds to absorption with thermalization loss.

where E_g is the band gap of the semiconductor. Therefore, nonabsorption of photons having less energy than the band gap of the semiconductor and loss of excess energy of the absorbed photons having more energy than the band gap are two important energy losses. Thermalization of hot carriers, created by photons having energy greater than the band gap of the semiconductor, causes a loss in photovoltage because higher band gap materials give higher V_{oc} than the lower band gap materials. In lower band gap materials V_{oc} is very low and most of the energy is wasted in the thermalization process. The problems of spectral mismatch and thermalization losses in single-junction solar cells could be minimized to some extent by tandem structures incorporating semiconductors of different band gaps [71,72]. A tandem structure effectively utilizes the photonic energy to give higher PCE but the absorption spectra of subcells should be complementary to each other, otherwise thermalization losses will not be reduced like in a homogeneous tandem cells (incorporating the same materials in subcells). Minimization of thermalization losses of photo energy in tandem structures maximizes V_{oc}.

3.5.1.4 Losses due to Nonuniform Optical Density Distribution

When light enters in a solar cell it penetrates through different layers to get absorbed in the active layer. When the active layer thickness is less than its absorption length, not all the photons entering the active layer are absorbed. If the back contact is reflecting, then the nonabsorbed light is reflected back into the active layer that interferes with the incident light. Interference between incident and back-reflected lights causes variation in the optical electric field in the device. For example, Brabec et al. calculated the number of photons absorbed in a P3HT:PCBM solar cell with structure glass (1 mm)/ITO (140 nm)/PEDOT:PSS (50 nm)/P3HT:PCBM (x nm)/Al (100 nm) as a function of P3HT:PCBM layer thickness and the results are shown in Figure 3.24 [73]. Different curves are for different internal quantum efficiencies (IQEs). The depths in the curves are because of destructive interference between incident and back-reflected light that is larger at thicknesses smaller than the coherence of the light. Calculations show that for 100% IQE, the J_{sc} will be 15.2 mA/cm^2 for 5000 nm of P3HT:PCBM thickness. Considering 100% external quantum efficiencies (EQEs) the maximum possible J_{sc} comes out to be 18.7 mA/cm^2. The dots on the graphs correspond to the current densities for a realistic thickness of 400 nm of P3HT:PCBM. Where the theoretical limit of J_{sc} for P3HT:PCBM is ~19 mA/cm^2, the practically achievable J_{sc} in real devices will be ~10–12 mA/cm^2. However, it is practically impossible for the EQE to be

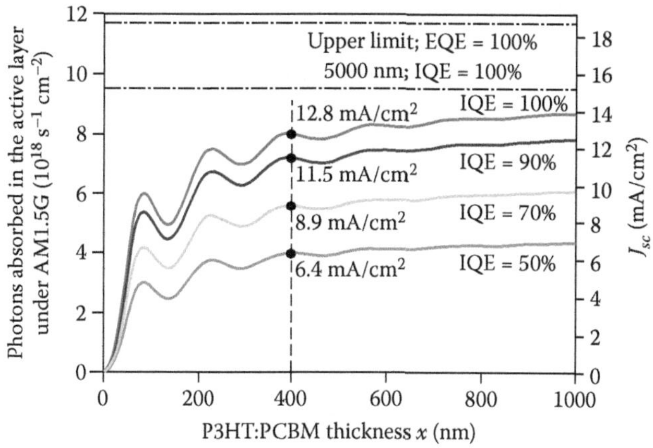

FIGURE 3.24
The number of photons absorbed in the active layer of glass (1 mm)/ITO (140 nm)/PEDOT:PSS (50 nm)/P3HT:PCBM (x nm)/Al (100 nm). The right-hand side axis represents the corresponding J_{sc} at various internal quantum efficiencies. (Modified with permission from G. Dennler, M. C. Scharber, and C. J. Brabec, 2009. *Adv. Mater.* 21: 1323.)

100% due to several loss factors discussed earlier and the practical current always remains less than the theoretical limit. For precise measurement of light absorption, one should calculate the distribution of optical electromagnetic field throughout the device. This calculation is carried out by the transfer-matrix formulation, which incorporates both the absorption and reflection in each layer of the device [31,74].

The effect of optical interference in thin-film solar cells is very important. For maximum absorption of light, the active layer should possess maximum light intensity. The effect of optical interference can be better understood from Figure 3.25a, which schematically

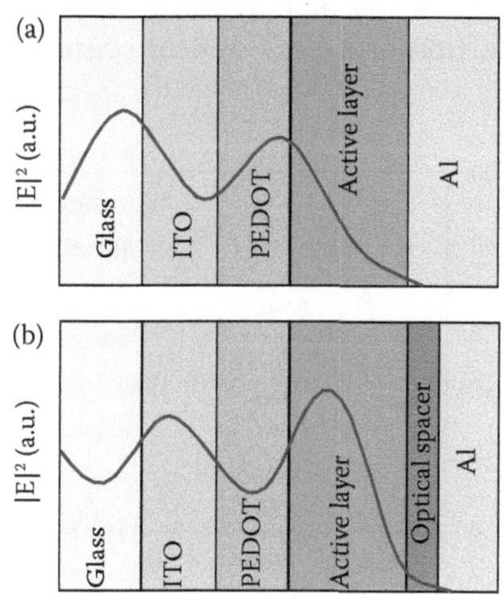

FIGURE 3.25
Schematic representation of the spatial distribution of the squared optical electric field $|E|^2$ of interfered light in (a) glass/ITO/PEDOT/active layer/Al and (b) glass/ITO/PEDOT/active layer/optical spacer/Al devices. (Modified with permission from J. Y. Kim et al., 2006. *Adv. Mater.* 18: 572–576.)

represents the spatial distribution of squared optical electric field strength of interfered light [75] in an OSC. The intensity of light is zero at the metallic back electrode and a relatively large fraction of active layer falls in the dead zone that significantly reduces photogeneration of charge carriers. This problem can be overcome by redistribution of light intensity by increasing the film thickness, but an increment in film thickness will introduce an additional series resistance that would result in lower fill factor. Alternately, application of an optical spacer is an effective approach to redistribute the light intensity in the cell. Figure 3.25b schematically shows the redistribution of squared optical electric field intensity with the use of an optical spacer.

Selection of the optical spacer for OSCs is an important issue. A good optical spacer for normal OSCs should possess: (a) good electron-accepting property with a conduction band edge below the LUMO level of acceptor, (b) good electron-transport property, (c) conduction band edge above the Fermi level of the electron-collecting cathode, and (d) high optical transparency to the incident light. Kim et al. used TiO_x as an optical spacer for P3HT:PCBM solar cells where a thin film of TiO_x was deposited on the top of the active layers via the solution-based sol–gel process [75]. Compared to a similar device having no optical spacer, an increase of ~50% in PCE was observed for the cell having a TiO_x optical spacer. Figure 3.26 shows the IPCE spectra of P3HT:PCBM solar cells fabricated with and without a TiO_x optical spacer. The device without TiO_x exhibited a maximum IPCE of ~60% at 500 nm, whereas the device with a TiO_x optical spacer exhibited ~40% increase over the entire IPCE spectrum. Application of a TiO_x optical spacer increased the PCE from 2.3% to 5.0%. This increase was a result of the redistribution of the optical intensity, with maximum in the P3HT:PCBM layer. Enhanced optical intensity caused enhanced absorption in the active layer that gave enhanced photocurrent.

It is worth mentioning that the use of an optical spacer is not always beneficial and it may also have negative effects on the cell performance. With extensive investigation, Gilot et al. presented theoretical and experimental studies on the effect of an optical spacer on the performance of P3HT:PCBM solar cells [76]. Investigations were performed on ITO/PEDOT:PSS/P3HT:PCBM/optical spacer/LiF/Al solar cells. An optical spacer was a thin layer of ZnO nanoparticles. J_{sc} was observed to oscillate with the active layer thickness, and for a device having no optical spacer, maximum optical electric fields were calculated to be at 80 and 210 nm. Insertion of a ZnO optical spacer shifted the intensity maxima

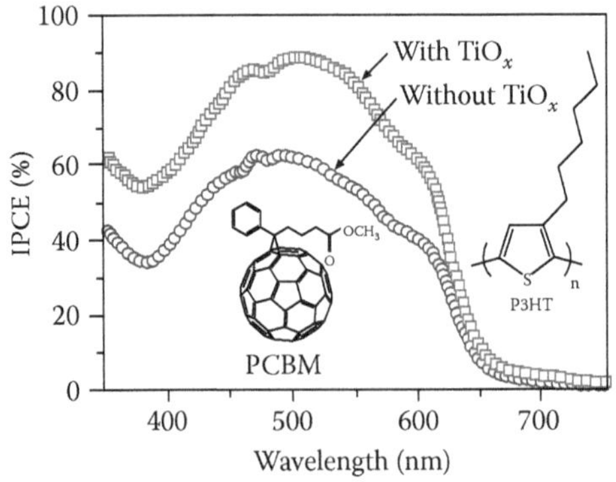

FIGURE 3.26
IPCE spectra of P3HT:PCBM solar cells with and without incorporation of TiO_x optical spacer before back electrode. (Modified with permission from J. Y. Kim et al., 2006. *Adv. Mater.* 18: 572–576.)

toward lower thicknesses. Calculations showed almost twofold increase in J_{sc} for 40 nm of active layer with 40 nm of optical spacer. However, for the thicker films ranging from 70 to 130 nm, the insertion of an additional optical spacer predicted deterioration in the cell performance. This was because when the active layer was already in the most effective position without an optical spacer, the insertion of an optical spacer would shift it away from the maxima zone leading to reduced J_{sc}. To verify these predicted results, Gilot et al. performed detailed experiments and observed an excellent agreement between theory and experiments. For thin films, insertion of an optical spacer was observed to significantly enhance the absorption, whereas for thicker films the absorption was less pronounced, and depending on the active layer thickness it might either increase or decrease. Therefore, enhancement in the cell performance by the use of an optical spacer is a subject of optimization. Application of optical spacers will be more pronounced for the solar cells where film thickness is limited by low charge carrier mobilities.

3.5.2 Electrical Losses

Like the optical losses, electrical losses are also very crucial in deciding the ultimate performance of solar cells. Electrical losses reduce the photoenergy conversion through several energy losses and recombination processes. Due to electrical losses not all the photogenerated charge carriers get extracted out of the cell. As the generation of excess charge carriers via photoabsorption is a deviation from the thermal equilibrium condition, the excess charge carries are subjected to recombination. The recombination takes place at different interfaces, contacts, and bulk of the solar cell. If the charge carriers are extracted out of the cell and allowed to recombine by passing through an external circuit we get a photocurrent. Recombination of charge carriers within the cell itself does not contribute to the photocurrent, therefore the internal recombination should be minimized. Recombination of charge carriers is determined by the electronic properties of the constituting materials and the active layer nanoscale morphology. Organic semiconductors are amorphous in nature and possess lots of charge carrier traps, which work as recombination centers. There are several ways for recombination of charge carriers, and it cannot be avoided but it can certainly be minimized. Charge carrier recombination not only reduces the photocurrent but also photovoltage and fill factor (*FF*). *FF* determines the maximum power generated by the cell, and as a result the energy conversion efficiency is reduced. Some of the important electrical loss mechanisms are discussed next.

3.5.2.1 Exciton Recombination Losses

Due to high binding energy in organic semiconductors, the excitons have a tendency to recombine at room temperature if some other breaking force is not available. Recombined excitons do not contribute to energy conversion, therefore recombination of excitons should not be allowed to happen. In single-material solar cells, the necessary force for exciton dissociation is provided by the energy offset at the metal–organic interface, but before that happens most of the excitons get quenched in bulk due to their low diffusion lengths or at the interface via surface states. Therefore, high exciton binding energy and low exciton diffusion lengths were among the major factors limiting efficiency of single component solar cells. The problem of exciton quenching was fixed to some extent by the donor–acceptor concept, where the necessary dissociation force is provided by the energy offset at the donor–acceptor interface, however, the excitons still had to diffuse to the interface to get dissociated. The excitons far from the interface do not make it and get recombined in the semiconductor. The problem of exciton recombination was further minimized in bulk-heterojunction structures where there are numerous donor–acceptor interfaces distributed throughout the bulk and the separation of two interfaces is within the diffusion length of excitons. Therefore, almost all the excitons generated in the active layer reach the donor–acceptor interface and get dissociated there.

3.5.2.2 Losses at the Donor–Acceptor Interface

Energy offset at the donor–acceptor interface is very crucial as it causes excitons to dissociate and also decides the photovoltage of the cell. When the energy offset between LUMOs of donor–acceptor materials is more than the exciton binding energy and there is an interfacial potential barrier to prevent hole transfer from the donor's HOMO to the acceptor's HOMO, the exciton dissociation efficiency at the interface comes out to be 100%. Exciton dissociation, at the donor–acceptor interface, happens ultrafast within ~10^{-15} s. Usually the exciton binding energy in organic semiconductors is 0.3–0.5 eV, but for most donor–acceptor combinations the LUMO energy offset (ΔE) is quite large. The extra offset energy is clearly a waste in photonic energy that is reflected in reduced photovoltage. For example, Figure 3.27a schematically shows the energy levels of the P3HT:PCBM donor–acceptor combination. The exciton binding energy in P3HT is ~0.5 eV [77], whereas ΔE is about 1.1 eV. The energy difference between the LUMO of the acceptor and the HOMO of the donor controls the V_{oc} in OSCs, and for larger differences higher V_{oc}s are observed [78,79]. For the P3HT:PCBM combination this HOMO–LUMO difference is around 1.1 eV and could be raised to 1.7 using another acceptor without any loss to the exciton dissociation efficiency. The increase in HOMO–LUMO difference would increase V_{oc} and as a result the PCE. For the P3HT:PCBM combination, there is clearly a loss of about 0.6 eV in electron energy, which could be utilized in increasing the photovoltage using some other acceptor with suitable energy levels. It is worth mentioning that the reduction in LUMO offset below the exciton binding energy might increase the V_{oc}, but it will reduce the probability of exciton dissociation and as a result will reduce the photocurrent. There could also be the possibility of exciton transfer from donor to acceptor, which might happen when exciton energy in the acceptor is lower than that in the donor. This energy transfer causes lots of energy loss and should always be avoided. Therefore, the respective energy levels of donor–acceptor materials are very important and should be

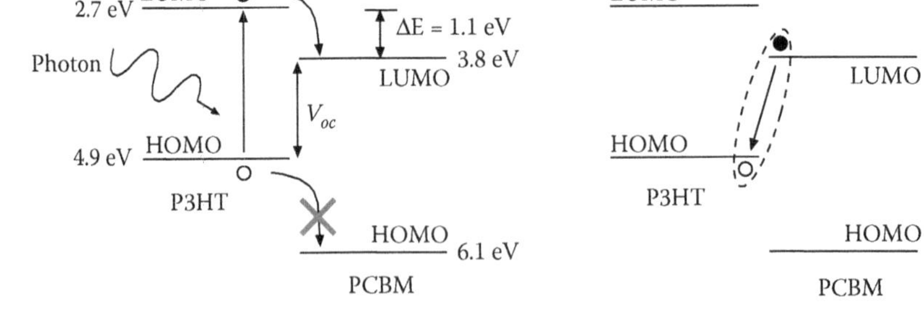

FIGURE 3.27
(a) Schematic energy level diagram of P3HT:PCBM blend and representation of exciton dissociation at the interface. (b) Recombination of charge carriers at the donor–acceptor interface via CTSs.

optimized to get maximum photonic conversion. Several new donor–acceptor materials have been synthesized that resulted in high-performance solar cells.

Recombination of charge carriers via charge transfer states (CTSs) at the donor–acceptor interface is another important loss mechanism in OSCs. After separation at the donor–acceptor interface, the electrons and holes are still bound together through coulombic attraction force and form the CTSs. The CTSs are electronic states that are formed after exciton dissociation and are also called polaron pairs, exciplex, bound electron–hole pairs, and bound radial pairs [80–83]. Under some conditions, the electrons may get back transferred from acceptor to donor by regeneration of the donor singlet exciton or transition to the triplet exciton state via electron back transfer. The energy of CTSs determines V_{oc}, and higher the energy of CTSs the higher the V_{oc}. CTS is a crucial intermediate step between exciton dissociation and charge generation. CTSs can return the charge carriers to the ground state and recombine them again if they encounter. The recombination of CTSs can be radiative or nonradiative. Recombination of charge carriers via CTSs is shown schematically in Figure 3.27b. The CTS energy is obviously less than the donor exciton energy. For photocurrent generation the CSTs need to be dissociated into free charge carriers. The necessary force for this purpose is provided by the built-in electric field of the cell and it should be sufficient enough to break apart the CTSs. Once CTSs are broken, the electrons move toward the cathode and the holes move toward the anode for collection. It has also been proposed that for charge separation, the energies of electronic states are more important than LUMO level offset and a LUMO offset of 0.3 eV would be sufficient enough for efficient dissociation of excitons in OSCs [84–87].

3.5.2.3 Recombination Losses in Bulk

After dissociation of CTSs, the charge carriers transport toward their respective electrode. The transportation happens through drift and diffusion processes. Most organic semiconductors have low charge carrier mobilities (μ) and, according to the Einstein relation, low charge carrier mobilities result in small diffusion coefficients ($D = \mu(kT/q)$). The diffusion coefficient and lifetime (τ) determine the charge carrier diffusion lengths ($L_{Diff} = \sqrt{D\tau}$). Low charge carrier mobilities also result in short drift lengths ($L_{Drift} = \mu\tau E_{eff}$, where E_{eff} is the effective electric field in the cell). When the distance to be transported by the charge carriers is more than their drift and diffusion lengths, the charge carriers recombine before they reach the electrodes. Additionally, the organic semiconductors have numerous traps that work as charge recombination sites and lots of charge carriers get trapped before they reach the electrodes.

3.5.2.4 Collection Losses at Metal Electrodes

After facing lots of recombination possibilities, the surviving charge carriers reach the metal–organic interface. Ideally all of them should collect but not all of them are able. The collection efficiency depends on surface states created at the metal–organic interface by imperfect film growth or deposition of hot metal vapors. Additionally, the metal–organic interfaces require being perfectly ohmic for efficient collection. Nonohmic contact and surface states reduce the number of charge carriers being collected at the electrodes. The surviving charge carriers finally collect at the electrodes to give photocurrent. The recombination losses caused by various factors are represented by parasitic series (R_s) and shunt (R_p) resistances in the equivalent circuit diagram. R_s represents the difficulties in charge collection and is shown connected in series, whereas R_p represents charge recombination and is shown connected in parallel to the cell [88]. The low R_s corresponds to high charge extraction capability, whereas low R_p corresponds to large recombination losses. R_s represents overall resistive contribution by bulk and contacts of the cell, whereas R_p represents the leakage and shunting contributions. R_s and R_p affect all the photovoltaic parameters and for maximum energy conversion R_s should be minimum with maximum R_p.

3.5.3 Donor–Acceptor Ratio

The donor–acceptor ratio in a BHJ layer is an important parameter that controls its morphological, charge transport, and exciton dissociation properties. For example, the best performance of MDMO-PPV, poly(3-octylthiophene-2,5-diyl-co-3-decyloxythipohene-2,5-diyl) (P3OT), and MEH-PPV solar cells was achieved when they were blended with PCBM in 1:4 wt% ratio [89–91], however, for the P3HT:PCBM blend different wt% ratios of typically 1:1 [92,93], 1:0.8 [75,94], and 1:0.6 [95] showed the best performance. Van Duren et al. extensively studied the nanoscale morphology of MDMO-PPV:PCBM blends using atomic force microscopy, transmission electron microscopy (TEM), and dynamic time of flight secondary ion mass spectroscopy (TOF-SIMS) for different weight percent ratios of MEH-PPV and PCBM [96]. The matrix exhibited a fairly homogeneous composite up to 50 wt% ratio of PCBM, however, a phase separation was observed somewhere between 60 and 75 wt% ratio of PCBM. Figure 3.28 shows TEM images of some of the MDMO-PPV:PCBM blends with different weight percent ratios of PCBM. The phase separation increased with PCBM concentration. Dark regions in the TEM images correspond to PCBM-rich regions. Absence of phase separation and percolating pathways for low PCBM concentrations exhibited poor charge transportation and were trapped in the active layer. Phase separation is important for charge transportation and reduction of bimolecular recombination, but large phase separation and formation of big size phases would cause reduced donor–acceptor interfacial area and reduced PCE. That is why for 90 wt% PCBM the efficiency was lower than that for 80 wt% PCBM [96].

The only drawback of BHJ solar cells is that for collection of the charge carriers at the respective electrodes, there must be percolating pathways for electron and hole transportation. The donor–acceptor ratio determines the width of separated phases that should

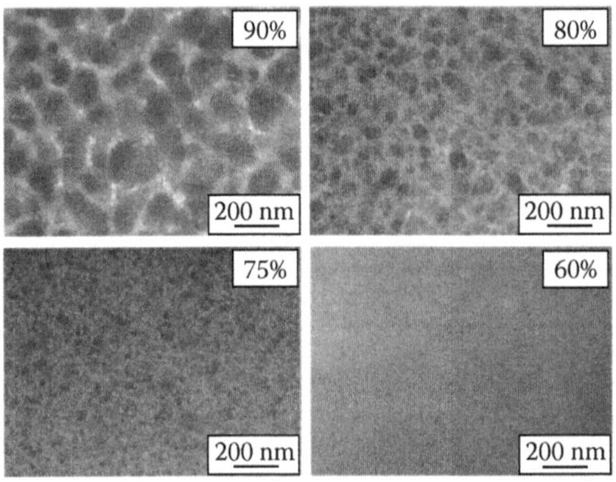

FIGURE 3.28
Transmission electron microscopy images of MDMO-PPV:PCBM blend for different wt% ratio of PCBM. (Modified with permission from J. K. J. van Duren et al., 2004. *Adv. Funct. Mater.* 14: 425–434.)

be less than the drift and diffusion lengths of the charge carriers. The optimum ratio of donor–acceptor materials is also decided by their charge carrier mobilities. A major portion of the incident light is absorbed in donor molecules and a certain minimum amount of acceptor molecules is required to dissociate them efficiently. Suppose the electron mobility in the acceptor is more than the hole mobility in the donor, then after dissociation at the donor–acceptor interface the electrons will move more rapidly toward the cathode than the holes toward the anode. The slowly moving holes would create a space charge in the device that would not only retard the next extracting holes but also reduce the collection of negatively charged electrons due to coulombic attraction. Overall the space charge reduces charge collection and results in reduced *FF*. To reduce the space charge effect in the active layer, the number of holes should be reduced and this can be done by reducing the amount of donor molecules in the blend. Though the reduced amount of donor molecules in such a case would give lower photocurrent, the improved *FF* would result in better PCE. Since electron mobility in PCBM is more than the hole mobility in MDMO-PPV, the optimum performance is obtained for a lower concentration of MDMO-PPV [91]. Similarly, for the cases where electron mobility in the acceptor is lower than the hole mobility in the donor, the optimum performance is achieved for a lower concentration of acceptor in the blend [92]. When the electron and hole mobilities are comparable to each other, the optimum performance is achieved for nearly the same amount of donor–acceptor materials in the blend [92,93]. Therefore for given donor–acceptor materials, the optimum ratio will depend on their charge carrier mobilities. It is important to note that the electronic properties of organic materials change from batch to batch. Even the same materials synthesized in same laboratory but in different batches may have different properties. The correct donor–acceptor ratio for given materials is a matter of optimization, and that is why P3HT from different laboratories has shown different optimum ratios with PCBM [75,92–95]. When P3HT chains are stacked together they show high crystallinity with a strong interchain overlap of wave functions. With a high degree of crystallinity in P3HT, the devices exhibit high performance. The conformational effects like crystallinity are controlled by choice of solvents and processing conditions. But finding of best weight ratio of donor–acceptor materials and optimization of processing conditions is really difficult, as it is found by trial and error method.

3.5.4 Active Layer Nanoscale Morphology

For efficient energy conversion in OSCs, the most important parameter is nanoscale morphology of the active layers [48,97–99]. The main objective over the last decade has been to understand and optimize the processing conditions to achieve excellent nanoscale morphology in OSCs, which is mandatory to achieve high efficiencies. PCBM exhibits high solubility in common organic solvents and forms nanocrystallites in polymer matrix. For polymer blends, the morphology of thin films can be controlled by optimizing the donor–acceptor ratio, active layer thickness, deposition method, curing process, and using appropriate solvents [48,98,99]. A mixed-solvent approach has also been successfully used to improve morphology of the active layers [19,100–103]. Yao et al. investigated the effect of the mixed-solvent approach on morphology of P3HT:PCBM thin films [100]. The mixed solvent possessed dichlorobenzene with a minute amount of 1,8-octanedithiol (ODT) as the additive. The mixed solvent caused ordered fibril crystalline domains of P3HT that were absent in the films processed without ODT. The ordered crystalline structure of P3HT led to enhanced performance of solar cells. Liang et al. employed the mixed-solvent approach for a PTB7:$PC_{71}BM$ blend [19]. DIO (1–8 diiodoctane) was used as an additive solvent with

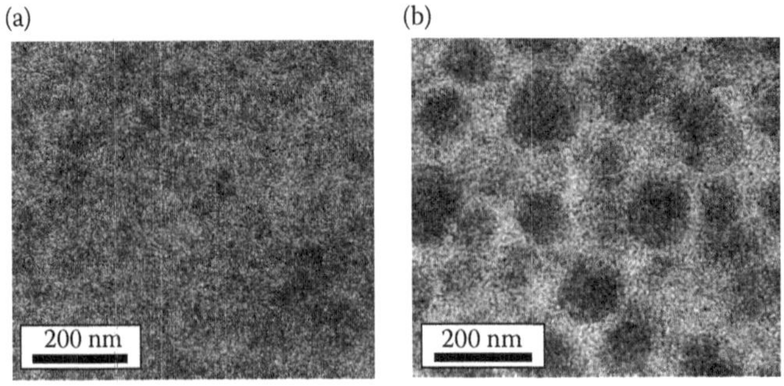

FIGURE 3.29
TEM images of PTB7:PC71BM films prepared from chlorobenzene (a) with and (b) without DIO. (Modified with permission from Y. Liang et al., 2010. *Adv. Mater.* 22: E135–E138.)

chlorobenzene. Where the cell from chlorobenzene only exhibited a PCE of 3.92%, the cell with mixed solvent resulted in a PCE of 7.4% [19]. The improved performance was attributed to the improved nanoscale morphology of the active layer induced by the mixed solvent. Figure 3.29 shows TEM images of PTB7:PC_{71}BM blend films prepared from chlorobenzene with and without DIO. Li et al. [99] proposed a solvent-annealing approach to improve the film morphology. The controlled rate of solidification during film growth controls the self-organization of polymer chains that results in better morphology [104]. Interaction of polymer blend with substrate is also important for controlled film morphology.

Block copolymers having both the donor and acceptor moieties on the same molecule were expected to give defined and stable phase separation, but due to low charge carrier mobilities and other losses, they could not show high performance. For example, Zhang et al. synthesized a block copolymer P3HT-*b*-poly(perylene diimide acrylate) composed of a P3HT donor and a poly(perylene diimide acrylate) acceptor [105]. This copolymer exhibited a PCE of only 0.49%. There are other reports as well on such systems, but the efficiency remained very low [106–108]. To control the nanoscale morphology of active layers, a nanoparticle approach to polymers has also been employed [109]. Active layer morphology affects photon absorption, exciton dissociation, and charge extraction, therefore achievement of optimum active layer morphology is the key for high-performance solar cells. Since there are so many parameters that affect the film morphology, it is bit difficult to predict and achieve optimum conditions for best morphology. Therefore, lots of experiments are performed by playing with all possible parameters to achieve optimum morphology.

3.6 Prediction of Possibly Achievable Efficiencies

Modeling and simulation of electronic devices is an important tool that can help in finding the optimum device parameters without performing too many experiments. It helps in finding the possible routes to improve device performance. Since there are so many parameters that control the performance of OSCs, exact calculation of the maximum possible efficiency is very difficult. Still, there have been some attempts to predict approximate achievable efficiency in OSCs. Considering optimum energy levels, band gap, and balanced charge carrier mobilities, Koster et al. made an early prediction that the PCE in single BHJ OSCs can approach 11% [110]. For the efficiency predictions, the most studied P3HT:PCBM system was taken to begin with. As discussed earlier, in the P3HT:PCBM system the extra ~0.6 eV of LUMO offset energy is dissipated into heat that is reflected in reduced V_{oc}, and the LUMO energy offset can be reduced to ~0.5 eV without compromising the electron transfer efficiency. By modulating the LUMO of the donor within the limit for efficient electron transfer, Koster et al. predicted an efficiency of ~6.6%, whereas by modulating the LUMO of the acceptor only, an efficiency of more than 8% was predicted [110]. Assuming the hole and electron mobilities to be 2×10^{-3} cm^2/Vs, with a polymer band gap of 1.9 eV and modulating the LUMO offset energy to 0.5 eV with an active layer thickness of 200 nm, they predicted an achievable efficiency of ~11%. It was just a prediction for given electron–hole mobilities, however, with better charge carrier mobilities and reducing the energy losses higher PCEs were possible.

Actually the maximum efficiency does not depend on the absolute positions of energy levels, but it is the energy offset and band gap of materials that actually do matter. If it is assumed that the solar emission intensity is completely absorbed by the materials, a band gap of 1.1 eV would be required to cover ~70% of AM1.5G solar photon flux, but most of the materials used earlier for OSCs possessed band gaps larger than 1.9 eV (P3HT band gap ~1.9 eV), which covered only up to 30% of photon flux. Therefore, to break the efficiency limitations, the possible solution was found to design and synthesize new low band gap materials with appropriate HOMO, LUMO energy levels, and high charge carrier mobilities. Keeping this idea in mind, several groups designed and synthesized low band gap polymers especially in the range 1.4–1.9 eV, which proved really beneficial for efficient energy conversion [19,68,101,111–117]. Molecular structures of some such materials are shown in Figure 3.30. These materials have shown very high PCEs in stand-alone single as well as tandem solar cells. Realization of high-performance solar cells with advanced materials led Scharber et al. to predict ~15% PCE with 75% *FF* and 80% EQE for single BHJ solar cells based on polymers having 1.45 eV band gap [118].

The maximum efficiency achieved so far for single-junction OSCs has gone up to 10.8% and it was reported by Liu et al. [21]. They used three donor polymers—poly[(5,6-difluoro-2,1,3-benzothiadiazol-4,7-diyl)-alt-(3,3‴-di(2-octyldodecyl) 2,2′;5′,2″;5″,2‴-quaterthiophen-5,5‴-diyl)] (PffBT4T-2OD), poly[(2,1,3-benzothiadiazol-4,7-diyl)-alt-(4′,3″-difluoro-3,3‴-di(2-octyldodecyl)-2,2′;5′,2″;5″,2‴-quaterthiophen-5,5‴-diyl)] (PBTff4T-2OD), and poly[(naphtho[1,2-c:5,6-c′]bis[1,2,5] thiadiazol-5,10-diyl)-alt-(3,3‴-di (2-octyldodecyl)-2,2′;5′,2″;5″,2‴-quaterthiophen-5,5‴-diyl)] (PNT4T-2OD)—and a range of fullerene derivatives as acceptors. Molecular structures of some of the fullerene derivatives are shown in Figure 3.31. The polymers exhibited more than 10% efficiencies with some suitable fullerene derivatives. The solar cells were prepared in inverted structure on ITO-coated glass substrates and exhibited very high *FF*s and EQEs. ZnO and MoO_3 were used as the electron-transport layer and anode interlayer, respectively. The performance of optimized solar cells incorporating different polymers and fullerene derivatives is shown in Table 3.1.

The initial predictions for maximum achievable efficiency in tandem configuration were made by Dennler et al. and up to 15% of efficiency was found achievable [119]. For this prediction, a tandem structure was considered that possessed two subcells connected in series. Both the top and bottom cells were considered to be made of a BHJ structure of PCBM and a donor. The two subcells were considered to be connected by a perfectly transparent layer such that the LUMO of PCBM of the bottom cell was perfectly aligned with the HOMO of the donor of the top cell. V_{oc} of the tandem cell was considered to be the exact sum of that of the two subcells, and V_{oc} of each cell was calculated from the difference

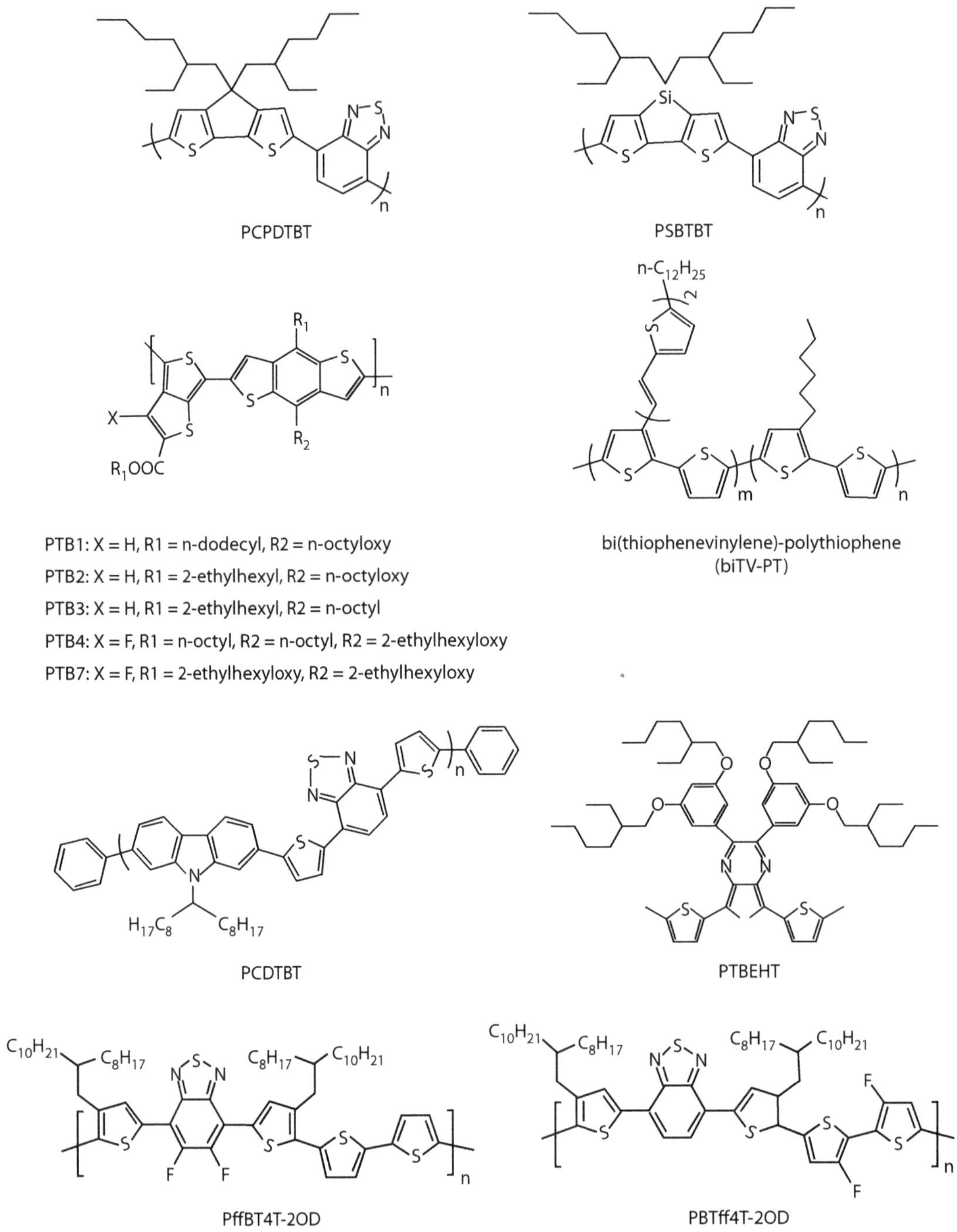

FIGURE 3.30
Molecular structures of some low band gap polymers: PCPDTBT, PSBTBT, PTB4, PTB7, PCDTBT, biTV-PT, PTBEHT, PffBT4T-2OD, and PBTff4T-2OD.

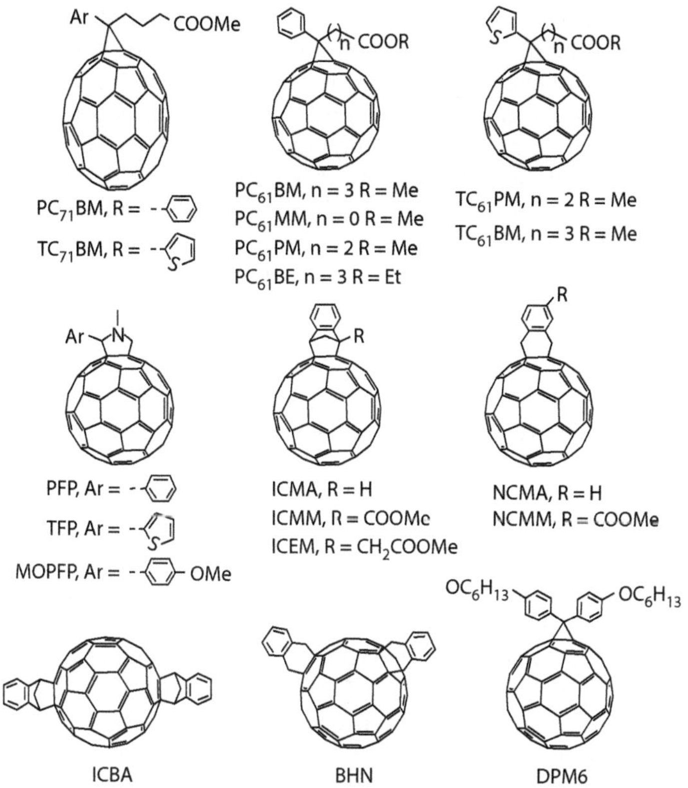

FIGURE 3.31
Molecular structure of some of the fullerene derivatives used as acceptors in OSCs.

TABLE 3.1
Photovoltaic Parameters of High-Performance Solar Cells

Active Layer	J_{sc} (mA/cm^2)	V_{oc} (V)	*FF*	η (%)
PffBT4T-2OD:TC_{71}BM	18.8	0.77	0.75	10.8
PffBT4T-2OD:PC_{71}BM	18.4	0.77	0.74	10.5
PffBT4T-2OD:PC_{61}PM	17.7	0.77	0.76	10.4
PffBT4T-2OD:ICMA	16.4	0.78	0.77	9.8
PffBT4T-2OD:TC_{61}PM	17.4	0.75	0.74	9.7
PffBT4T-2OD:PC_{61}BM	17.1	0.77	0.73	9.6
PBTff4T-2OD:PC_{71}BM	18.2	0.77	0.74	10.4
PBTff4T-2OD:TC_{71}BM	18.7	0.76	0.68	9.7
PBTff4T-2OD:PC_{61}PM	18.6	0.76	0.69	9.6
PNT4T-2OD:PC_{71}BM	19.8	0.76	0.68	10.1

Source: Data from Y. Liu et al., 2014. *Nat. Commun.* 5: 5293.

between the HOMO of the donor and the LUMO of the acceptor in their respective subcells [119]. Each subcell was considered to absorb light from 350 nm to the wavelength corresponding to the band gap of the donors. The EQE of each cell was considered to be some constant varying from 0% to 65% and IQE was taken to be 85% with *FF* equal to 0.65. Taking the absorption loss into account, the photocurrent densities of the two subcells were calculated. The current of the tandem cell was considered to be the smaller current of single cells. For calculations, the band gap of each donor was varied from 1 to 3.2 eV and the LUMO level of the donor varied from 3 to 4 eV as the LUMO level of PCBM lies at 4.3 eV. Using these parameters the highest efficiency achievable was calculated for stand-alone top, stand-alone bottom, and the tandem cells. P3HT and PCPDTBT are two very important donors that have shown high PCEs, therefore calculations were made considering these polymers as one of the donors in the subcells. Considering P3HT:PCBM as the bottom cell, the efficiency of the tandem cell was calculated for variable donors in the top cell mixed with PCBM and the maximum achievable efficiency was calculated to be around 9%. Importantly, the stand-alone top cell, itself of 9% tandem solar cell, could give an efficiency of 11%. Therefore, a tandem solar cell does not always give the efficiency more than that of corresponding subcells. But when individual donor materials perform below their capabilities, the tandem configurations could be very beneficial.

For the second case, PCPDTBT:PCBM was considered to be in the top cell, and the donor of the bottom cell was varied and considered mixed with PCBM. The ultimate efficiency in this case was observed to go up to 12%. For the highest efficiency achievable in tandem solar cell incorporating PCBM in both the top and bottom cells, the LOMO levels of the donor materials in both subcells were considered to be at 4 eV and the band gaps of the donors were varied. The efficiency of almost 15% seemed to be achievable with the bottom donor having a band gap of 1.6 eV and the top donor having a band gap of 1.3 eV [119]. It was pointed out that 15% was not the maximum possible value of efficiency in OSCs, as these calculations were made considering PCBM as the acceptor and taking constant EQE up to 65%. Also the benefits of donors having less absorption overlap had been neglected. Choice of another acceptor with different properties could give more efficient solar cells.

A prediction made by Brabec and colleagues suggests that even up to 21% of efficiency is achievable in tandem OSCs [120]. This prediction was made using recent knowledge and the experimentally feasible values of parameters. Advanced materials and devices have shown very high *FF* and EQEs, which have resulted in high prediction values. Detailed investigations based on the transfer formalism matrix for individual material combinations were performed. Following the same approach as demonstrated by Dennler et al. [119], Brabec and his colleagues considered IQE equal to 100%, an *FF* of 75%, and EQE of 80%, and predicted an achievable efficiency of 21% in tandem configuration. The results of optical simulations are shown in Figure 3.32 as a function of the band gap of donors in the bottom and top cells. There have been a number of reports on efficient tandem OSCs [27,69,121–127] that will be discussed in detail in the next section; for instance, in 2013 Yang's group reported double-junction tandem solar cells having 10.6% PCE [121]. A novel low band gap polymer, poly[2,7-(5,5-bis-(3,7-dimethyloctyl)-5*H*-dithieno[3,2-b:2′,3′-d] pyran)-alt-4,7-(5,6-difluoro-2,1,3-benzothia diazole) (PDTP-DFBT), with band gap of 1.38 eV was synthesized that gave an efficiency of 7.9% in the stand-alone single cell with the $PC_{71}BM$ acceptor. This polymer enabled 10.6% PCE in the tandem configuration when the bottom cell was based on a P3HT:ICBA (indene-C_{60} bisadduct) blend and the top cell was based on a PDTP-DFBT:$PC_{61}BM$ blend. This cell exhibited a J_{sc} of 10.1 mA/cm^2, V_{oc} of 1.51 V, and *FF* of 68.5%. In 2014 Yang's group achieved 11.5% PCE in a triple-junction tandem OSC [69]. To keep the current mismatch minimum among the constituent subcells, they performed an optical simulation for a balanced rate of photon absorption in the subcells. Using the simulated studies the triple-junction solar cells were fabricated in inverted configuration where the bottom cell was based on the blend of P3HT (band gap 1.9 eV) with ICBA, the middle cell was based on the blend of poly[4,8-bis(5-(2-ethylhexyl)thiophen-2-yl)benzo[1,2-b:4,5-b′]dithiophene-co-3-fluorothieno[3,4-b]

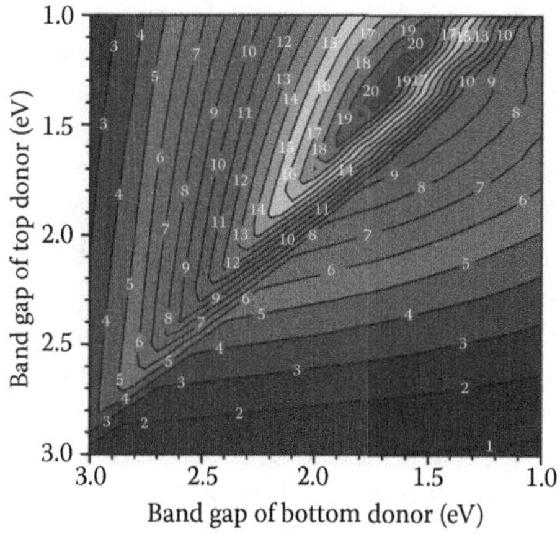

FIGURE 3.32
Efficiency curves calculated by Li et al. for tandem OSCs made of two subcells. The curve profiles vary with the band gaps of donors of both the top and bottom subcells. (Modified with permission from N. Li et al., 2014. *Adv. Energy Mater.* 4: 1400084.)

thiophene-2-carboxylate] (PTB7-Th, band gap 1.58 eV) with $PC_{71}BM$, and the top cell was based on the blend of PDTP-DFBT (band gap 1.38 eV) with $PC_{71}BM$. PEDOT:PSS/ZnO layers were used as interconnecting layers. An unoptimized triple-junction cell exhibited an efficiency of 8.74%, whereas the optimization in thicknesses of the absorbing and interconnecting layers resulted in 11.5% PCE.

3.7 A Review on Recent Developments in BHJ OSCs

3.7.1 Developments in Single BHJ OSCs

Since the early report by Heeger et al. in 1995 on single BHJ solar cells, many efforts, which include various treatments, improved device architectures, new advanced materials, improved morphology, and suitable solvents with additives, have been employed to improve the performance of BHJ OSCs. In 2003, Padinger et al. [128] fabricated polymer solar cells based on P3HT and PCBM donor–acceptor materials. A postproduction thermal treatment at 75°C improved the cell efficiency from 0.4% to 2.5% and the postproduction thermal treatment at 75°C in the presence of an externally applied voltage of 2.7 V resulted in an overall efficiency of 3.5%. This improvement in efficiency was attributed to improved nanoscale morphology and better charge carrier mobilities in the active layer upon post production thermal treatment. This effect was stimulated further in the presence of an external electric field. Later on in 2005, Heeger et al. achieved 5.0% PCE in a simple P3HT:PCBM BHJ solar cell with postproduction heat treatment at 150°C [97]. This high efficiency was because of high regioregularity of P3HT and improved nanoscale morphology. High regioregularity imparts very high hole mobility and absorption of photons at longer wavelengths. Along with regioregularity, the polydispersity and molecular weights of P3HT have also been observed to have significant effect on the device performance [62,129,130].

To minimize the spectral mismatch with the solar spectrum, several new low band gap polymers were synthesized, which resulted in rather efficient solar cells. Though the initial efforts were not very successful for high performance solar cells, later efforts showed the performance of low band gap materials could be improved a lot [19,116]. One of the most important low band gap polymers is poly[2,6-(4,4-bis-(2-ethylhexyl)-4*H*-cyclopenta[2,1-b;3,4-b]-dithiophene)-alt-4,7-(2,1,3-benzothiadiazole)] (PCPDTBT) [101]. PCPDTBT (band gap = 1.46 eV) improved light harvesting and extended the absorption spectrum up to ~900 nm. To formulate the solar cells, PCPDTBT was incorporated with $PC_{71}BM$ because, compared to $PC_{61}BM$, $PC_{71}BM$ possesses enhanced absorption in the visible range of the solar spectrum. The addition of 1,8-octanedithiol in a chlorobenzene solution of PCPDTBT:$PC_{71}BM$ resulted in significant change in its absorption spectrum. It happened because of the strong interaction of 1,8-octanedithiol with PCPDTBT chain that resulted in improved local structural ordering and better film morphology. The optimum devices had an average efficiency of 5.5% under 100 mW/cm^2 AM1.5G irradiance. Low HOMO level is the only drawback of PCPDTBT, which does not allow V_{oc} to be more than 0.6–0.7 V with PCBM. The current strategy to overcome this problem is the design and synthesis of novel polymers with high HOMO and acceptors with low LUMO energy levels. Kim et al. employed PCPDTBT along with P3HT in tandem configuration where $PC_{61}BM$ and $PC_{70}BM$ were used, respectively, as electron acceptors for both the polymers [68]. This cell exhibited a maximum PCE of 6.5%. The combination of PCPDTBT and P3HT resulted in a relatively broad coverage of the solar spectrum and gave an IPCE of >35% from 750 to 800 nm and >32% below 440 nm.

Although the efforts toward synthesis of new efficient low band gap polymers have been quite successful, but there is still room for new materials with better performance. Like, for example, Liang et al. [111] synthesized a series of new low band gap polymers based on alternating ester substitutes thieno [3,4-b] thiophene and benzodithiophene units. These polymers are called PTB polymers. These polymers exhibit a band gap of ~1.7 eV and very good hole mobilities. The physical properties of these polymers were fine-tuned to optimize their photovoltaic properties. V_{oc} of the cells was increased by adjusting the LUMO and HOMO energy levels of the polymers. The cell prepared by one of the polymers named PTB4 (HOMO at −5.12 eV and LUMO at −3.31 eV) as the donor and $PC_{61}BM$ as the acceptor exhibited a PCE of ~6%. DIO was mixed with 1,2-dichlorobenzene in 3% volume ratio, which improved the film morphology with finely distributed polymer/fullerene interpenetrating networks, giving high-performance solar cell. Liang et al. developed a new polymer from the same class of PTB polymers and named it PTB7 [19]. This polymer exhibits excellent photovoltaic properties compared to PTB4. The solar cell prepared by the blend of PTB7 and $PC_{71}BM$ in o-dichlorobenzene exhibited a PCE of 6.22% and the efficiency increased to 7.18% when o-dichlorobenzene mixed with DIO (3% v/v) was used as the solvent. However, the same combination exhibited an efficiency of 3.9% with chlorobenzene solvent and mixing 3% of DIO in chlorobenzene resulted in higher current density and the efficiency increased to 7.4%. This dramatic enhancement in the efficiency was due to improved active layer morphology. It was clearly seen that the PTB7:$PC_{71}BM$ cell exhibits very high photon harvesting and, as a result, very high EQE. A review on low band gap polymers has recently been reported by Bundgaard et al. [131]. As an alternate to PCBM, various new acceptor materials that absorb visible light more strongly have also been synthesized and used with P3HT [132–134]. Also, various derivatives of C_{60} and C_{70} have been synthesized to improve the processability and morphology in blends with different conjugated polymers [78,117].

In 2011, He et al. fabricated polymer solar cells based on PCDTBT and PTB7 incorporating an alcohol/water soluble conjugated polymer poly[(9,9-bis(3′-(*N*,*N*-dimethylamino) propyl)-2,7-fluorene)-alt-2,7-(9,9-dioctylfluorene)] (PFN) as the cathode interlayer and achieved a PCE of 6.79% and 8.37% for the respective polymers [135]. In both the cases $PC_{71}BM$ was used as the electron acceptor. The controlled devices of PCDTBT and PTB7 without a PFN interlayer exhibit efficiencies of 4.02% and 7.13%, respectively. Incorporation of PFN resulted in significant improvement in J_{sc}, V_{oc}, and *FF* leading to enhanced efficiency. Application of PFN in place of a Ca cathode interlayer resulted in better cell performance, and the performance improved further when PFN was incorporated in association with a Ca interlayer. It was observed that a PFN interlayer develops an interface dipole, which enhances the built-in electric field. Improvement in the built-in electric field not only enhances V_{oc} but also J_{sc} by reducing recombination losses. A PFN interlayer improved charge transport properties across the device and reduced space charge buildup that resulted in better *FF*.

These high-efficiency solar cells were achieved in normal structure, which incorporated PEDOT:PSS on ITO as the anode buffer layer with Ca/Al or Ba/Al cathodes. Use of PEDOT:PSS as the anode buffer layer and highly sensitive and unstable low work functions metals like Ca and Ba as the cathode buffer layer make OSCs very unstable [136], therefore inverted geometry was proposed, which incorporates high work function metals and oxide-semiconductor-based buffer layers. Inverted geometry uses Au or Ag electrodes as the anode and ITO as the cathode. Both the electrodes could be used with metal oxide buffer layers like MoO_3 for anode and ZnO for cathode giving air-stable solar cells. Moreover, the inverted geometry also has the beauty that all the materials including top electrodes can be processed in large area by printing techniques, making polymer solar cell manufacturing more cost effective. Initial efforts on inverted solar cells gave low PCEs, but now there are several reports in the literature on highly efficient inverted solar cells [20,24,137–139]. Even today's record-breaking single BHJ solar cell is based on inverted geometry [21]. Yang et al. prepared inverted polymer solar cells based on the BHJ network of PBDT-DTNT with $PC_{71}BM$ and achieved a PCE of 8.4% under AM1.5G irradiation. PBDT-DTNT is an important low band gap polymer, which has absorption from 400 to 800 nm with a band gap of 1.48 eV. The solar cells incorporated PFN-Br as the cathode interlayer. Without the PFN-Br interlayer the PCE was only 6.1%. Solar cells were prepared in structures ITO/ZnO/PBDT-DTNT:$PC_{71}BM$/MoO_3/Ag and ITO/ZnO/PFN-Br/PBDT-DTNT:$PC_{71}BM$/MoO_3/Ag. Thin films of ZnO (~40 nm), PFN-Br (~5 nm), and PBDT-DTNT:$PC_{71}BM$ (~250 nm) were prepared by spin cast method from their solutions, whereas MoO_3 (8 nm) and Ag (60 nm) were deposited by thermal evaporation in a vacuum chamber. PFN-Br on ZnO had a great impact on the solar cell performance. To explore the effect of PFN-Br, detailed investigations were carried out and from ultraviolet photoelectron spectroscopy (UPS) analysis and it was found that a dipole is formed between PFN-Br and ZnO layers that resulted in a higher built-in field and V_{oc}. Additionally, the incorporation of PFN-Br suppressed bimolecular recombination and enhanced IPCE response throughout the spectrum.

In another report by Yang et al., they used polyethyleneimine (PEI) as the cathode interlayer in inverted structure and achieved a high PCE [22]. PTB7:$PC_{71}BM$ blend was the light-absorbing medium and solar cells were prepared on ITO-coated glass substrates in ITO/TiO_x/PEI/PTB7:$PC_{71}BM$/MoO_3/Ag structure. The solar cell with the TiO_x/PEI electron-transport layer (ETL) exhibited an average PCE of 8.72% (9.08% for champion cell), which was, respectively, 7% and 7.38% with PEI and TiO_x only transport layers. Like PFN and PFN-Br, PEI also improved active layer contact by reducing contact resistance/contact roughness and improving the electron mobility. Spontaneous orientation of the ethylamine group within PEI develops a permanent dipole at TiO_x/active layer interface that leads to reduced work-function of TiO_x and enhanced solar cell performance. Reduced

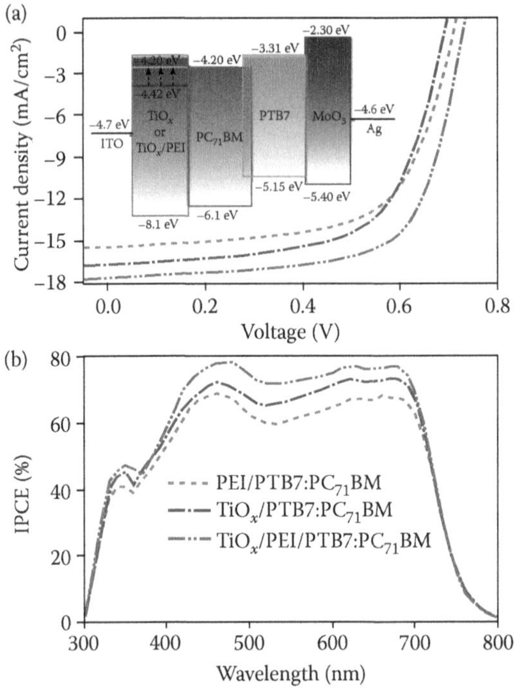

FIGURE 3.33
(a) *J–V* characteristics of ITO/ETL/PTB7:$PC_{71}BM/MoO_3$/Ag solar cells under 100 mW/cm² irradiance of AM1.5G illumination. (b) IPCE spectra of the corresponding solar cells with different ETLs. Inset shows the schematic energy level diagram of the inverted solar cells. (Redrawn with permission from D. Yang et al., 2014. *J. Mater. Chem. A* 2: 17281–17285.)

PL intensity of TiO_x/PTB7 after incorporation of PEI interlayer indicated reduced recombination and enhanced separation of charge carriers in PTB7. TiO_x/PEI extracts out the electrons more efficiently from PTB7 than TiO_x only. Figure 3.33 shows the *J–V* characteristics and IPCE spectra of ITO/ETL/PTB7:$PC_{71}BM/MoO_3$/Ag solar cells where ETLs are TiO_x, PEI, and TiO_x/PEI. He et al. prepared an inverted polymer solar cell using PFN as the cathode interlayer and PTB7:$PC_{71}BM$ blend as the light-absorbing medium and achieved a remarkable 9.2% PCE certified by the National Center of Supervision and Inspection on Solar Photovoltaic Products Quality of China (CPVT) [20]. The solar cells were prepared in structure ITO/PFN/PTB7:$PC_{71}BM/MoO_3$/Ag without any metal oxide electron-transport layer.

Earlier, it was believed that the inverted polymer solar cells would have lower efficiency than the normal ones with same photoactive materials, but the work of He et al. showed that optically and electrically optimized inverted solar cells would have not only higher efficiencies but also high stability with ease in production in large area [20]. Where the normal ITO/PEDOT:PSS/PTB7:$PC_{71}BM$/Ca/Al solar cell showed a maximum PCE of 8.24%, the inverted ITO/PFN/PTB7:$PC_{71}BM/MoO_3$/Ag solar cell exhibited 9.15% PCE. The normal solar cells had a J_{sc} of 15.4 mA/cm², V_{oc} of 0.759 V, and *FF* of 70.6%; on the other hand, the inverted solar cell had a J_{sc} of 17.2 mA/cm², V_{oc} of 0.74 V, and *FF* of 72%. From one-dimensional transfer-matrix formalism (TMF) for optical absorption it was found that inverted structure harvested more photons from incident light with reduced bimolecular recombination than those harvested by normal device. Therefore, the inverted solar cell exhibited higher J_{sc} than the normal one. XPS studies showed that application of PFN on ITO reduces its work function from 4.7 to 4.1 eV. It had also been shown in previous studies that a dipole developed at the ITO/PFN interface changes the ITO work function and energy level alignment toward better photovoltaic performance [135]. Reduced work function of ITO makes it better cathode with ohmic contact for inverted solar cells.

The solar cells were stored in air and also tested for their long-term stability. The normal cell lost half of its initial efficiency in 10 days, but the inverted solar cell retained 95% of its initial efficiency up to 62 days of air exposure. This showed that inverted solar cells are not only efficient but also maintain high stability in ambient air. Inverted solar cells were also prepared using ZnO in place of PFN, but PFN-based solar cells exhibited better performance. The chemical structures of PFN, PFN-Br, and PEI materials used for the cathode interlayer in the polymer solar cell are shown in Figure 3.34.

Most of the high-efficiency solar cells were based on a few selected polymers like PTB7 and PBDT-DTNT with $PC_{71}BM$ acceptor. Photovoltaic performance of these polymers has been found acceptor sensitive, and replacement of $PC_{71}BM$ by some other fullerene derivatives resulted in reduced efficiency around 6%–7% [140]. Therefore, for high PCE the polymers needed properties as decided by $PC_{71}BM$ and this limits variety in polymers. This creates a problem especially for tandem solar cells where perfectly matching polymer/acceptor combinations are required. For rapid development of these solar cells, the flexibility in selection of donors and acceptors materials is a must. Or the polymers should be such that the variation of fullerene derivatives does not affect the polymer:fullerene morphology. Liu et al. used three different polymers and achieved more than 10% PCE

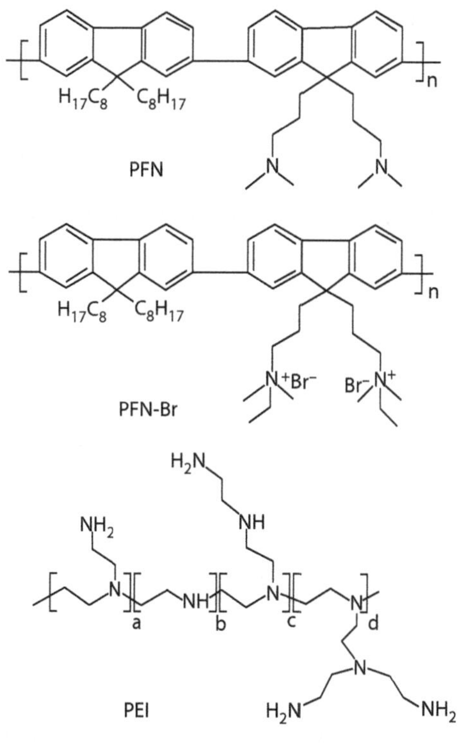

FIGURE 3.34
Chemical structures of PFN, PFN-Br, and PEI electron-transport materials.

using a number of fullerenes [21]. The chemical structures of the polymers and fullerenes used by Liu et al. are shown in Figures 3.30 and 3.31. The photovoltaic parameters of different solar cells prepared by Liu et al. are tabulated in Table 3.1. The solar cell based on the PffBT4T-2OD:TC$_{71}$BM blend showed a PCE of 10.8%, which is the maximum efficiency reported so far in single BHJ solar cells. The solar cells exhibited very high *FF* up to 77%, which could be attributed to small but pure and highly crystalline regions in polymer. Even the nontraditional fullerenes gave very high efficiency ranging from 8.6% to 10.8%. High efficiency form nontraditional fullerenes reduced our dependence on PC$_{71}$BM. The active layer morphology was insensitive to variation in fullerenes and was controlled by temperature-dependent aggregation behavior of the blends. Liu et al. noticed that the alkyl chains like 2-octyldodecyl on quaterthiophene (like in PffBT4T-2OD) cause temperature-dependent aggregation behavior in polymers. This allows the polymer solution to be processed at elevated temperatures for controlled aggregation with high crystallinity and robust morphology. This behavior could be exploited during device preparation to achieve high performance. Figure 3.35a shows the ultraviolet–visible (UV–vis) absorption spectra of PffBT4T-2OD in thin film and dichlorobenzene solution at different temperatures. When the temperature of PffBT4T-2OD solution was reduced from 85°C to 25°C the absorption spectra exhibited a red shift. At high temperature the polymer is well dissolved and forms no aggregates, but at low temperatures occurrence of a strong 0–0 transition peak at ~700 nm represents strong aggregation. The absorption spectrum of solution at 25°C is almost similar to that of the film, which was highly crystalline. It means the polymer is already aggregated at low temperatures, but if it is cast from warm solution and then allowed to cool during the film formation, the rate of aggregation and film morphology can be controlled to achieve high performance. Even the spin rate for film formation during spin coating showed an important effect on film crystallinity. From x-ray diffraction (XRD) studies it was observed that π–π ordering of the polymer decreases with an increase in spin rate.

The aggregation of polymer depends on solution concentration, temperature, film drying time, and aggregation kinetics. Low spin rates give the polymer sufficient time to aggregate and form crystalline regions. The substrate temperature was also observed to affect film aggregation and crystallinity. Fast spin rate and high substrate temperatures

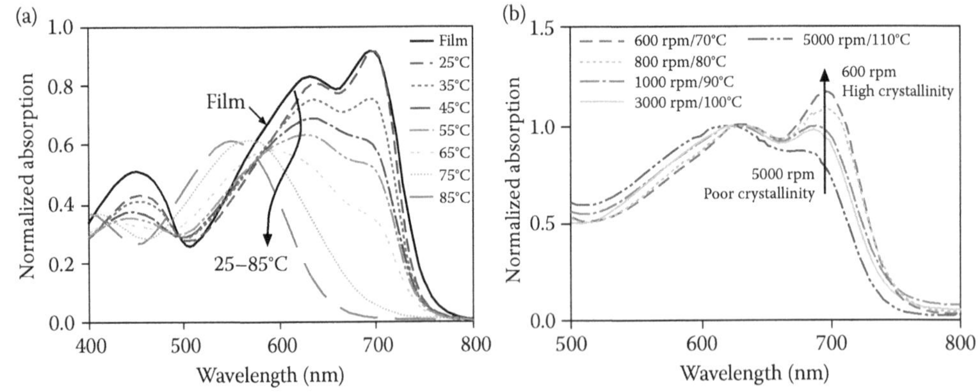

FIGURE 3.35
(a) UV–vis absorption spectra of PffBT4T-2OD in thin film and solution at different temperatures. (b) UV–vis absorption spectra of PffBT4T-2OD films obtained at different spin rates and substrate temperatures. (Reproduced with permission from Y. Liu et al., 2014. *Nat. Commun.* 5: 5293.)

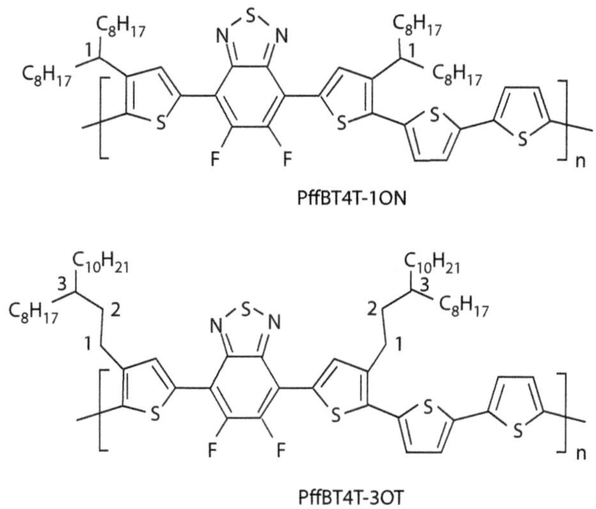

FIGURE 3.36
Chemical structures of PffBT4T-1ON and PffBT4T-3OT. The two polymers have alkyl chains branched at 1 and 3 carbon atoms.

decreased 0–0 transition peak in UV–vis absorption spectra indicating reduced crystallinity as shown in Figure 3.35b. The solar cells prepared at high spin rates and high substrate temperatures exhibited reduced hole mobilities and as a result reduced efficiency. Rate of aggregation during film formation controls the film morphology or in other terms the electrical and optical properties of the film. The aggregation of PffBT4T-2OD was insensitive to the presence of fullerenes. The temperature-dependent aggregation of polymers like PffBT4T-2OD is caused by their branched alkyl chain and their position on quaterthiophene. Liu et al. prepared a number of other polymers having the same backbone as PffBT4T-2OD but with the alkyl chains branched at the first and third chain carbon atoms. These polymers were named PffBT4T-1ON and PffBT4T-3OT, and their structures are shown in Figure 3.36 [21]. Interestingly PffBT4T-1ON did show aggregation neither at 85°C nor at 25°C and formed films of very poor crystallinity giving very poor PCE. On the other hand, PffBT4T-3OT had high aggregation at both the 85°C and 25°C leading to quick gelation of the solution even before casting. The alkyl chains and their position in PffBT4T-1ON caused too much steric hindrance, whereas the alkyl chains in PffBT4T-3OT cause very little steric hindrance leading to quick aggregation and making film processing very difficult. The alkyl chains branched at second position in PffBT4T-2OD gives optimal tradeoff giving controllable aggregation. For more on the effect of alkyl chain branching on polymer properties and solar cells, see Liu et al. [21].

3.7.2 Developments in Tandem BHJ OSCs

Though the concept of tandem structure in OSCs was used by Hiramoto et al. in 1990 [60], but to discuss the major developments made in this area I start with the tandem OSCs prepared by Drechsel et al. in 2005 [32]. Drechsel et al. staked two *p-i-n* solar cells where each subcell had the blend of ZnPc and C_{60} as the active medium [32]. The *p*-type and *n*-type transport layers in both the subcells were the F4TCNQ-doped MeOTPD and rhodamine-B-doped C_{60}, respectively. The two *p-i-n* cells were connected by an ultrathin layer of Au. The Au interlayer forms a quasi-ohmic tunneling contact to the *p*-doped MeO-TPD hole-transport layer of the back cell and to the *n*-doped C_{60} electron-transport layer of the front cell. The electrons from the front cell combine with holes from the back cell at the Au interlayer. This tandem structure is different from those with undoped transport layers as both the layers adjacent to the interlayer are wide band gap doped layers and work as spacers between the photoactive layers of each subcell. With the optimized thickness, the tandem *p-i-n* cell exhibited a PCE of 3.8%, which was almost double of that achieved with the single *p-i-n* cell [32]. The tandem *p-i-n* solar cell exhibited higher *FF* and low recombination losses compared to that of the single *p-i-n* cell. Xue et al. [61] demonstrated an efficiency of ~5.7% by staking two CuPc/C_{60} hybrid planar-mixed heterojunction cells in series. The two subcells were connected by a thin layer of Ag nanoclusters buried in 5 nm of 4,4′,4″-tris(3-methyl-phenyl-phenyl-amino)triphenylamine (m-MTDATA) doped with 5 mol% of 2,3,5,6-tetrafluoro-7,7,8,9-tetracyano quinodimethane (F4TCNQ). PTCBI and BCP were used as exciton blocking layers for the front and back cells, respectively. To get maximum absorption of incident light, the tandem cell was prepared in a way that longer wavelength light absorbed in the front cell and shorter wavelength light absorbed in the back cell. The V_{oc} of the cell was almost twice that of the single cell and reached ~1.2 V under high illumination intensity.

Shrotriya et al. [66] fabricated two solar cells with semitransparent LiF/Al/Au cathode and ITO/PEDOT:PSS anode. In both devices, the active layer was the blend of MEH-PPV and PCBM. The cells were stacked in series and parallel. For series connection, the devices were kept in front of each other such that the light transmitted by the front cell was absorbed by the back cell. For serially connected devices, the V_{oc} of the tandem cell was doubled, whereas J_{sc} and *FF* remained more or less unchanged. For parallel-connected devices the J_{sc} was doubled, whereas V_{oc} and *FF* remained almost unchanged. As a result of stacking, the PCEs for the stacked structures were 2.4 ± 0.2% (for series connection) and 2.5 ± 0.1% (for parallel connection) compared to ~1.2 ± 0.1% for the individual cells. Efforts were also made to make tandem solar cells incorporating both the small molecular and polymer materials, but the efficiency remained very low, even lower than the individual subcell [28,63,65]. Achieving high efficiency from tandem OSCs had been a tough task for a long time, as it required photoactive materials that not only harvests photons efficiently but are also compatible to each other. In 2007, Kim et al. reported a tandem OSC that employed two subcells of polymers having complementary absorption spectra. The front cell was based on a PCPDTBT:PC_{61}BM blend, whereas the back cell was based on a P3HT:PC_{71}BM blend. The two cells were connected through a thin TiO_x layer and the optimized tandem solar cell showed a PCE of 6.5%. The schematic structure and working mechanism of this cell with an energy level diagram is shown in Figure 3.16.

For high V_{oc} and large photocurrent from tandem OSCs, the energy levels of polymers are required to be fine-tuned for minimum energy loss. The materials were required to have high charge carrier mobilities with fine phase separation in the active layer. Because a large part of the incident light was already absorbed by the front cell, the back cell produced low current. Therefore, to minimize the current mismatch, the back cell is required to generate a high current, which had been a real challenge from existing low band gap polymers. Design and synthesis of new low band gap polymers, generating high current from low-energy photons, was urgently needed for efficient tandem solar cells. In 2012, Yang and his collaborators designed and synthesized a novel low band gap conjugated polymer poly{2,6′-4,8-di(5-ethylhexylthienyl)benzo[1,2-*b*;3,4-b]dithiophene-alt-5-dibutyloctyl-3,6-bis(5-bromothiophen-2-yl)pyrrolo[3,4-*c*]pyrrole-1,4-dione} (PBDTT-DPP)

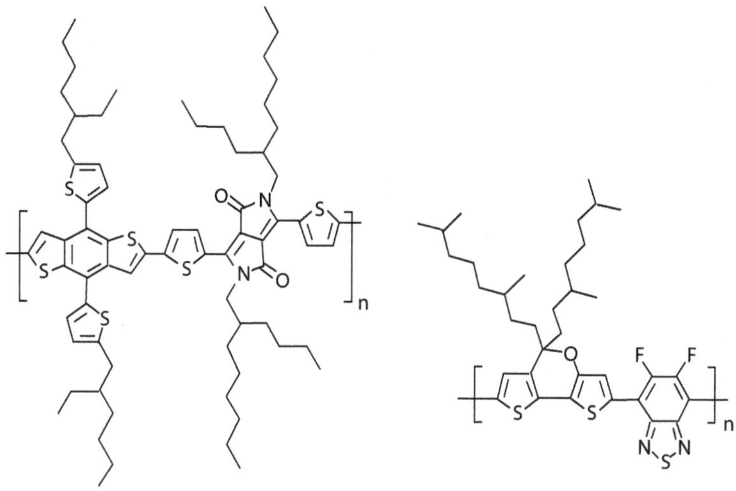

FIGURE 3.37
Chemical structures of PBDTT-DPP and PDTP-DFBT low band gap polymers.

(band gap ~1.44 eV) and achieved 8.62% PCE (certified by the National Renewable Energy Laboratory [NREL]) in tandem structure in association with P3HT (band gap ~1.9 eV) [124]. The absorption spectrum of PBDTT-DPP ranges from 500 to 850 nm, and the two polymers covered the solar spectrum from 350 to 850 nm with a very small spectral overlap. The chemical structure of PBDTT-DPP is shown in Figure 3.37. PBDTT-DPP had a good hole mobility of 3.1×10^{-4} cm^2/V and the single-junction solar cells prepared with PBDTT-DPP:$PC_{71}BM$ blend only exhibited a PCE of 6.5% both in normal and inverted architectures. Therefore, the tandem solar cell was prepared in inverted structure because of its simplicity and high stability. The tandem solar cell was prepared in structure ITO/ZnO/P3HT:ICBA/PEDOT:PSS/ZnO/PBDTT-DPP:$PC_{71}BM$/MoO_3/Ag, where PEDOT:PSS was used as the intermediate bridging layer. ZnO nanoparticle film was used as the electron-transport layer and prepared from its solution by spin coating. In the front cell the active layer of P3HT:ICBA (1:1 wt%) was spin coated from its 1.8% blend solution in 1,2-dichlorobenzene on the top of ZnO film. Then PEDOT:PSS was spin coated and annealed at 150°C for 10 min. After that ZnO film was spin coated on PEDOT:PSS layer and annealed at 150°C for 10 min. In the back cell PBDTT-DPP:$PC_{71}BM$ (1:2 wt%) was spin coated from its 8 mg/ml solution in 1,2-dichlorobenzene on the top of PEDOT:PSS film. Finally, to complete the cell, 15 nm of MoO_3 and 100 nm of Ag were sequentially deposited by thermal evaporation in vacuum. Whereas the optimized single-junction P3HT:ICBA solar cell had a J_{sc} of 9.56 mA/cm^2, V_{oc} of 0.85 V, *FF* of 70.2%, and PCE of 5.7%, the PBDTT-PDD:$PC_{71}BM$ solar cell had a J_{sc} of 13.5 mA/cm^2, V_{oc} of 0.74 V, *FF* of 65.1%, and PCE of 6.5%. The tandem cell had a certified J_{sc} of 8.26 mA/cm^2, V_{oc} of 1.56 V, *FF* of 66.8%, and PCE of 8.62% under 100 mW/cm^2 irradiance of AM1.5G illumination. These solar cells also showed very high stability with almost no degradation up to 30 days of testing.

In 2013, Yang and his collaborators further designed and synthesized another novel polymer poly[2,7-(5,5-bis-(3,7-dimethyloctyl)-5*H*-dithieno[3,2-b:2′,3′-d]pyran)-alt-4,7-(5,6-difluoro-2,1,3-benzothiadiazole)] (PDTP-DFBT) with a band gap of 1.38 eV and absorption spectrum spanning from ~500 to 900 nm [141]. The chemical structure of PDTP-DFBT is also shown in Figure 3.37. This polymer had a high hole mobility and deep HOMO level such that the single-junction solar cell itself with $PC_{71}BM$ acceptor gave 7.9% PCE. The performance of PDTP-DFBT was also tested with $PC_{61}BM$ but a slightly lower PCE of 7.1% was achieved in single BHJ solar cell. This polymer led to an NREL-certified 10.6% PCE in tandem structure with P3HT. The two polymers had complementary absorption spectra covering the solar spectrum from 350 to 900 nm with very little overlap. Tandem solar cells were prepared in inverted structure in the configuration ITO/ZnO/P3HT:ICBA/ PEDOT:PSS/ZnO/PDTP-DFBT:PCBM/MoO_3/Ag, where high band gap polymer P3HT was blended with ICBA to formulate the front cell, and the low band gap polymer PDTP-DFBT was blended with $PC_{61}BM$ or $PC_{71}BM$ to constitute the back cell. Fabrication method of this tandem cell was the same as mentioned earlier for PBDTT-DPP-based tandem solar cells. The photovoltaic performance of different single-junction and tandem solar cells are shown in Table 3.2. The two single BHJ solar cells of PDTP-PFBT exhibited different PCE mainly because of different J_{sc} due to different absorption coefficients of $PC_{61}BM$ and $PC_{71}BM$. V_{oc}s of the two tandem solar cells were exactly equal to the sum of those of corresponding single BHJ solar cells, indicating perfect interconnection of the front and back cells. Interestingly, where PDTP-DFBT exhibited higher PCE with $PC_{71}BM$ in single BHJ solar cells, it showed higher efficiency in tandem structure with the $PC_{61}BM$ acceptor. To understand this behavior, Yang et al. measured the absorption spectra of different polymer:fullerene combinations, namely, P3HT:ICBA, PDTP-DFBT:$PC_{61}BM$, and PDTP-DFBT:$PC_{71}BM$, which are shown in Figure 3.38 [141]. Blending of $PC_{61}BM$ with PDTP-DFBT enhanced the absorption spectrum slightly in the region 300–400 nm with slight overlap with that of P3HT:ICBA, whereas blending of $PC_{71}BM$ with PDTP-DFBT enhanced the absorption spectrum in the entire visible region with significant overlap with absorption spectrum of P3HT:ICBA. Different spectral overlaps of front and back solar cells affect the current matching (large overlap would cause lower current in the back cell) and therefore the efficiency of tandem solar cells. Due to larger overlap in absorption spectra of the front and back cell, the back PDTP-DFBT:$PC_{71}BM$ cell generated lower current leading to lower efficiency. However, it gave higher efficiency in the single-junction cell due to broader absorption spectrum. It shows that the maximum efficiency in OSCs is not only a matter of efficient materials but also of device optimization. Therefore, selection of suitable fullerene derivatives to match the currents from the front and back cells could be an important tool to improve the performance of tandem devices.

TABLE 3.2

Photovoltaic Parameters of Single-Junction and Tandem Solar Cells Based on P3HT and PDTP-DFBT

	J_{sc} (mA/cm^2)	V_{oc} (V)	*FF* (%)	PCE (%)
Single-Junction Devices				
P3HT:ICBA	10.3	0.84	71.1	6.1
PDTP-DFBT:$PC_{61}BM$	15.4	0.70	66.2	7.1
PDTP-DFBT:$PC_{71}BM$	17.8	0.68	65.0	7.9
Tandem Devices				
P3HT:ICBA/PDTP-DFBT:$PC_{61}BM$	10.1	1.53	68.5	10.6
P3HT:ICBA/PDTP-DFBT:$PC_{71}BM$	9.8	1.51	69.2	10.2

Source: Data from J. You et al., 2013. *Nat. Commun.* 4: 1446.

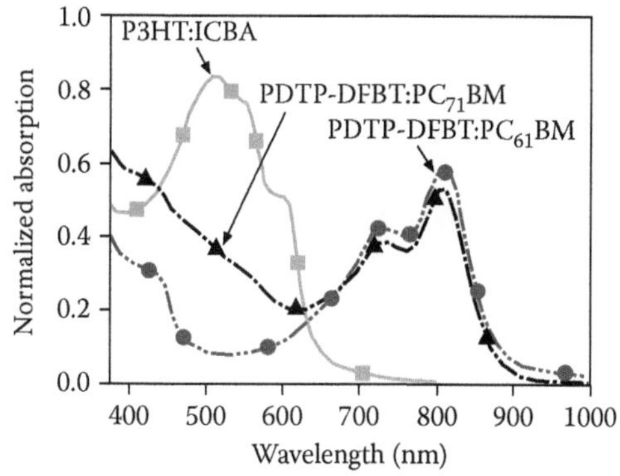

FIGURE 3.38
Absorption spectra of P3HT:ICBA, PDTP-DFBT:$PC_{61}BM$, and PDTP-DFBT:$PC_{71}BM$ blends. (Modified from J. You et al., 2013. *Nat. Commun.* 4: 1446.)

Later in 2014, Yang et al. reported a triple-junction tandem OSC having a record 11.5% PCE from polymeric systems [69]. For this purpose, they selected three polymers with a large difference in their band gaps: P3HT (E_g ~ 1.9 eV), PTB7 (E_g ~ 1.58 eV), and PDTP-DFBT (E_g ~ 1.38 eV) satisfying the condition $E_{g1} > E_{g2} > E_{g3}$. P3HT was used as the front absorber, whereas PTB7 and PDTP-DFBT were used as the middle and the back absorbers, respectively. P3HT absorbs higher energy photons (>1.9 eV) and leaves lower energy photons to be absorbed by lower band gap PTB7 and PDTP-DFBT polymers. The combination of three complementary polymers utilizes the excess thermalization energy and harvests the incident photons more efficiently. As a result of improved photon harvesting, the triple-junction solar cell exhibited high efficiency with V_{oc} exceeding 2 V. Figure 3.39a shows the schematic structure of the triple-junction solar cell. To improve the visible responses, P3HT was blended with ICBA (1:1 wt%), whereas PTB7 and PDTP-DFBT were blended with $PC_{71}BM$ in 1:1.5 and 1:2 wt% ratios, respectively. P3HT:ICBA and PDTP-DFBT:$PC_{71}BM$ solutions were prepared in dichlorobenzene, whereas PTB7:$PC_{71}BM$ solution was prepared in chlorobenzene with 3% 1,8-diiodooctane and deposited by spin coating. LBG in Figure 3.39 stands for the low band gap polymer PDTP-DFBT. Figure 3.39b shows the refractive indices (n) and extinction coefficients (k) of the three active layers P3HT:ICBA, PTB7:$PC_{71}BM$, and PDTP-DFBT:$PC_{71}BM$. The three polymers have complementary absorption spectra covering the solar spectrum from 400 to 950 nm.

It can be seen from the energy level diagram of the tandem cell that the HOMO energy level of P3HT exists at 5.0 eV, and to make perfect contact for hole recombination with electrons of the next cell, PEDOT:PSS having the same work function of 5.0 eV was used as the interconnecting layer of front and middle cells. Since PTB7 has slightly deep HOMO of 5.2 eV, to overcome the mismatch in the energy levels of PTB7 and PEDOT:PSS a thin layer of WO_3 was inserted between PTB7 and PEDOT:PSS to constitute an efficient interlayer for the middle and back cells. To minimize the current mismatch among front, middle, and back cells, the authors first investigated the performance of three polymers in single-junction configurations [69]. The photovoltaic performance of these solar cells is tabulated in Table 3.3. The solar cells were prepared in inverted structure ITO/ZnO/active layer/anode. For P3HT:ICBA-based single-junction solar cells, PEDOT:PSS/Al was used as the anode, whereas for those based on PTB7:$PC_{71}BM$ and PDTP-DFBT:$PC_{71}BM$ the anode was WO_3/Al. Due to improved photon harvesting by lower band gap polymers, the J_{sc}

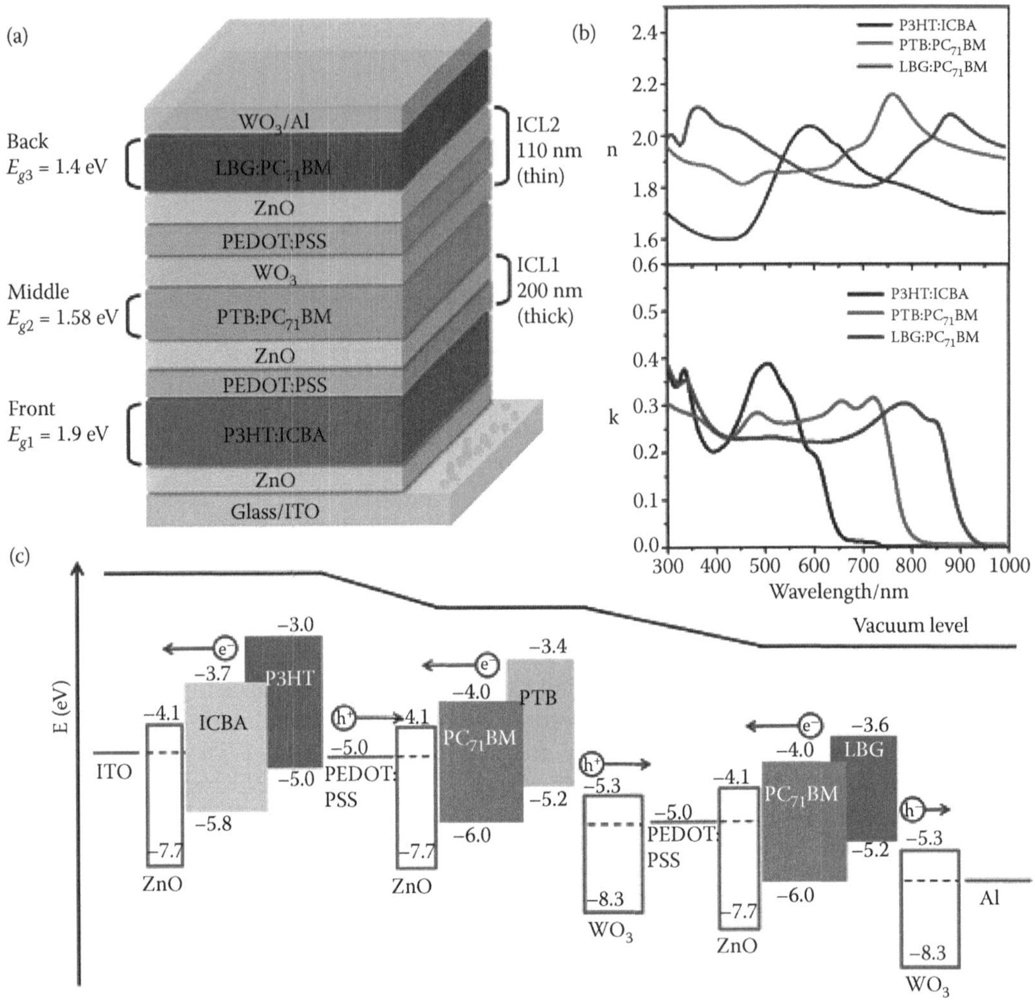

FIGURE 3.39
(a) Schematic structure of the triple junction tandem polymer solar cell in inverted architecture. (Data from C. C. Chen et al., *Adv. Mater.* 26 (2014): 5670–5677.) Here LBG stands for the low band gap polymer PDTP-DFBT. (b) Refractive indices (n) and extinction coefficient (k) of the three light-absorbing mediums P3HT:ICBA, PTB:PC71BM, and PDTP-DFBT:PC_{71}BM. (c) Schematic energy level diagram of the triple junction solar cell. (Reproduced with permission from C. C. Chen et al., 2014. *Adv. Mater.* 26: 5670–5677.)

increased with reduction in band gap, which was perfect and consistent with the requirement for tandem configuration. To understand the behavior in stacking, the authors prepared a number of double-junction solar cells; P3HT:ICBA/PTB7:PC_{71}BM (bottom part of the triple-junction cell), PTB7:PC_{71}BM/PDTP-DFBT:PC_{71}BM (top part of the triple-junction cell), and P3HT:ICBA/PDTP-DFBT:PC_{71}BM (middle part of the triple junction missing) and their photovoltaic performance are shown in Table 3.3. Since P3HT and PDTP-DFBT had complementary absorption spectra, the maximum photocurrent from the front cell of the P3HT:ICBA/PDTP-DFBT:PC_{71}BM double-junction tandem cell matched with that of the back cell that gave nearly ideal current in tandem configuration. J_{sc} in the double-junction PTB7:PC_{71}BM/PDTP-DFBT:PC_{71}BM solar cell was less than that of larger band

TABLE 3.3
Photovoltaic Parameters of Single-Junction, Double-Junction, and Triple-Junction Solar Cells

	J_{sc} (mA/cm²)	V_{oc} (V)	*FF* (%)	PCE (%)
Single-Junction Devices				
P3HT:ICBA	9.93	0.84	69.23	5.79
PTB7:$PC_{71}BM$	14.51	0.78	67.81	7.68
PDTP-DFBT:$PC_{71}BM$	17.49	0.70	64.35	7.82
Double-Junction Tandem Devices				
P3HT:ICBA/PTB7:$PC_{71}BM$	8.81	1.58	68.83	9.58
PTB7:$PC_{71}BM$/PDTP-DFBT:$PC_{71}BM$	11.30	1.42	66.67	10.70
P3HT:ICBA/PDTP-DFBT:$PC_{71}BM$	9.50	1.53	67.65	9.83
Triple-Junction Tandem Devices				
P3HT:ICBA/PTB7:$PC_{71}BM$/PDTP-DFBT:$PC_{71}BM$ (unoptimized)	6.95	2.22	56.64	8.74
P3HT:ICBA/PTB7:$PC_{71}BM$/PDTP-DFBT:$PC_{71}BM$ (optimized)	7.63	2.28	66.39	11.55

Source: Data from C. C. Chen et al., 2014. *Adv. Mater.* 26: 5670–5677.

gap PTB7 solar cells and that was because of partial absorption overlap between PTB7 and PDTP-DFBT. Similarly, the double-junction P3HT:ICBA/PTB7:$PC_{71}BM$ solar exhibited lower photocurrent than that of P3HT solar cell due to overlap of P3HT and PTB7 absorption spectra. It shows that the spectral overlap of subcells in a tandem device is very critical and to minimize this effect the materials should have a large band gap difference of more than 0.5 eV. Yang et al. performed optical simulation and showed that by minimizing the effect of optical interference one can achieve efficient power conversion from sequential stacking of polymers like P3HT, PTB7, and PDTP-DFBT, which have band gap differences of only 0.3 and 0.2 eV. Where the unoptimized triple-junction solar cells exhibited an efficiency of 8.74%, the optimization in film thicknesses of different interconnecting, transporting, and light-absorbing layers led to 11.55% PCE.

References

1. C. K. Chiang, C. R. Fincher, Jr., Y. W. Park, A. J. Heeger, H. Shirakawa, E. J. Louis, S. C. Gau, and A. G. MacDiarmid, 1977. Electrical conductivity in doped polyacetylene, *Phys. Rev. Lett.* 39: 1098.
2. C. W. Tang, 1986. Two-layer organic photovoltaic cell, *Appl. Phys. Lett.* 48: 183.
3. C. W. Tang and S. A. VanSlyke, 1987. Organic electroluminescent diodes, *Appl. Phys. Lett.* 51: 913.
4. H. Koezuka, A. Tsumara, and T. Ando, 1987. Field-effect transistor with polythiophene thin film, *Synth. Met.* 18: 699–704.
5. J. H. Burroughes, C. A. Jones, and R. H. Friend, 1988. New semiconductor device physics in polymer diodes and transistors, *Nature* 335: 137.
6. G. Horowitz, D. Fichou, X. Z. Peng, Z. Xu, and F. Garnier, 1989. A field-effect transistor based on conjugated alpha-sexithienyl, *Sol. State Commun.* 72: 381–384.
7. S. Shao, J. Liu, J. Bergqvist, S. Shi, C. Veit, U. Würfel, Z. Xie, and F. Zhang, 2013. In situ formation of MoO_3 in PEDOT:PSS matrix: A facile way to produce a smooth and less hygroscopic hole transport layer for highly stable polymer bulk heterojunction solar cells, *Adv. Energy Mater.* 3: 349–355.

8. F. Liu, S. Shao, X. Guo, Y. Zhao, and Z. Xie, 2010. Efficient polymer photovoltaic cells using solution-processed MoO_3 as anode buffer layer, *Sol. Energy Mater. Sol. Cells* 94: 842–845.
9. J. Jung, D. L. Kim, S. H. Oh, and H. J. Kim, 2012. Stability enhancement of organic solar cells with solution-processed nickel oxide thin films as hole transport layers, *Sol. Energy Mater. Sol. Cells* 102: 103–108.
10. C. K. Cho, W. J. Hwang, K. Eun, S. H. Choa, S. I. Na, and H. K. Kim, 2011. Mechanical flexibility of transparent PEDOT:PSS electrodes prepared by gravure printing for flexible organic solar cells, *Sol. Energy Mater. Sol. Cells* 95: 3269–3275.
11. M. D. Irwin, D. B. Buchholz, A. W. Hains, R. P. H. Chang, and T. J. Marks, 2008. p-Type semiconducting nickel oxide as an efficiency-enhancing anode interfacial layer in polymer bulk-heterojunction solar cells, *PNAS* 105: 2783–2787.
12. V. Shrotriya, G. Li, Y. Yao, C.-W. Chu, and Y. Yang, 2006. Transition metal oxides as the buffer layer for polymer photovoltaic cells, *Appl. Phys. Lett.* 88: 073508–073510.
13. K. Takanezawa, K. Tajima, and K. Hashimoto, 2008. Efficiency enhancement of polymer photovoltaic devices hybridized with ZnO nanorod arrays by the introduction of a vanadium oxide buffer layer, *Appl. Phys. Lett.* 93: 063308–063310.
14. Y. Kinoshita, R. Takenaka, and H. Murata, 2008. Independent control of open-circuit voltage of organic solar cells by changing film thickness of MoO_3 buffer layer, *Appl. Phys. Lett.* 92: 243309–2433011.
15. D. Y. Kim, J. Subbiah, G. Sarasqueta, F. So, H. Ding, and G. Y. Irfan, 2009. The effect of molybdenum oxide interlayer on organic photovoltaic cells, *Appl. Phys. Lett.* 95: 093304–093306.
16. L. Chen, P. Wang, F. Li, S. Yu, and Y. Chen, 2012. Efficient bulk heterojunction polymer solar cells using PEDOT/PSS doped with solution-processed MoO_3 as anode buffer layer, *Sol. Energy Mater. Sol. Cells* 102: 66–70.
17. L. Yang, H. Xu, H. Tian, S. Yin, and F. Zhang, 2010. Effect of cathode buffer layer on the stability of polymer bulk heterojunction solar cells, *Sol. Energy Mater. Sol. Cells* 94: 1831–1834.
18. C. C. Chen, L. Dou, R. Zhu, C. H. Chung, T. B. Song, Y. B. Zheng, S. Hawks, G. Li, P. S. Weiss, and Y. Yang, 2012. Visibly transparent polymer solar cells produced by solution processing, *ACS Nano* 6: 7185–7190.
19. Y. Liang, Z. Xu, J. Xia, S. T. Tsai, Y. Wu, G. Li, C. Ray, and L. Yu, 2010. For the bright future-bulk heterojunction polymer solar cells with PCE of 7.4%, *Adv. Mater.* 22: E135–E138.
20. Z. He, C. Zhong, S. Su, M. Xu, H. Wu, and Y. Cao, 2012. Enhanced power-conversion efficiency in polymer solar cells using an inverted device structure, *Nat. Photon.* 6: 591–595.
21. Y. Liu, J. Zhao, Z. Li, C. Mu, W. Ma, H. Hu, K. Jiang, H. Lin, H. Ade, and H. Yan, 2014. Aggregation and morphology control enables multiple cases of high-efficiency polymer solar cells, *Nat. Commun.* 5: 5293.
22. D. Yang, P. Fu, F. Zhang, N. Wang, J. Zhang, and C. Li, 2014. High efficiency inverted polymer solar cells with room-temperature titanium oxide/polyethylenimine films as electron transport layers, *J. Mater. Chem. A* 2: 17281–17285.
23. S. H. Liao, H. J. Jhuo, Y. S. Cheng, and S. A. Chen, 2013. Fullerene derivative-doped zinc oxide nanofilm as the cathode of inverted polymer solar cells with low band gap polymer (PTB7-Th) for high performance, *Adv. Mater.* 25: 4766–4771.
24. T. Yang, M. Wang. C. Duan, X. Hu, L. Huang, J. Peng, F. Huang, and X. Gong, 2012. Inverted polymer solar cells with 8.4% efficiency by conjugated polyelectrolyte, *Energy Environ. Sci.* 5: 8208–8214.
25. A. K. Gosh and T. Feng, 1973. Rectification, space-charge limited current, photovoltaic and photoconductive properties of Al/tetracene/Au sandwich cell, *J. Appl. Phys.* 44: 2781–2788.
26. A. Hadipour, B. De Boer, J. Wildman, F. B. Kooistra, J. C. Hummelen, M. G. R. Turbiez, M. M. Wienk, R. A. J. Janssen, and P. W. M. Blom, 2006. Solution-processed organic tandem solar cells, *Adv. Funct. Mater.* 16: 1897–1903.
27. S. Sista, M. H. Park, Z. Hong, Y. Wu, J. Hou, W. L. Kwan, G. Li, and Y. Yang, 2010. Highly efficient tandem polymer photovoltaic cells, *Adv. Mater.* 22: 380–383.

28. J. Gilot, M. M. Weink, and R. A. J. Janssen, 2007. Double and triple junction polymer solar cells processed from solution, *Appl. Phys. Lett.* 90: 143512.
29. J. Gilot, M. M. Weink, and R. A. J. Janssen, 2010. Optimizing polymer tandem solar cells, *Adv. Mater.* 22: E67–E71.
30. S. R. Forrest, W. Y. Yoon, L. Y. Leu, and F. F. So, 1989. Optical and electrical properties of isotype crystalline molecular organic heterojunctions, *J. Appl. Phys.* 66: 5908–5914.
31. P. Peumans, A. Yakimov, and S. R. Forrest, 2003. Small molecular weight organic thin-film photodetectors and solar cells, *J. Appl. Phys.* 93: 3693–3723.
32. J. Drechsel, B. Männig, F. Kozlowski, M. Pfeiffer, K. Leo, and H. Hoppe, 2005. Efficient organic solar cells based on a double *p-i-n* architecture using doped wide-gap transport layers, *Appl. Phys. Lett.* 86: 244102.
33. K. L. Mutolo, E. I. Mayo, B. P. Rand, S. R. Forrest, and M. E. Thompson, 2006. Enhanced open-circuit voltage in subphthalocyanine/C60 organic photovoltaic cells, *J. Am. Chem. Soc.* 128: 8108–8109.
34. H. Kumar, P. Kumar, R. Bhardwaj, G. D. Sharma, S. Chand, S. C. Jain, and V. Kumar, 2009. Broad spectral sensitivity and improved efficiency in CuPc/Sub-Pc organic photovoltaic devices, *J. Phys. D* 42: 15103.
35. B. P. Rand, J. Xue, F. Yang, and S. R. Forrest, 2005. Organic solar cells with sensitivity extending into the near infrared, *Appl. Phys. Lett.* 87: 233508.
36. K. Suemori, T. Miyata, M. Yokoyama, and M. Hiramoto, 2004. Organic solar cells protected by very thick naphthalene tetracarboxylic anhydride films, *Appl. Phys. Lett.* 85: 6269–6271.
37. J. H. Schön, C. Kloc, E. Bucher, and B. Batlogg, 2000. Efficient organic photovoltaic diodes based on doped pentacene, *Nature* 403: 408–410.
38. K. Schulze, C. Uhrich, R. Schuppel, K. Leo, M. Pfeiffer, E. Brier, E. Reinold, and P. Bauerle, 2006. Efficient vacuum deposited organic solar cells based on a new low band gap oligothiophene and fullerene C60, *Adv. Mater.* 18: 2872–2875.
39. Y. Hirose, A. Kahn, V. Aristov, P. Soukiassian, V. Bulovic, and S. R. Forrest, 1996. Chemistry and electronic properties of metal–organic semiconductor interfaces: Al, Ti, In, Sn, Ag, and Au on PTCDA, *Phys. Rev. B* 54: 13748–13758.
40. P. Peumans, V. Bulovic, and S. R. Forrest, 2000. Efficient photon harvesting at high optical intensities in ultrathin organic double-heterostructure photovoltaic diodes, *Appl. Phys. Lett.* 76: 2650–2652.
41. P. Peumans and S. R. Forrest, 2001. Very-high-efficiency double-heterostructure copper phthalocyanine/C60 photovoltaic cells, *Appl. Phys. Lett.* 79: 126–128.
42. J. Xue, S. Uchida, B. P. Rand, and S. R. Forrest, 2004. 4.2% efficient organic photovoltaic cells with low series resistances, *Appl. Phys. Lett.* 84: 3013–3015.
43. M. Y. Chan, C. S. Lee, S. L. Lai, M. K. Fung, F. L. Wong, H. Y. Sun, K. M. Lau, and S. T. Lee, 2006. Efficient organic photovoltaic devices using a combination of exciton blocking layer and anodic buffer layer, *J. Appl. Phys.* 100: 94506.
44. M. Y. Chan, S. L. Lai, M. K. Fung, C. S. Lee, and S. T. Lee, 2007. Doping-induced efficiency enhancement in organic photovoltaic devices, *Appl. Phys. Lett.* 90: 23504.
45. G. Yu and A. J. Heeger, 1995. Charge separation and photovoltaic conversion in polymer composites with internal donor/acceptor heterojunctions, *J. Appl. Phys.* 78: 4510–4515; G. Yu, J. Gao, J. Hummelen, F. Wudl, and A. J. Heeger, 1995. Polymer photovoltaic cells: Enhanced efficiencies via a network of internal donor–acceptor heterojunctions, *Science* 270: 1789–1791.
46. J. J. M. Halls, A. C. Walsh, N. C. Greenham, E. A. Marseglia, R. H. Friend, S. C. Moratti, and A. B. Holmes, 1995. Efficient photodiodes from interpenetrating polymer networks, *Nature* 376: 498–500.
47. J. Xue, S. Uchida, B. P. Rand, and S. R. Forrest, 2005. A hybrid planar-mixed molecular heterojunction photovoltaic cell, *Adv. Mater.* 17: 66–71.
48. S. E. Shaheen, C. J. Brabec, N. S. Sariciftci, F. Padinger, T. Fromhert, and J. C. Hummelen, 2001. 2.5% efficient organic plastic solar cells, *Appl. Phys. Lett.* 78: 841–843.

49. P. Peumans, S. Uchida, and S. R. Forrest, 2003. Efficient bulk heterojunction photovoltaic cells using small-molecular weight organic thin films, *Nature* 425: 158–162.
50. S. Uchida, J. Xue, B. P. Rand, and S. R. Forrest, 2004. Organic small molecule solar cells with a homogeneously mixed copper phthalocyanine: C60 active layer, *Appl. Phys. Lett.* 84: 4218–4220.
51. M. Hiramoto, H. Fujiwara, and M. Yokohama, 1991. Three layered organic solar cell with a photoactive interlayer of co-deposited pigments, *Appl. Phys. Lett.* 58: 1062–1064.
52. M. Hiramoto, H. Fujiwara, and M. Yokohama, 1992. *p-i-n* like behavior in three-layered organic solar cells having a co-deposited interlayer of pigments, *J. Appl. Phys.* 72: 3781–3787.
53. J. Huang, M. Pfeiffer, A. Werner, J. Blochwitz, K. Leo, and S. Liu, 2002. Low-voltage organic electroluminescent devices using *pin* structures, *Appl. Phys. Lett.* 80: 139–141.
54. B. Maennig, M. Pfeiffer, J. Blochwitz, A. Nollau, X. Zhou, P. Simon, and K. Leo, 2001. Controlled *p*-type doping of polycrystalline and amorphous organic layers: Self-consistent description of conductivity and field effect mobility by a microscopic percolation mode, *Phys. Rev. B* 64: 195208.
55. A. G. Werner, F. Li, K. Harada, M. Pfeiffer, T. Fritz, and K. Leo, 2003. Pyronin B as a donor for *n*-type doping of organic thin films, *Appl. Phys. Lett.* 82: 4495–4497.
56. D. Gebeyehu, M. Pfeiffer, B. Maennig, J. Drechsel, A. Werner, and K. Leo, 2004. Highly efficient *p-i-n* type organic photovoltaic devices, *Thin Solid Films* 451–452: 29–32.
57. J. Rostalski and D. Meissner, 2000. Monochromatic versus solar efficiencies of organic solar cells, *Sol. Energy Mater. Sol. Cells* 61: 87–95.
58. B. Maennig, J. Drechsel, D. Gebeyehu, P. Simon, F. Kozlowski, A. Werner, F. Li et al., 2004. Organic *p-i-n* solar cells, *Appl. Phys. A* 79: 1–14.
59. R. R. King, D. C. Law, K. M. Edmondson, C. M. Fetzer, G. S. Kinsey, H. Yoon, R. A. Sherif et al., 2007. Multijunction solar cells with over 40% efficiency and future directions in concentrator PV, *22nd European Photovoltaic Solar Energy Conference*, 11–15.
60. M. Hiramoto, M. Suezaki, and M. Yokoyama, 1990. Effect of thin gold interstitial-layer on the photovoltaic properties of tandem organic solar cell, *Chem. Lett.* 19: 327.
61. J. Xue, S. Uchida, B. P. Rand, and S. R. Forrest, 2004. Asymmetric tandem organic photovoltaic cells with hybrid planar-mixed molecular heterojunctions, *Appl. Phys. Lett.* 85: 5757–5759.
62. W. Ma, J. Y. Kim, K. Lee, and A. J. Heeger, 2007. Effect of the molecular weight of poly3-hexylthiophene) on the morphology and performance of polymer bulk heterojunction solar cells, *Macromol. Rapid Commun.* 28: 1776–1780.
63. G. Dennler, H. J. Prall, R. Koeppe, M. Egginger, R. Autengruber, and N. S. 2006. Sariciftci, Enhanced spectral coverage in tandem organic solar cells, *Appl. Phys. Lett.* 89: 73502.
64. A. Yakimov and S. R. Forrest, 2002. High photo-voltage multiple heterojunction organic solar cells incorporating interfacial metallic nano-clusters, *Appl. Phys. Lett.* 80: 1667–1669.
65. A. Colsmann, J. Junge, C. Kayser, and U. Lemmer, 2006. Organic tandem solar cells comprising polymer and small molecule subcells, *Appl. Phys. Lett.* 89: 203506.
66. V. Shrotriya, E. H. En Wu, G. Li, Y. Yao, and Y. Yang, 2006. Efficient light harvesting in multiple-device stacked structure for polymer solar cells, *Appl. Phys. Lett.* 88: 64104.
67. K. Kawano, N. Ito, T. Nishimori, and J. Sakai, 2006. Open circuit voltage of stacked bulk heterojunction organic solar cells, *Appl. Phys. Lett.* 88: 73514.
68. J. Y. Kim, K. Lee, N. E. Coates, D. Moses, T. Q. Nguyen, M. Dante, and A. J. Heeger, 2007. Efficient tandem polymer solar cells fabricated by all-solution processing, *Science* 317: 222–225.
69. Chun-Chao Chen, Wei-Hsuan Chang, Ken Yoshimura, Kenichiro Ohya, Jingbi You, Jing Gao, Zirou Hong, and Yang Yang, 2014. An efficient triple-junction polymer solar cell having a PCE exceeding 11%, *Adv. Mater.* 26: 5670–5677.
70. D. Meyerhofer, 1978. Characteristics of resist films produced by spinning, *J. Appl. Phys.* 49: 3993.
71. J. You, L. Dou, K. Yoshimura, T. Kato, K. Ohya, T. Moriarty, K. Emery et al., 2013. A polymer tandem solar cell with 10.6 % PCE, *Nat. Commun.* 4: 1446.
72. J. You, L. Dou, Z. Hong, G. Li, and Y. Yang, 2013. Recent trends in polymer tandem solar cells research, *Prog. Polym. Sci.* 38: 1909–1928.

73. G. Dennler, M. C. Scharber, and C. J. Brabec, 2009. Polymer fullerene bulk-heterojunction solar cells, *Adv. Mater.* 21: 1323.
74. L. A. A. Pettersson, L. S. Roman, and O. Inganas, 1999. Modeling photocurrent action spectra of photovoltaic devices based on organic thin films, *J. Appl. Phys.* 86: 487–496.
75. J. Y. Kim, S. H. Kim, H. H. Lee, K. Lee, W. Ma, X. Gong, and A. J. Heeger, 2006. New architecture for high-efficiency polymer photovoltaic cells using solution-based titanium oxide as an optical spacer, *Adv. Mater.* 18: 572–576.
76. J. Gilot, I. Barbu, M. M. Weink, and R. A. J. Janssen, 2007. The use of ZnO as optical spacer in polymer solar cells: Theoretical and experimental study, *Appl. Phys. Lett.* 91: 113520.
77. K. Kanai, T. Miyazaki, H. Suzuki, M. Inaba, Y. Ouchi, and K. Seki, 2010. Effect of annealing on the electronic structure of poly3-hexylthiophene) thin film, *Phys. Chem. Chem. Phys.* 12: 273–282.
78. C. J. Brabec, A. Cravino, D. Meissner, N. S. Sariciftci, T. Fromherz, M. T. Rispens, L. Sanchez, and J. C. Hummelen, 2001. Origin of the open circuit voltage of plastic solar cells, *Adv. Funct. Mater.* 11: 374–380.
79. L. J. A. Koster, V. D. Mihailetchi, R. Ramaker, and P. W. M. Blom, 2005. Light intensity dependence of open-circuit voltage of polymer: Fullerene solar cells, *Appl. Phys. Lett.* 86: 123509.
80. V. Dyakonov and E. Frankevich, 1998. On the role played by polaron pairs in photophysical processes in semiconducting polymers, *Chem. Phys.* 227: 203–217.
81. A. C. Morteani, P. Sreearunothai, L. M. Herz, R. H. Friend, and C. Silva, 2004. Exciton regeneration at polymeric semiconductor heterojunctions, *Phys. Rev. Lett.* 92: 247402.
82. M. C. Scharber, N. A. Schultz, N. S. Sariciftci, and C. J. Brabec, 2003. Optical and photocurrent detected magnetic resonance studies on conjugated polymer/fullerene composites, *Phys. Rev. B* 67: 085202.
83. D. Veldman, Ö. Ipek, S. C. J. Meskers, J. Sweelssen, M. M. Koetse, S. C. Veenstra, J. M. Kron, S. S. van Bavel, J. Loos, and R. A. J. Janssen, 2008. Compositional and electric field dependence of the dissociation of charge transfer excitons in alternating polyfluorene copolymer/fullerene blends, *J. Am. Chem. Soc.* 130: 7721–7735.
84. M. A. Green, K. Emery, Y. Hishikawa, W. Warta, and E. Dunlops, 2012. Solar cell efficiency tables version 39), *Prog. Photovolt.: Res. Appl.* 20: 12–20.
85. R. A. J. Janssen and J. Nelson, 2013. Factors limiting device efficiency in organic photovoltaics, *Adv. Mater.* 25:1847–1858.
86. T. M. Clarke and J. R. Durrant, 2010. Charge photo-generation in organic solar cells, *Chem. Rev.* 110: 6736–6767.
87. J. D. Servaites, M. A. Ratner, and T. J. Marks, 2011. Organic solar cells: A new look at traditional models, *Energy Environ. Sci.* 4: 4410–4422.
88. S. M. Sze, 1981. *Physics of Semiconductor Devices*, John Wiley & Sons, New York.
89. V. D. Mihailetchi, L. J. A. Koster, P. W. M. Blom, C. Melzer, B. De Boer, J. K. J. van Duren, and R. A. J. Janssen, 2005. Compositional dependence of the performance of polyp-phenylene vinylene):methanofullerene bulk heterojunction solar cells, *Adv. Funct. Mater.* 15: 795–801.
90. C. Shi, Y. Yao, Y. Yang, and Q. Pei, 2006. Regioregular copolymers of 3-alkoxythiophene and their photovoltaic application, *J. Am. Chem. Soc.* 128: 8980–8986.
91. S. C. Jain, T. Aernout, A. K. Kapoor, V. Kumar, W. Geens, J. Poortmans, and R. Mertens, 2005. *I–V* characteristics of dark and illuminated PPV-PCBM blends solar cells, *Synth. Met.* 148: 245–250.
92. Y. Kim, S. A. Choulis, J. Nelson, D. D. C. Bradley, S. Cook, and J. R. Durrant, 2006. Device annealing effect in organic solar cells with blends of regioregular poly3-hexylthiophene) and soluble fullerene, *Appl. Phys. Lett.* 86: 63502.
93. R. De Bettignies, J. Leroy, M. Firon, and C. Sentein, 2006. Accelerated lifetime measurements of P3HT:PCBM solar cells, *Synth. Met.* 156: 510–513.
94. M. Y. Chiu, U. S. Jeng, M. S. Su, and K. H. Wei, 2010. Morphologies of self-organizing regioregular conjugated polymer/fullerene aggregates in thin film solar cells, *Macromolecules* 43: 428–432.

95. D. H. Wang, H. K. Lee, D. G. Choi, J. H. Park, and O. O. Park, 2009. Solution-processable polymer solar cells from a poly3-hexylthiophene)/[6,6]-phenyl C_{61}-butyric acid methyl ester concentration graded bilayers, *Appl. Phys. Lett.* 95: 43505.
96. J. K. J. van Duren, X. Yang, J. Loos, C. W. T. Bulle-Lieuwma, A. B. Sieval, J. C. Hummelen, and R. A. J. Janssen, 2004. Relating the morphology of polyp-phenylenevinylene)/methanofullerene blends to solar-cell performance, *Adv. Funct. Mater.* 14: 425–434.
97. W. Ma, C. Yang, X. Gong, K. Lee, and A. J. Heeger, 2005. Thermally stable, efficient polymer solar cells with nanoscale control of the interpenetrating network morphology, *Adv. Funct. Mater.* 15: 1617–1622.
98. V. D. Mihailetchi, H. Xie, B. De Bore, L. M. Popescu, J. C. Hummelen, P. W. M. Blom, and L. J. A. Koster, 2006. Origin of the enhanced performance in poly3-hexylthiophene): [6,6]-phenyl C61-butyric acid methyl ester solar cells upon slow drying of the active layer, *Appl. Phys. Lett.* 89: 12107.
99. G. Li, V. Shrotriya, H. Huang, Y. Yao, T. Moriarty, K. Emery, and Y. Tang, 2005. High-efficiency solution processable polymer photovoltaic cells by self-organization of polymer blends, *Nat. Mater.* 4: 864–868.
100. Y. Yao, J. H. Hou, Z. Xu, G. Li, and Y. Yang, 2008. Effects of solvent mixtures on the nanoscale phase separation in polymer solar cells, *Adv. Funct. Mater.* 18: 1783–1789.
101. J. Peet, J. Y. Kim, N. E. Coates, W. L. Ma, D. Moses, A. J. Heeger, and G. C. Bazan, 2007. Efficiency enhancement in low-band gap polymer solar cells by processing with alkane dithiols, *Nat. Mater.* 6: 497–500.
102. J. Peet, C. Soci, R. C. Coffin, T. Q. Nguyen, A. Mikhailovsky, D. Moses, and G. C. Bazan, 2006. Method for increasing the photoconductive response in conjugated polymer/fullerene composites, *Appl. Phys. Lett.* 89: 252105.
103. C. V. Hoven, X. D. Dang, R. C. Coffin, J. Peet, T. Q. Nguyen, and G. C. Bazan, 2010. Improved performance of polymer bulk heterojunction solar cells through the reduction of phase separation via solvent additives, *Adv. Mater.* 22: E63–E66.
104. G. Li, Y. Yao, H. C. Yang, V. Shrotriya, G. Yang, and Y. Yang, 2007. "Solvent annealing" effect in polymer solar cells based on poly(3-hexylthiophene) and methanofullerenes, *Adv. Funct. Mater.* 17: 1636–1644.
105. Q. Zhang, A. Cirpan, T. P. Russell, and T. Emrick, 2009. Donor–acceptor poly(thiophene-block-perylene diimide) copolymers: Synthesis and solar cell fabrication, *Macromolecules* 42: 1079–1082.
106. R. Guntner, U. Asawapirom, M. Forster, G. Schmitt, B. Stiller, B. Tiersch, A. Falcou, H. G. Nothofer, and U. Scherf, 2002. Conjugated polyfluorene/polyaniline block copolymers-improved synthesis and nanostructure formation, *Thin Solid Films* 417: 1–6.
107. M. Sommer and M. Thelakkat, 2006. Synthesis, characterization and application of donor–acceptor block copolymers in nano-structured bulk heterojunction solar cells, *Euro. Phys. J. Appl. Phys.* 36: 245–249.
108. S. S. Sun, 2003. Design of a block copolymer solar cell, *Sol. Energy Mater. Sol. Cells* 79: 257–264.
109. T. Kietzke, D. Neher, K. Landfester, R. Montenegro, R. Guntner, and U. Scherf, 2003. Novel approaches to polymer blends based on polymer nanoparticles, *Nat. Mater.* 2: 408–412.
110. J. L. A. Koster, V. D. Mihailetchi, and P. W. M. Blom, 2006. Ultimate efficiency of polymer/fullerene bulk heterojunction solar cells, *Appl. Phys. Lett.* 88: 93511.
111. Y. Liang, D. Feng, Y. Wu, S. T. Sai, G. Li, C. Ray, and L. Yu, 2009. Highly efficient solar cell polymers developed via fine-tuning of structural and electronic properties, *J. Am. Chem. Soc.* 131: 7792–7799.
112. D. Muhlbacher, M. Scharber, M. Morana, Z. Zhu, D. Waller, R. Gaudiana, and C. J. Brabec, 2006. High photovoltaic performance of a low-band gap polymer, *Adv. Mater.* 18: 2884–2889.
113. M. M. Wienk, M. G. R. Turbiez, M. P. Struijk, M. Fonrodona, and R. A. J. Janssen, 2006. Low-band gap poly(di-2-thienylthienopyrazine): Fullerene solar cells, *Appl. Phys. Lett.* 88: 153511.
114. F. Zhang, W. Mammo, L. M. Andersson, S. Admassie, M. R. Andersson, and O. Inganäs, 2006. Low-band gap alternating fluorene copolymer/methanofullerene heterojunctions in efficient near-infrared polymer solar cells, *Adv. Mater.* 18: 2169–2173.

115. Y. Zou, D. Gendron, R. Badrou-Aich, A. Najari, Y. Tao, and M. Leclerc, 2009. A high-mobility low-band gap poly(2,7-carbazole) derivative for photovoltaic applications, *Macromolecules* 42: 2891–2894.
116. N. Blouin, A. Michaud, and M. Leclerc, 2007. A low-band gap poly(2,7-carbazole) derivative for use in high-performance solar cells, *Adv. Mater.* 19: 2295–2300.
117. F. B. Kooistra, J. Knol, F. Kastenberg, L. M. Popescu, W. J. H. Verhees, J. M. Kroon, and J. C. Hummelen, 2007. Increasing the open circuit voltage of bulk-heterojunction solar cells by raising the LUMO level of the acceptor, *Org. Lett.* 9: 551–554.
118. M. Scharber and N. Sariciftci, 2013. Efficiency of bulk-heterojunction organic solar cells, *Prog. Polym. Sci.* 38: 1929–1940.
119. G. Dennler, M. C. Scharber, T. Ameri, P. Denk, K. Forberich, C. Waldauf, and C. J. Brabec, 2008. Design rules for donors in bulk-heterojunction tandem solar cells: Towards 15% energy-conversion efficiency, *Adv. Mater.* 20: 579–583.
120. N. Li, D. Baran, G. D. Spyropoulos, H. Zhang, S. Berny, M. Turbiez, T. Ameri, F. C. Kreb, and C. J. Brabec, 2014. Environmentally printing efficient organic tandem solar cells with high fill factor: A guideline towards 20% PCE, *Adv. Energy Mater.* 4: 1400084.
121. J. You, L. Dou, K. Yoshimura, T. Kato, K. Ohya, T. Moriarty, K. Emery et al., 2013. A polymer tandem solar cell with 10.6% power conversion efficiency, *Nat. Commun.* 4: 1446.
122. Y. Liu, C. C. Chen, Z. Hong, J. Gao, Y. M. Yang, H. Zhou, L. Dou, G. Li, and Y. Yang, 2013. Solution-processed small-molecule solar cells: Breaking the 10% power conversion efficiency, *Sci. Reports* 3: 3356.
123. J. Jo, J. R. Pouliot, D. Wynands, S. D. Collins, J. Y. Kim, T. L. Nguyen, H. Y. Woo, Y. Sun, M. Leclerc, and A. J. Heeger, 2013. Enhanced efficiency of single and tandem organic solar cells incorporating a diketopyrrolopyrrole-based low-bandgap polymer by utilizing combined ZnO/polyelectrolyte electron-transport layers, *Adv. Mater.* 25: 4783.
124. L. T. Dou, J. You, J. Yang, C. C. Chen, Y. He, S. Murase, T. Moriarty, K. Emery, G. Li, and Y. Yang, 2012. Tandem polymer solar cells featuring a spectrally matched low band gap polymer, *Nat. Photon.* 6: 180–185.
125. L. T. Dou, J. Gao, E. Richard, J. You, C. C. Chen, K. C. Cha, Y. He, G. Li, and Y. Yang, 2012. Systematic investigation of benzodithiophene and diketopyrrolopyrrole-based low-band gap polymers designed for single junction and tandem polymer solar cells, *J. Am. Chem. Soc.* 134: 10071–10079.
126. J. Yang, R. Zhu, Z. Hong, Y. He, A. Kumar, Y. Li, and Y. Yang, 2011. A robust inter-connecting layer for achieving high performance tandem polymer solar cells, *Adv. Mater.* 23: 3465–3470.
127. C. H. Chou, W. L. Kwan, Z. Hong, L. M. Chen, and Y. Yang, 2011. Metal-oxide interconnection layer for polymer tandem solar cells with an inverted architecture, *Adv. Mater.* 23: 1282–1286.
128. F. Padinger, R. S. Rittberger, and N. S. Sariciftci, 2003. Effects of postproduction treatment on plastic solar cells, *Adv. Funct. Mater.* 13: 85–88.
129. R. C. Hiorns, R. De Bettignies, J. Leroy, S. Bailly, M. Firon, C. Sentein, A. Khoukh, H. Preudhomme, and C. D. Lartigau, 2006. High molecular weights, polydispersities, and annealing temperatures in the optimization of bulk-heterojunction photovoltaic cells based on poly(3-hexylthiophene) or poly(3-butylthiophene), *Adv. Funct. Mater.* 16: 2263–2273.
130. A. Zen, J. Pflaum, S. Hirschmann, W. Zhuang, F. Jaiser, U. Asawapirom, J. P. Rabe, U. Scherf, and D. Neher, 2004. Understanding the origin of the 535 nm emission band in oxidized poly(9,9-dioctylfluorene): The essential role of inter-chain/inter-segment interactions. *Adv. Funct. Mater.* 14: 757–764.
131. E. Bundgaard and F. C. Krebs, 2007. Low band gap polymers for organic photovoltaics, *Sol. Energy Mater. Sol. Cells* 91: 954–985.
132. R. Y. C. Shin, T. Kietzke, S. Sudhakar, A. Dodabalapur, Z. K. Chen, and A. Sellinger, 2006. N-type conjugated materials based on 2-vinyl-4,5-dicyanoimidazoles and their use in solar cells, *Chem. Mater.* 19: 1892–1894.
133. J. Li, F. Dierschke, J. Wu, A. C. Grimsdale, and K. Mullen, 2006. Poly(2,7-carbazole) and perylene tetracarboxydiimide: A promising donor/acceptor pair for polymer solar cells, *J. Mater. Chem.* 16: 96–100.

134. N. Camaioni, G. Ridolfi, V. Fattori, L. Favaretto, and G. Barbarelle, 2005. Branched thiophene-based oligomers as electron acceptors for organic photovoltaics, *J. Mater. Chem.* 15: 2220–2225.
135. Z. He, C. Zhong, X. Huang, W. Y. Wong, H. Wu, L. Chen, S. Su, and Y. Cao, 2011. Simultaneous enhancement of open-circuit voltage, short-circuit current density, and fill factor in polymer solar cells, *Adv. Mater.* 23: 4636–4643.
136. M. Jørgensen, K. Norman, and F. C. Krebs, 2008. Stability/degradation of polymer solar cells, *Sol. Energy Mater. Sol. Cells* 92: 686–714.
137. H. Choi, J. S. Park, E. Jeong, G. H. Kim, B. R. Lee, S. O. Kim, M. H. Song, H. Y. Woo, and J. Y. Kim, 2011. Combination of titanium oxide and a conjugated polyelectrolyte for high performance inverted type organic optoelectronic devices, *Adv. Mater.* 23: 2759.
138. Z. Tan, W. Zhang, Z. Zhang, D. Qian, Y. Huang, J. Hou, and Y. Li, 2012. High-performance inverted polymer solar cells with solution-processed titanium chelate as electron-collecting layer on ITO electrode, *Adv. Mater.* 24: 1476–1481.
139. C. E. Small, S. Chen, J. Subbiah, C. M. Amb, S. W. Tsang, T. H. Lai, J. R. Reynolds, and F. So, 2012. High-efficiency inverted dithienogermole-thienopyrrolodione-based polymer solar cells, *Nat. Photon.* 6: 115.
140. M. Wang, X. Hu, P. Liu, W. Li, X. Gong, F. Huang, and Y. Cao, 2011. Donor–acceptor conjugated polymer based on naphtho[1,2-c:5,6-c]bis[1,2,5]thiadiazole for high-performance polymer solar cells. *J. Am. Chem. Soc.* 133: 9638–9641.
141. J. You, L. Dou, K. Yoshimura, T. Kato, K. Ohya, T. Moriarty, K. Emery et al., 2013. A polymer tandem solar cell with 10.6% PCE, *Nat. Commun.* 4: 1446.

4

Device Physics and Modeling

4.1 Introduction

Understanding the physics of organic solar cells (OSCs) is a must for their rapid development and successful implementation in energy production. Along with high efficiency, the stability and reliability in different environmental conditions are highly important issues that should be well understood for the success of this technology. The photovoltaic effect, generation of photocurrent and photovoltage, and calculation of power conversion efficiency (PCE) in OSCs were discussed in previous chapters. In this chapter, I shall present an in-depth analysis of device functioning, the charge transport mechanism, and effects of surrounding conditions on the cell performance. Energy conversion by an OSC can be broken down into four important processes: (1) exciton generation via light absorption, (2) exciton diffusion, (3) exciton dissociation, and (4) charge carrier transportation and collection. The overall PCE of a solar cell is decided by individual efficiencies of these processes and the product of these efficiencies gives the external quantum efficiency (EQE). Therefore, a clear fundamental understanding of these processes is necessary for the design and development of efficient and stable OSCs. Device modeling is an important tool to interpret the device behavior and find out the optimum conditions for high performance; therefore some of the models developed to interpret the behavior of OSCs will also be discussed here.

4.2 Exciton Generation, Diffusion, and Dissociation

Due to absorption, the intensity of incident light decreases exponentially in the absorbing medium according to $I(\lambda, x) = I_0(\lambda)\exp(-\alpha(\lambda)x)$, where $\alpha(\lambda)$ is absorption coefficient of the absorbing medium. Considering that each absorbed photon generates one exciton, the exciton generation rate can be given by Equation 1.12. The exciton density decreases exponentially and due to ultrafast dissociation at the donor–acceptor interface, the density decreases to zero there. The concentration gradient of excitons makes them migrate to the donor–acceptor interface. This concept applies to the excitons generated in both the donor and acceptor phases. Exciton migration in organic semiconductors can happen in three ways known as cascade or radiative energy transfer, Dexter energy transfer, and Förster energy transfer, which are shown schematically in Figure 4.1. In cascade energy transfer the photoexcited molecule emits a photon, which is subsequently absorbed by the next molecules resulting in exciton transfer from one molecule to other. This process continues and the excitons migrate over a long range comparable to absorption length. This energy transfer requires significant overlap between emission and absorption spectra

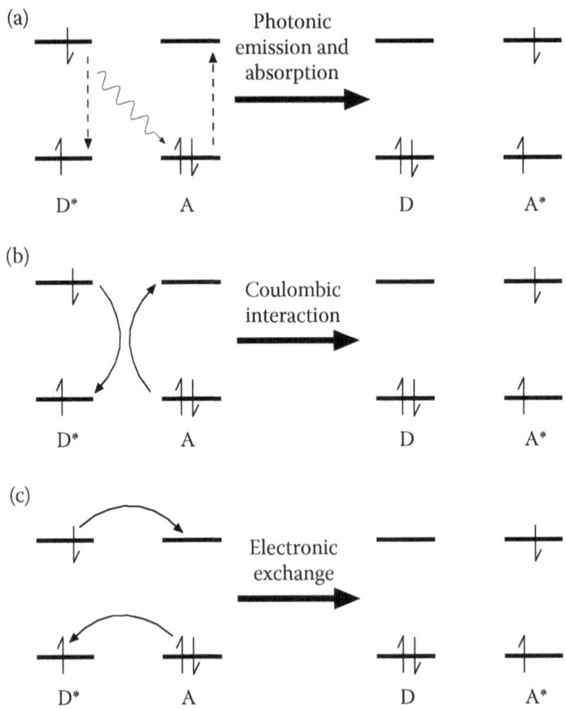

FIGURE 4.1
Schematic representation of (a) cascade, (b) Förster, and (c) Dexter energy transfer between exciton donor (D) and exciton acceptor (A) molecules. "*" represents the exited state.

of the material. However, Förster energy transfer is a nonradiative energy transfer that involves dipole–dipole interaction between the two molecules. There is no direct exchange of electrons and it happens via coulombic coupling of electromagnetic fields of the dipoles. Förster energy transfer can be taken as emission and subsequent absorption of a virtual photon, and it requires a spectral overlap between excited and the nearby ground-state molecule. The rate of Förster energy transfer (k_F) is written as [1]

$$k_F(d) = \frac{1}{\tau}\left(\frac{R_0}{d}\right)^6, \quad (4.1)$$

where d is the intermolecular spacing, τ is the exciton lifetime, and R_0 is the Förster radius for energy transfer given by

$$R_0^6 = \frac{9\eta_{PL}k^2}{128\pi^5 n^4}\int \lambda^4 F_D(\lambda)\sigma_A(\lambda)d\lambda. \quad (4.2)$$

Here η_{PL} is the PL efficiency of the excited state, k^2 is the dipole orientation factor, n is the refractive index at the wavelength where the spectral overlap integral is maximized, λ is the wavelength, F_D is the normalized fluorescence of the exciton donor, and σ_A is the absorption cross section of the exciton acceptor. Förster energy transfer is a long-range energy transfer and occurs over a length scale of 1–10 nm. As the intermolecular distance decreases the coulombic interaction between the molecules increases and the dipole–dipole coupling becomes stronger. In case of Dexter energy transfer, it involves orbital–orbital interactions and direct exchange of electrons takes place between the molecules. For electronic exchange, the electronic densities of both the exciton exchanging molecules should overlap. The rate of Dexter energy transfer k_D is written as [2]

$$k_D(d) = KJe\left(-\frac{2d}{L}\right), \tag{4.3}$$

where K corresponds to specific orbital interaction, J is a spectral overlap integral normalized for the extinction coefficient of the ground-state molecule, and L is the van der Waals radius. Dexter energy transfer is a short-range energy transfer on the length scale of 0.1–1 nm. It can happen between both the radiative and nonradiative states. The triplet excitons, which are nonradiative, migrate via Dexter energy transfer.

Unfortunately due to short lifetime (nanoseconds), the excitons are able to diffuse to a very short distance only. For efficient energy conversion, all of the generated excitons should diffuse to the donor–acceptor interface that requires the donor–acceptor domain size to be on the order of exciton diffusion length. Understanding of origin of short diffusion lengths and development of new methods to enhance this parameter is of high importance. The materials with high diffusion lengths will not only be beneficial for the simple bilayer structures but also for design and processing of bulk-heterojunction (BHJ) solar cells in terms of relaxing the constraints on domain size. The simple models, developed to describe the exciton diffusion, consider it as a nearest-neighbor hopping event [3], whereas in complex models the dispersive nature of energy transfer in energetic, spatial, and orientational disordered systems has been considered [4–7]. Considering no second-order processes like annihilation and fission, the exciton diffusion is modeled using the continuity equation [8]

$$\frac{\partial n}{\partial t} = D\nabla^2 n(r_n) + G(r_n) - \frac{n(r_n)}{\tau} \tag{4.4}$$

where n is the exciton density, r_n is the position corresponding to density n, τ is the exciton lifetime, and G is the exciton generation rate. The first term corresponds to diffusion, the second term corresponds to generation, and the last term corresponds to recombination of excitons. Diffusion length, L, is related to lifetime, τ, as $L = \sqrt{D\tau}$, where D is the diffusion coefficient.

There are several reports in the literature that describe the methods for calculation of exciton diffusion coefficient and diffusion lengths in organic semiconductors [8,9–20]. These methods are either spectroscopic or charge-carrier-based techniques. Spectroscopic techniques involve photon detection, whereas charge carrier techniques involve detection of charge carriers. Surface quenching [12–14], volume quenching [15,16], time-resolved microwave conductivity [20], surface photoconductivity measurements [21], and exciton–exciton annihilation [17,18] are some of the important methods employed to study the exciton diffusion in organic materials. Surface quenching is the most frequently used spectroscopic method for this purpose where the organic semiconductor is deposited on a suitable quencher. The quencher is a material that readily dissociates the excitons at the interface by means of charge transfer. For most of the polymers, the fullerene derivatives are used as quenchers. However, fullerene derivatives interdiffuse into polymers and lose the distinct interface that gives overestimated values of diffusion lengths [11]. Deposition on the quencher causes a loss in fluorescence that can be quantified by comparing the fluorescence of two identical films: one with and the other without exciton quencher. The loss in fluorescence depends on exciton diffusion coefficient and film thickness. For precise calculation of exciton diffusion coefficients, accurate measurement of film thickness is very important. Surface quenching experiments are performed either in steady state or in transient state to measure the exciton diffusion coefficient and lifetime [9,13].

Luer et al. performed oxygen-induced fluorescence quenching studies on P3HT and found the exciton diffusion coefficient to be 5×10^{-4} cm^2s^{-1}, corresponding to a diffusion length of about 4 nm [15]. Shaw et al. performed time-resolved measurements on fluorescence of P3HT and exciton diffusion length of about 8.5 nm was calculated [9]. In the studies performed by Shaw et al., TiO_2 was chosen as the quencher to form a well-defined distinct interface with P3HT. Film thicknesses of the P3HT were calculated with the help of ellipsometry technique where the effects of reflection and interferences were fully taken into account. P3HT films were grown on fused silica substrates and 80 nm TiO_2-coated fused silica substrates. P3HT films were optically excited with 400 nm pulses of 100 fs duration at a frequency of 80 MHz, and the fluorescence was recorded over 600–730 nm with time-resolved decay using the Hamamatsu C6860 streak camera in synchroscan mode. A ratio of quenched fluorescence to that without a quencher was calculated. For exciton–exciton annihilation the laser output was modified to give 400 nm pulses of 100 fs duration at a frequency of 50 kHz. The photogenerated excitons would definitely decay after their lifetime, but presence of the quencher accelerates the exciton quenching. For 6.5 nm thick film, the decay in fluorescence was significantly faster for the film deposited on TiO_2 compared to that on bared fused silica substrate. For both cases, fluorescence decay occurred fast initially and slowly afterward. The emission of 6.5 nm film on TiO_2, quenched completely after ~250 ps. This was a direct indication of exciton quenching by diffusion to P3HT/TiO_2 interface. A similar trend was observed for thicker films as well but the extent of quenching decreased with an increase in P3HT film thickness that supported the phenomenon of quenching by diffusion to the interface. Figure 4.2 shows the photoluminescence decay in P3HT films of 6.5 and 32 nm on bare fused silica and TiO_2-coated fused silica substrates. The value of the exciton diffusion coefficient was calculated by fitting the fluorescence decay to the one-dimensional diffusion equation

$$\frac{\partial n(z,t)}{\partial t} = D\frac{\partial^2 n(z,t)}{\partial z^2} - k(t)n(z,t) \tag{4.5}$$

where $n(z,t)$ is the exciton distribution throughout the P3HT film as a function of time, D is the exciton diffusion coefficient, and $k(t)$ is the rate of fluorescence decay without a

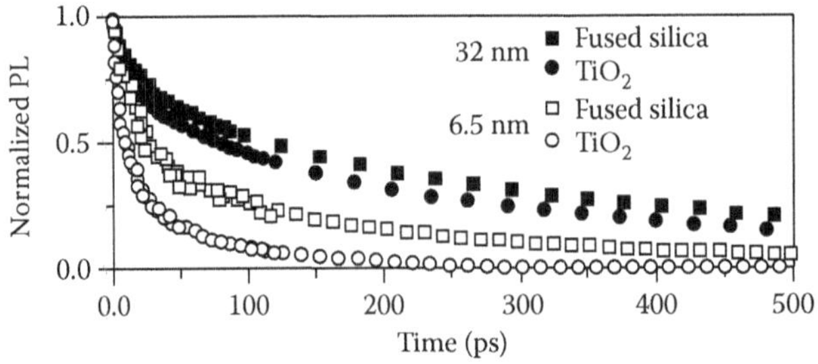

FIGURE 4.2
Transient decay in photoluminescence of P3HT films on fused silica and TiO_2-coated fused silica substrates. (Modified from P. E. Shaw, A. Ruseckas, and I. D. W. Samuel, 2008. *Adv. Mater.* 20: 3516.)

quencher. Due to near normal incidence the reflections were negligible and the distribution of excitons in P3HT film was considered to be $n(z,0) = n(0)e^{-\alpha z}$. The fluorescence decay rate would be uniform throughout the P3HT film when there is no quencher. As the excitons dissociate ultrafast at the interface with an infinite quencher, the boundary condition in the presence of a quencher was taken to be $n(L,t) = 0$, where L was P3HT thickness. $k(t)$ was set for each polymer film and then applied to decay on TiO_2 and the value of D was obtained with curve fitting. A single value of D of $1.8 \pm 0.3 \times 10^{-3}$ cm^{2}s^{-1} was obtained from the fitting of all curves.

Reports on different values of diffusion coefficients and diffusion lengths in P3HT can be attributed to different measurement techniques and the film properties. Transient absorption is another spectroscopy technique to measure diffusion length in organic semiconductors by the exciton–exciton annihilation process [22,23]. The film is excited with a monochromatic light pulse followed by a variable wavelength pulse, and the differential transmission of the film is measured on a picosecond (ps) time scale. By fitting the transient absorption decay over a range of exciton densities, the diffusion coefficient is determined. In time-resolved microwave conductivity measurements the organic films are coated on a comparatively more conducting quencher so that a microwave cavity is formed and a change in conductivity of photogenerated charge carriers in quencher results in reflection of differential microwave power [20]. The time-dependent change in reflected power from the cavity, due to change in conductivity of the quenching film, is measured that leads to the measurement of diffusion length. In surface photoconductivity measurements, two electrodes are deposited on the film to monitor photoconductivity [21]. Surface photoconductivity depends on photogenerated excitons, their diffusion to the surface, and then their dissociation by some means. This technique is generally applied to single crystals, where absorption is anisotropic for different azimuthal angles and gives different penetration lengths. The differential surface conductivity gives diffusion length, as the smaller diffusion length will give large differential and vice versa.

The optical absorption length in organic semiconductors is on the length scale of 100 nm, and the exciton diffusion lengths of a few nanometers in that case limits the photoconversion efficiency. Therefore, a balance between absorption length and exciton diffusion length is required for efficient power conversion. Though the BHJ concept has shown overwhelming exciton diffusion to the interface as the domain size of constituent materials is less than the diffusion length of excitons, but for bilayer structures it is still a serious issue.

4.3 Device Operation Mechanism

The operation mechanism of different kinds of OSCs were discussed in Section 3.3 of Chapter 3. However, in general, the energy conversion in OSCs includes light absorption, exciton generation and dissociation into charge carriers, charge carrier recombination, transportation, and collection at the corresponding electrodes. Photon harvesting in OSCs encounters a number of optical and electrical losses and only the photogenerated charge carriers, which manage to survive and extract out of the device actually contribute to electricity. Only a fraction of incident light is absorbed because some of the incident photons are reflected back from the solar cell and some are left from being absorbed. The absorbed photons generate excitons that need to dissociate into free electrons and holes.

For dissociation, excitons need to diffuse to the donor–acceptor interface where the LUMO (lowest unoccupied molecular orbital) level offset between donor and acceptor materials provides necessary force to dissociate. Due to low diffusion lengths, the excitons in the vicinity of the donor–acceptor interface dissociate efficiently, and those far from the interface recombine without contributing to photocurrent. The problem of inefficient exciton dissociation was solved in BHJ structures where donor–acceptor materials are mixed together such that the donor–acceptor phase widths are less than the exciton diffusion lengths. Therefore, almost all the excitons diffuse to the donor–acceptor interface giving ~100% exciton dissociation. Exciton dissociation takes place by transfer of electrons to the LUMO of acceptor. The transfer of holes from donor to accepter is prevented by higher HOMO (highest occupied molecular orbital) energy of the donor. Higher HOMO energy of the donor also helps in dissociation of excitons generated in the acceptor. Exciton dissociation gives electrons in the acceptor phase and holes in the donor phase around the interface. Because of low charge carrier mobilities in organic materials, even after dissociation of excitons the charge carriers accumulate near the heterojunction and formulate a space charge, which further controls the photovoltaic performance. The separated charge carriers at the donor–acceptor interface are still bound together by coulombic force and form charge transfer states (CTSs). CTSs are electronic states of a polaron pair of coulombically bound negative and positive polarons and sometimes are also known as exciplex, bound electron–hole pairs, interfacial charge pairs, geminate pairs, charge transfer excitons, and bound intermolecular radial pairs [24–27]. Some of the CTSs may also recombine and the recombination may be radiative or nonradiative.

CTSs are a result of intermolecular overlap between donor and acceptor molecules. CTSs have direct effects on the photovoltaic performance by affecting the short-circuit current density (J_{sc}) and open-circuit voltage (V_{oc}). The maximum value of V_{oc} is determined by the energy of CTSs. However, it will be important to know whether the maximum photocurrent and maximum V_{oc} can be achieved simultaneously. Optimization of the PCE by tuning the properties of CTSs is a difficult task but should be carefully optimized. The creation of CTSs in OSCs is evident by the generation of photocurrent due to absorption of photons with energy well below the HOMO–LUMO gap of the active layer [28,29]. Vandewal et al. reported the photocurrent generation due to CTSs absorption in polymer–fullerene solar cells [30]. Tvingstedt et al. reported a weak electroluminescence from several polymer–fullerene BHJ solar cells [31], and the emitted light was significantly different from that of the pure materials in the blend. Because of the radiative charge transfer complex recombination, a separate photoluminescence peak was observed in the PPV:PCBM combination [32,33], whereas in the P3HT:PCBM blend a peak was observed as the lower energy shoulder in donor photoluminescence [34]. This emission was from the interfacial CTS recombination. Emission from the solar cells is an important tool for the characterization of CTSs in OSCs. These results showed the possibility of absorption of photons having energy less than the band gap of the organic semiconductor via CTSs. As the energy of CTSs determines the V_{oc}, it should be tuned by selecting material combinations such that it is lower than but as close as possible to the exciton energy. Also the losses due to triplet excitons decay should be avoided. It is quite obvious that the fundamental understanding of the processes involving CTSs is essential for optimization of the performance of OSCs. Deibel et al. discussed in detail the role of CTSs on the performance of OSCs [35]. More details on the role of CTSs in OSCs can be found in other articles [35–38].

The CTSs require an additional force to break apart and the necessary force is provided by an internal built-in electric field of the cell. The probability of CTSs dissociation

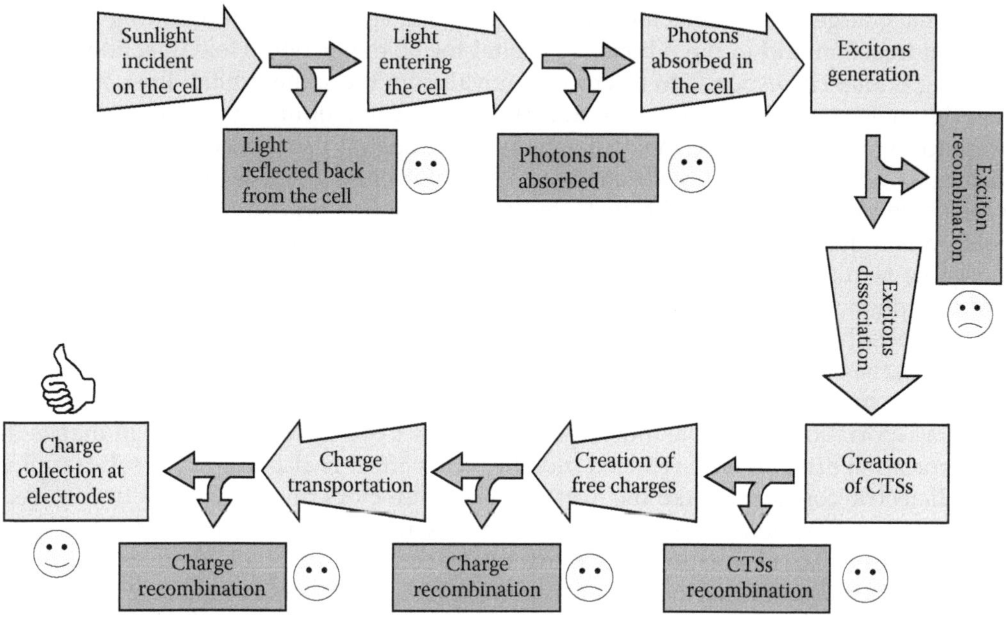

FIGURE 4.3
Scheme of photon harvesting by OSCs.

is described by the Onsager–Braun model [39]. After dissociation of CTSs, the separated electrons and holes transport through respective channels (electrons in acceptor channels and holes in donor channels) to collect at the respective electrodes. The separated charge carriers may recombine or get trapped in localized states during transportation, which results in reduced photocurrent. The surviving charge carriers extract out of the cell via the drift–diffusion process and formulate a photocurrent. Charge transport usually happens via the hopping phenomenon from one localized state to another. As a whole, the extracted photocurrent does not solely depend upon photogeneration of excitons but also on the exciton diffusion, exciton dissociation, and charge transport properties of donor/acceptor materials. As a whole, the possible processes happening in an OSC from light incidence to collection of charge carriers are shown schematically in Figure 4.3.

4.4 Charge Carrier Transport in Organic Semiconductors

Transportation of charge carriers through organics plays a crucial role in deciding the device performance, therefore knowledge of charge transport mechanisms in organic semiconductors is quite imperative not only for fundamental academic interest but also for reduced losses and further development in organic devices. To study the charge transport phenomenon in individual organic semiconductors, special types of devices are prepared where semiconductors are sandwiched between two electrodes and charge carriers are injected from one side and allowed to transport to the other side. The injection is done by electrical or optical excitation. Injection of an electron or a hole causes an additional charge on the molecule that is accommodated in the π* antibonding orbital for electron injection and in the π bonding orbital for hole injection. Hole injection in the π bonding orbital corresponds to electron removal from the same orbital. In such a condition the molecule is in excited state and the charge's surrounding is distorted because presence of a charge on the molecule changes the spatial distribution of electrons in σ molecular orbitals that causes a change in the molecular bond length. The charge on a molecule in combination with distorted surrounding is known as polaron. In a way, polaron is a quasi-particle where electronic charge is surrounded by phonon cloud. In inorganic semiconductors, due to rigid lattice structure no polaron is formed or the electron–phonon interaction is quite weak. In organic semiconductors the holes are actually positively charged polarons, whereas electrons are negatively charge polarons. Therefore, the transportation of electrons and holes in organic semiconductors is actually not the transportation of electrons and holes but the transportation of polarons. For polaronic transportation the whole mustered energy has to be transported, which makes the transportation difficult with reduced mobility. This effect is also known as self-trapping. Low dielectric constants make the organic semiconductors to electron–electron correlation and electron–phonon coupling, which are absent in high dielectric constant materials. The charge carriers or polarons move in the semiconductor via drift and diffusion processes. Diffusion causes local displacement of charge carriers around an average position, whereas drift causes displacement in average position. In organic films, drift usually dominates the charge transport.

To interpret the experimental observations different groups proposed different charge transport mechanisms that made the charge carrier transport in organic semiconductors a debatable issue. There were the conflicts on the nature of current flow in organic devices, whether it was injection limited or bulk limited. Later it was realized that if the injecting contacts are ohmic (no injection barrier), the charge transport would be bulk limited, and for high injection barriers the injection-limited phenomena would become important [40]. For any device, the most important aspect is boundary conditions, that is, the contacts, that are responsible for charge injection and extraction. Undoped organic semiconductors have negligible intrinsic charge carriers and the current flow happens due to charge carriers injected from the contacts or photogeneration.

4.4.1 Injection-Limited Charge Transport

Fowler–Nordheim (FN) tunneling and thermionic emission are the two most important interface-/injection-limited phenomena. FN theory, also known as field emission, is based on the quantum mechanical tunneling of charge carriers from the metal to the energy band of the semiconductor in the presence of a high electric field. According to FN theory, the presence of a large electric field at the surface of a metallic cathode reduces the width of the energy barrier and helps the charge carriers to tunnel into the semiconductor. Once the charge carriers are inside the semiconductor, they are free to move within the energy band and are swept immediately to the counter electrode. The calculation of current flown is based on the WKB (Wigner, Kramers, Brilliuon) approximation yielding the following relation between the tunneling current (J_{FN}) and the electric field F,

$$J_{FN} = \frac{A^* q^2 F^2 \exp\left(-(2\alpha\phi_B^{3/2}/3qF)\right)}{\phi_B \alpha^2 k_B^2}, \tag{4.6}$$

where $\alpha = 4\pi m^{*1/2}/h$, A^* is the FN coefficient, m^* is the effective mass of charge carriers, h is Planck's constant, ϕ_B is the injection barrier for charge carriers from the metal electrode to the semiconductor, and k_B is the Boltzmann constant. The FN mechanism has been extensively studied in metal-oxide-semiconductor (MOS) structures where it has shown to be a dominant mechanism. At low electric fields, the direct tunneling mechanism dominates the tunneling current.

In case of thermionic emission the charge carriers overcome of the injection barriers to inject into the semiconductor, and it is a thermal-assisted injection phenomenon. The thermionic emission is further supported by the Schottky effect. The Schottky effect results in induced image force lowering of the potential barrier for charge carrier emission from metal to the semiconductor under the influence of an electric field [41]. The current density due to thermionic emission (J_{RS}) of charge carriers over an injection barrier ϕ_B is given by the Richardson–Schottky (RS) equation

$$J_{RS} = AT^2 \exp\left(\frac{-\phi_B + \beta_{RS}F^{1/2}}{k_B T}\right)\exp\left(\frac{qV}{k_B T} - 1\right), \tag{4.7}$$

where $A = (4\pi q m^* k_B^2/h^3)$ and $\beta_{RS} = \sqrt{(q^3/4\pi\varepsilon\varepsilon_0)}$, m^* is the effective mass of charge carriers, h is Planck's constant, ε is the dielectric constant of the semiconductor, and ε_0 is the permittivity of free space.

Earlier studies by I. D. Parker showed that the current flow through MEH-PPV with finite (nonzero) injection Schottky barrier in organic light-emitting diodes (OLEDs) was determined by FN tunneling [42]. Kawabe et al. used FN formalism and imposed tunneling of charge carriers as one of the boundary conditions in organic devices [43]. Lupton and Samuel [44] found a reasonable fit of experimental *J–V* characteristics of MEH-PPV with the FN tunneling model, but only over a small portion of the applied voltage. The injection barrier height calculated by FN analysis was observed to be much smaller than that obtained by direct measurements of the built-in electric field and offset between metal work function and HOMO level of MEH-PPV. The fits of experimental data to the injection-limited thermionic emission theory broke down at low temperatures [44]. Abkowitz et al. proposed a model of thermally assisted tunneling into a trap-free transporting polymer and found good agreement with experimental data, where neither FN tunneling nor thermionic emission could account for the temperature and field dependence of the *J–V* characteristics [45]. Campbell and coworkers [46,47] studied a PPV derivative and concluded that in case of single-carrier injection, the injection-limited tunneling theory showed very poor agreement, whereas experimental data showed good agreement with the bulk-limited conduction. Later several models including unipolar and bipolar device models, time-dependent models, and single-layer and multilayer device models were also proposed to explain the experimental observations in organic devices [48–51].

4.4.2 Bulk-Limited Charge Transport

In bulk-limited conduction the current is controlled by bulk of the material, where transport phenomenon may be band transport or hopping transport. When the interaction between nearest neighbors is strong, like in molecular crystals, the band transport takes place. Band transport requires the bands to be wider than the energetic uncertainty of charge carriers. When there is large variation in site energy and interaction between neighbor molecules is quite weak, the hopping transport becomes dominant. In both transport phenomena, charge carrier mobility is an important fundamental parameter. In hopping transport the mobility is thermally activated, whereas in band transport it is thermally deactivated. Usually for better understanding of transport phenomenon, devices are prepared such that only one type of charge carriers is allowed to pass through the semiconductor. Selection of the transporting charge carriers is made by the choice of injecting and extracting electrodes. Depending upon the charge carriers passing through the semiconductor, the devices are either called electron-only or hole-only devices. Analyzing the behavior of current with applied electric field at different temperatures, the mechanism of charge transport in organic semiconductors is understood.

4.4.2.1 Band Transport

Due to low mobility of charge carriers in organic semiconductors, the injected charge carriers develop a space charge electric field for the next injecting charge carriers and control the current flow. Such a current is known as the space charge limited current (SCLC) and depends on the electric field and charge carrier distribution in the film. The SCLC was first explained by Mott and Gurney in trap-free insulators, when they found that the current would flow even in an insulator if the charge carriers could be injected [52]. Solving the Poisson and current-density equations in a trap-free insulator with the boundary conditions that the injecting contact is ohmic (or the injection barrier is zero), and diffusion is negligible, the following *J*–*V* relation was derived,

$$J = \frac{9}{8}\mu\varepsilon\varepsilon_0\frac{V^2}{d^3} \tag{4.8}$$

where μ is the charge carrier mobility, ε is the dielectric constant of the film, ε_0 is the permittivity of free space, and d is the film thickness. This equation was later known as Mott's trap-free V^2 law or Child's law. This law states that the current through a trap-free insulator depends quadratically on the applied voltage and inversely on the thickness cube. Later the charge transport in solids containing traps was first studied by Helfrich and Mark [53]. Traps are nothing but the energy states that are created in the forbidden energy gap of the material by chemical or physical defects/impurities. The charge carriers moving in the transport bands may get trapped in these states and no longer contribute to the current until they are released back into the transport states. One trap state can only trap one charge carrier at a time. The energetic distribution of traps can be at a single energy level, Gaussian or exponential. The single energy level distribution of traps is written as

$$h(E) = H_a\delta(E - E_t), \tag{4.9}$$

where H_a is total trap density, E_t is energy of the trap level, and δ is the Dirac delta function. For Gaussian traps the distribution is given by

$$h(E) = \frac{H_d}{(2\pi)^{1/2}\sigma_t}\exp\left(-\frac{(E - E_{tm})^2}{2\sigma_t^2}\right), \tag{4.10}$$

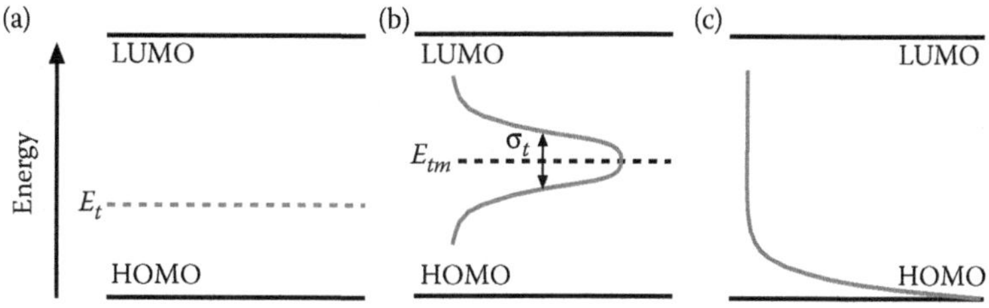

FIGURE 4.4
Schematic representation of (a) single level, (b) Gaussian, and (c) exponential distribution of traps in forbidden energy gap of an organic semiconductor.

where H_d is the total trap density, $(2\pi)^{1/2}$ is the normalization factor, σ_t is the standard deviation of Gaussian distribution, and E_{tm} is the trap energy at maximum trap density. On the other hand, exponential distribution of traps in energy is given by

$$h(E) = \frac{H_b}{k_B T_c} \exp\left(-\frac{E}{k_B T_c}\right), \tag{4.11}$$

where H_b is the total trap density, k_B is the Boltzmann constant, and T_c is the characteristic distribution temperature for exponential traps. All of these trap distributions are shown schematically in Figure 4.4. The coupled Poisson and current density equation are solved considering the aforementioned traps distributions to get current flow through the materials as a function of applied voltage [54,55]. Organic semiconductors have a large density of traps and if the trap states are deep and temperature is low, then the injected charge carriers get trapped in these states. Trapping of charge carriers influences the space charge and electric field distributions. At a given voltage only a fixed amount of total charge is supported in a sample therefore the number of free charge carriers is reduced considerably. As a result of reduced free charge carriers, the current flow through the semiconductor is reduced. All the traps below Fermi level are filled and all above it are empty. When all the traps get filled by charge carriers their effect becomes negligible and the semiconductor behaves as if there were no traps. Analyzing the nature of J–V characteristics in different voltage range at different temperatures, the nature of traps, trap density, and mobility of charge carriers can be calculated.

For single-carrier devices, for example, a hole-only device, when diffusion is negligible the current flow is determined by solution of the following coupled current density and Poisson's equations [54,55]:

$$J = q\mu_p p(x) F(x), \tag{4.12}$$

and

$$\frac{dF(x)}{dx} = \frac{q}{\varepsilon\varepsilon_0}(p(x) + p_t(x)), \tag{4.13}$$

where μ_p is the hole mobility, $p(x)$ is the free hole density, $p_t(x)$ is the trapped hole density, and $F(x)$ is the electric field. The diffusion current as calculated by Mott was negligible and therefore was omitted from Equation 4.12. Omission of diffusion current here is an assumption that makes the mathematical calculations easy. For electron-only devices, the corresponding equations for electrons are used to determine the current flow. The preceding coupled equations are solved using the boundary condition

$$V = \int_0^d F(x)\,dx, \tag{4.14}$$

where d is the film thickness. For any distribution, the filled density of trap (p_t) is determined from

$$p_t(x) = \int_{E_f(x)}^{\infty} h(E)\,dE \tag{4.15}$$

where $E_f(x)$ corresponds to the Fermi energy level at position x. For exponential distribution of traps, Equation 4.15 gives

$$p_t(x) = H_b \exp\left(-\frac{E_f(x)}{k_B T_c}\right). \tag{4.16}$$

At any point x, the free hole density is a unique function of applied voltage and is given by the Fermi–Dirac distribution

$$p(x) = N_v \frac{2}{\sqrt{\pi}} \int_0^{\infty} \frac{\eta^{1/2}}{1+\exp\left(\eta + (E_f(x)/k_B T)\right)}\,d\eta \tag{4.17}$$

where N_v is the effective density of states in valance band. For low carrier densities and/or high temperatures, the preceding equation reduces to

$$p(x) = N_v \exp\left(-\frac{E_f(x)}{k_B T}\right). \tag{4.18}$$

For an exponential distribution of traps with 0 K approximation, the number of traps that are filled will be given by integrating Equation 4.16 from $E_f(x)$ to ∞. Using Equation 4.18 $p_t(x)$ comes out to be

$$p_t(x) = H_b \left(\frac{p(x)}{N_v}\right)^{1/l}, \tag{4.19}$$

where $l = T_c/T$, where T_c is the characteristic temperature of the trap distribution and T is the absolute temperature. The energy $k_B T_c$ is the characteristic energy of the exponential trap distribution at which the density of the trap is $1/e$ of the trap density at band edge ($E = 0$). Using two key assumptions—(1) the electrodes make ohmic contact with the organics and (2) trapped carrier density $p_t(x) \gg$ free carrier density $p(x)$—Helfrich and Mark solved the coupled current-density and Poisson's equations, and for exponential distribution of traps the following J–V relation was derived:

$$J = q^{1-l}\mu_p N_v \left(\frac{2l+1}{l+1}\right)^{l+1} \left(\frac{l}{l+1}\frac{\varepsilon\varepsilon_0}{H_b}\right)^l \frac{V^{l+1}}{d^{2l+1}}. \tag{4.20}$$

The current remains SCLC in the presences of traps as the SCLC depends inversely on the film thickness. The charge transport in solids, containing different kinds of traps, is well documented in the books of Kao and Hwang [54] and Lampert and Mark [55]. In early work (in the 1970s and 1980s) the experiments were generally performed at relatively low voltages and these approximations were valid for a good approximation. In modern technology these two approximations put severe restrictions on the usefulness of this model.

4.4.2.2 Effects of Nonzero Schottky Barrier

Owing to the simple device geometry in organic devices, that is, metal contact/organic layer(s)/metal contact, a one-dimensional device model is generally sufficient. Using appropriate boundary conditions, the current density and Poisson's equations are solved to obtain current densities, carrier densities, electric field, and electrostatic potential. For most of the metal–organic Schottky contacts the contact barriers are not zero [56]. Since no theory of bulk-limited transport in organic materials existed for this case and the experimental data were always compared with the theory for zero Schottky barriers [40,46,47,54–59] until the concept of the nonzero injection barrier was introduced and the modified band transport theory was presented [56,60–62]. If the Schottky barriers are not very large, the current still remains SCLC. I present here a brief work done by us regarding SCLC transport with nonzero Schottky barriers.

Most organic semiconductors have been observed to have exponential distribution of traps, therefore here only exponential traps would be considered. However, similar analysis can be taken up to determine current flow for any type of trap distributions. For exponential distribution of traps assuming $p_t(x) \gg p(x)$, the J–V relation for the hole current of an organic Schottky diode with a finite injection barrier was derived to be [56]

$$J = q^{1-l}\mu_p N_v \left(\frac{2l+1}{l+1}\right)^{l+1} \left(\frac{l}{l+1}\frac{\varepsilon\varepsilon_0}{H_b}\right)^l \frac{V^{l+1}}{\left[(d+C)^{(2l+1)/(l+1)} - C^{(2l+1)/(l+1)}\right]^{l+1}}, \tag{4.21}$$

where C is a constant and independent of x and electric field $F(x)$ but it depends on J (or V). Using necessary boundary conditions, the value of C came out to be

$$C = J\left[\frac{1}{q\mu_p p(0)}\right]^{(l+1)/l} \left(\frac{\varepsilon\varepsilon_0}{qH_b}\right) \left(q\mu_p N_v\right)^{(1/l)} \left(\frac{l}{l+1}\right) \tag{4.22}$$

where $p(0)$ is hole density at the injecting contact at $x = 0$. The injection barriers directly control $p(0)$ according to the relation $p(0) \approx N_0 \exp(-\phi_B/k_BT)$, where N_0 is the electron density at the Fermi level of injecting metal electrode. For ohmic contact ($\phi_B = 0$), $p(0)$ becomes equal to N_0, which is infinity large ~10^{20} cm^{-3}, giving $C \approx 0$ and Equation 4.21 reduces to Equation 4.20. For $C > d$ the characteristics deviate considerably from Equation 4.20 and for $C \gg d$, Equation 4.21 gives

$$J = q\mu_p p(0)\frac{V}{d}. \tag{4.23}$$

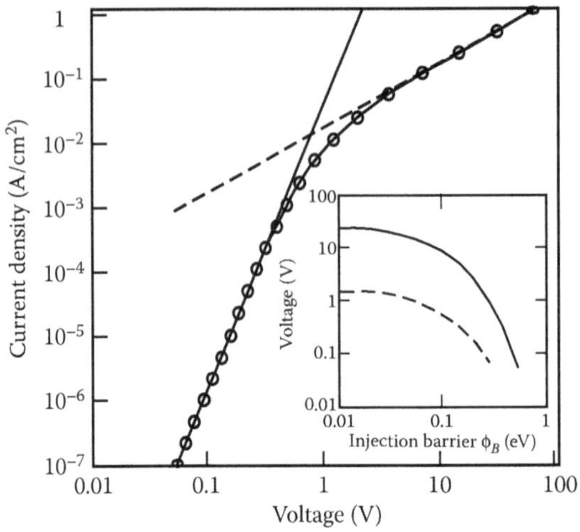

FIGURE 4.5
Plots of Equations 4.20 (solid line), 4.21 (solid curve with symbols), and 4.23 (dashed line) considering Schottky barrier of 0.12 eV and $N_0 = 10^{19}$ cm^{-3}, $T_c = 1000$ K, $T = 295$ K, $N_v = 2 \times 10^{19}$ cm^{-3}, $\mu_p = 1 \times 10^{-5}$ cm^2 V^{-1} s^{-1}, $H_b = 1 \times 10^{17}$ cm^{-3}, $\varepsilon = 3$, and $d = 100$ nm. Inset: Voltages corresponding to deviation from Equation 4.20 (dashed) and approaching Ohm's law (solid) (within 10% error) as a function of the injection barrier. (Reproduced with permission from P. Kumar et al., 2007. *Synth. Met.* 157: 905.)

This result is very important since it shows that for finite (nonzero) Schottky barriers and large values of C, the current changes from SCLC to ohmic current. The calculated J–V curves of Equations 4.20 and 4.21, for given values of parameters, are shown in Figure 4.5. The solid line corresponds to Equation 4.20 and the dashed line represents the ohmic region Equation 4.23. Symbols are for Equation 4.21 with an injection Schottky barrier of 0.12 eV. The plot of Equation 4.20 agrees with that of Equation 4.21 up to a certain value of applied voltage. At high voltages, the plot of Equation 4.21 deviates from that of Equation 4.20 and bends downward. Further increase in applied voltage makes the curve to approach asymptotically the Ohm's law. Calculations show that this result is independent of parameter values. The voltages (within 10% error) at which current deviates from Equation 4.20 and approaches Ohm's law are shown by dashed and solid curves, respectively, as a function of the injection barrier in the inset of Figure 4.5. As the injection barrier increases, both voltages decrease first slowly and then rapidly.

For detailed understanding of the aforementioned results, the electric field and charge carrier distributions were also calculated (see Figure 4.6) for different values of $p(0)$ at a given current density and given value of l [56]. For ohmic contacts, the injected charge carrier density at the injecting contact is infinitely large and decays gradually away from the contact and constitute a space charge. A similar trend is observed in the electric field distribution. As the $p(0)$ decreases (due to increase in the injection barrier) both the electric field and charge carrier density decrease and tend to become uniform. For uniform electric field and charge carrier distributions, the current would follow Ohm's law. Similarly, for a given finite-injected charge carrier density, the electric field and charge carrier distributions tend to become uniform at high current or voltage, leading to ohmic conduction as shown in Figure 4.5. Some hole-only organic Schottky diodes were prepared using MEH-PPV and P3HT, where ITO/PEDOT:PSS was used as the hole injecting electrode and Au

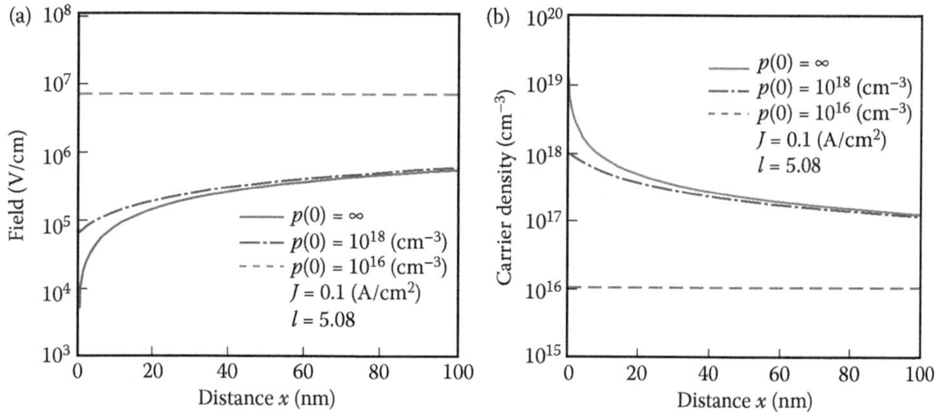

FIGURE 4.6
Distributions of (a) electric field and (b) charge carrier density in an organic diode for different values of $p(0)$ and given values of $J = 0.1$ A/cm^2 and $l = 5.08$. Values of other parameters considered are $N_0 = 10^{20}$ cm^{-3}, $H_b = 2 \times 10^{17}$ cm^{-3}, $N_v = 2 \times 10^{19}$ cm^{-3}, $\varepsilon = 3$, $\mu = 1 \times 10^{-5}$ cm^2 V^{-1} s^{-1}, and $d = 100$ nm. (Reprinted with permission from P. Kumar et al., Characteristics of a conducting organic diode with finite (non-zero) Schottky barrier, *J. Appl. Phys.* 100, 114506. Copyright 2006, American Institute of Physics.)

was used as the hole extracting electrode and *J*–*V* characteristics were measured at different temperatures. The experimental data was compared with Equation 4.21 at different temperatures and a good agreement was observed [56,60].

However, for more accurate analysis it would be interesting to discuss the case where the injection Schottky barrier is not zero and the approximation $p_t(x) \gg p(x)$ is also relaxed. For this case the coupled Poisson's and current density equations give

$$\int_0^d dx = \frac{\varepsilon\varepsilon_0}{q} \int_{F(0)}^{F(d)} \frac{1}{(J/q\mu_p F) + H_b (J/q\mu_p F N_v)^{1/l}} dF, \tag{4.24}$$

where $F(0)$ and $F(d)$ are the electric fields at the injecting contact ($x = 0$) and at the exit contact ($x = d$), respectively. $F(0)$ can be obtained from the condition of continuity,

$$J = q\mu_p p(0) F(0). \tag{4.25}$$

An analytical solution cannot be obtained in this case. Therefore, the complete *J*–*V* characteristics are obtained by solving Equation 4.24 numerically following the method given by Kumar et al. [63].

4.4.2.3 Trap-Filled Limit

At high applied voltages when the injected charge carrier density becomes very large, either all the traps are filled or their effect becomes negligible. The voltage at which the effect of traps become negligible is known as trap-filled voltage (V_{TFL}). There was a wide belief in the literature that once all the traps are filled, the current in an insulator would be governed by Mott's trap-free V^2 law and the V_{TFL} was determined as the voltage at which plots of Equation 4.20 and V^2 law intersect and was given by [54,55]

$$V_{TFL,lit} = q\frac{d^2}{\varepsilon\varepsilon_0}\left[\frac{9}{8}\frac{H_b{}^l}{N_v}\left(\frac{l+1}{l}\right)^l\left(\frac{l+1}{2l+1}\right)^{l+1}\right]^{l/(l-1)}. \quad (4.26)$$

The additional subscript "*lit*" corresponds to old literature. This approach to calculate V_{TFL} was proved wrong, as V^2 law can be achieved only when $p(x) \gg p_t(x)$ and in that case Equation 4.20 will not be valid [64].

If the injection contact is nonzero, only a fraction of the traps are filled and the V^2 law would be observed at much larger voltage than the voltage at which the maximum traps are filled. In other words, the occurrence of V^2 law depends upon $p(0)$. The Fermi level is determined by the density of free carriers, which cannot exceed $p(0)$ at a given temperature. To obtain the maximum number of traps that can be filled at any point x, the Fermi level E_f corresponding to $p(x) = p(0)$ was determined using Boltzmann statistics. And according to Equation 4.19 the maximum number of traps that can be filled (H_b') is given by

$$H_b' = H_b\left(\frac{p(0)}{N_v}\right)^{(1/l)}. \quad (4.27)$$

In principle one ought to use Fermi–Dirac statistics to calculate V_{TFL} but the use of Boltzmann's statistics was simple and it introduced a negligible error [61]. To calculate the error introduced by Boltzmann statistics, we calculated E_f from Equations 4.17 and 4.18 for different values of p. Figure 4.7 compares the Fermi energies calculated from Equations 4.17 (dashed curve) and 4.18 (solid curve) for different values of p. For these calculations,

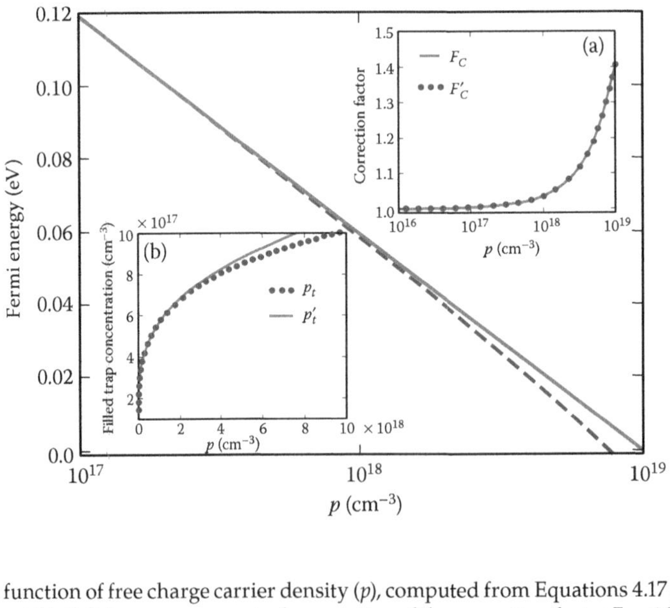

FIGURE 4.7
Fermi energies as a function of free charge carrier density (p), computed from Equations 4.17 (dashed curve) and 4.18 (solid curve). Inset (a): Solid curve represents the variation of the correction factor F_C with p and solid circles are for the empirical relation Equation 4.28. Inset (b): Variation of p_t and p_t' with p. Note that the difference in p_t and p_t' decreases with reduction in p.

N_v was considered to be 10^{19} cm^{-3}. It was seen that for $p > 1 \times 10^{18}$ cm^{-3} the Fermi energy (E_f) from Equation 4.17 was less than that from Equation 4.18. At very large carrier concentration, for example, $p = 7.5 \times 10^{18}$ cm^{-3}, the Fermi energy from Equation 4.18 was about 11 times of that from Equation 4.17. If E_f is not several kT away from the band edge, Equation 4.17 becomes necessary to determine E_f for a given concentration of free carriers. As we switch from Equation 4.18 through 4.17 the Fermi level E_f decreases for a given p, that is, the Fermi level comes closer to the band edge. Now it is obvious that the use of Fermi–Dirac statistics will give a larger value of p_t (see Equation 4.19). Therefore, to get the correct value of p_t we had to increase the free carrier concentration p. The advantage of this procedure was that we could still use Equation 4.19 provided we used the "fictitious" value of p as a mathematical artifice. We consider p_f as the new value of free carriers and defined F_C as the correction factor equal to p_f/p. The correction factor, F_C, could be calculated using a polynomial F_C' of the form

$$F_C' = \left[1 + \frac{p}{3N_v} + \left(\frac{p}{3N_v}\right)^2 - \left(\frac{p}{3N_v}\right)^3 + \left(\frac{p}{3N_v}\right)^4 - \left(\frac{p}{3N_v}\right)^5\right]. \tag{4.28}$$

F_C' was obtained by the numerical fitting of F_C. A comparison of computed calculations for F_C and F_C' for different values of p is shown in inset (a) of Figure 4.7. Note that this empirical relation is suitable only when the Fermi level is in the band gap or touches the valance band. Once the Fermi level enters the valance band, this relation does not remain valid. We could write the value of p_f in terms of p as

$$p_f = p\left[1 + \frac{p}{3N_v} + \left(\frac{p}{3N_v}\right)^2 - \left(\frac{p}{3N_v}\right)^3 + \left(\frac{p}{3N_v}\right)^4 - \left(\frac{p}{3N_v}\right)^5\right]. \tag{4.29}$$

Therefore, the correct value of p_t was obtained by replacing p with p_f in Equation 4.19. Let p_t' be the correct value of the filled traps. Then,

$$p_t' = H_b\left(\frac{p_f}{N_v}\right)^{1/l}. \tag{4.30}$$

In this way the calculated values of filled traps using Equations 4.19 and 4.30 for different values of p are shown in inset (b) of Figure 4.7. The values of the parameters used for these calculations are $T_c = 1200$ K, $T = 300$ K, $H_b = 10^{18}$ cm^{-3}, and $N_v = 10^{19}$ cm^{-3}. It was found that the use of Equation 4.18 introduced an error in p, which was the maximum when the Fermi level touched the valance band and it came out to be ~30%. The error decreased rapidly as the Fermi level moved away from the valance band. It was also observed that the error, introduced by the use of Equation 4.18, in the calculated filled traps increased with the free carriers until all the traps were actually filled. For free carrier density 7.67×10^{18} cm^{-3}, the error in p_t came out to be only 6.93%, which decreased with an increase in l (T_c/T). For $l = 20$ the error in p_t reduced to 1.35%. Although the correction in p_t at room temperature was small, the accuracy in application of Equation 4.19 improved rapidly as temperature decreased. At very high carrier concentrations when the Fermi level entered the valance band, no further error in p_t was observed as all the traps were filled and p_t had reached its saturated value H_b. Once all the traps were filled, *J–V* characteristics were not in error due to the use of Equation 4.19.

For the characteristics beyond the trap filled limit, the Poisson equation is given by

$$\frac{dF(x)}{dx} = \frac{q}{\varepsilon\varepsilon_0}\left(p(x) + H_b'\right), \tag{4.31}$$

We considered three-dimensionless distance (w), field (u), and potential (v) variables such as

$$u = \frac{H_b'}{p(x)} = \frac{qH_b'\mu_p}{J} F(x), \tag{4.32}$$

$$w = \frac{q^2 H_b'^2 \mu_p}{\varepsilon\varepsilon_0 J} x, \tag{4.33}$$

and

$$v = \frac{q^3 H_b'^3 \mu_p^2}{\varepsilon\varepsilon_0 J^2} V(x). \tag{4.34}$$

Integration of Poisson's Equation 4.31 and use of Equations 4.33 and 4.34 gave

$$w = u - \ln(1+u) + C, \tag{4.35}$$

where C is the constant of integration. The constant of integration C was introduced to account for the nonzero electric field at the junction. Using the boundary condition that at $x = 0$, $w = 0$ and $u = H_b'/p(0)$, C came out to be

$$C = \ln\left(1 + \frac{H_b'}{p(0)}\right) - \frac{H_b'}{p(0)}. \tag{4.36}$$

Integrating Equation 4.33 and using Equation 4.34 we got

$$v = \frac{u^2}{2} - \frac{1}{2}\left[\frac{H_b'}{p(0)}\right]^2 - u + \frac{H_b'}{p(0)} + \ln(1+u) - \ln\left(1 + \frac{H_b'}{p(0)}\right). \tag{4.37}$$

Subscript a was used to represent the values of w, u, and v at $x = d$. Now in terms of u_a, w_a, and v_a, Equations 4.35 and 4.37 could be written as

$$w_a = u_a - \ln(1+u_a) + \ln\left(1 + \frac{H_b'}{p(0)}\right) - \frac{H_b'}{p(0)}, \tag{4.38}$$

and

$$v_a = \frac{u_a^2}{2} - \frac{1}{2}\left[\frac{H_b'}{p(0)}\right]^2 - u_a + \frac{H_b'}{p(0)} + \ln(1+u_a) - \ln\left(1 + \frac{H_b'}{p(0)}\right). \tag{4.39}$$

Equations 4.32 and 4.34 gave the following relations for J and V:

$$J = \frac{q^2 H_b'^2 \mu d}{\varepsilon\varepsilon_0 w_a}, \tag{4.40}$$

and

$$V = \frac{qH_b' d^2}{\varepsilon\varepsilon_0}\frac{v_a}{w_a^2}. \tag{4.41}$$

The relation between J and V at lower and higher values of voltages could be obtained for higher and lower values of u_a, respectively. The value of u_a is given by $H_b'/p(d)$ (see Equation 4.32). For large values of u_a, v_a/w_a^2 tends to 0.5 and V tends to

$$V = V_{TFL}' = 0.5\frac{qH_b' d^2}{\varepsilon\varepsilon_0}. \tag{4.42}$$

This is the voltage at which the maximum possible traps are filled and we called it the actual trap-filled limit voltage (V_{TFL}'). Equations 4.40 and 4.41 gave the following explicit J–V relation:

$$\frac{J}{V} = \frac{d}{qH_b'\mu}\left(\frac{v_a}{w_a}\right) \tag{4.43}$$

Figure 4.8 shows the plot of Equation 4.43 (dashed curve) along with the plot of Equation 4.20 and the numerically calculated J–V curve using Equation 4.24. The dash-dotted straight line represents Equation 4.20, while the solid curve is the numerically calculated

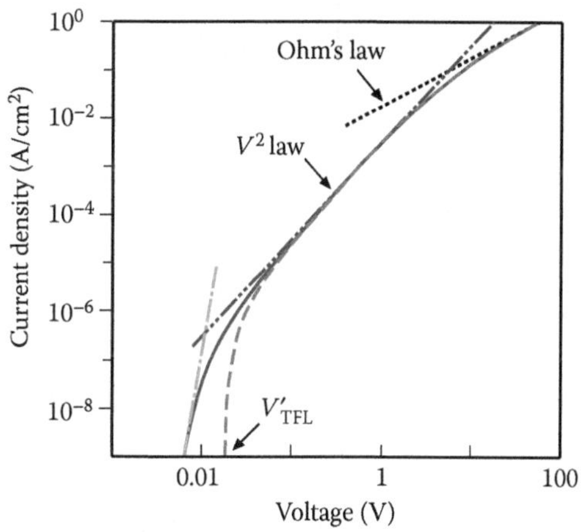

FIGURE 4.8
Computed J–V characteristics of an organic semiconductor in contact with a metal electrode. The dash-double dotted line represents the V^2 law and the dotted line is for Ohm's law. The dash-dotted line is the plot of Equation 4.20, and the dashed curve corresponds Equation 4.43. The solid curve was obtained by numerical calculations without using any approximation. (Reproduced from P. Kumar et al., 2009. *Eur. Phys. J. E* 28: 361–368 with kind permission of *The European Physical Journal* (EPJ).)

one without using any approximation. The dash-double dotted and dash-dotted lines represent the V^2 and Ohm's laws, respectively. The values of the parameters used in the calculations were $\mu = 10^{-5}$ cm^2 V^{-1} s^{-1}, $\varepsilon = 3$, $T = 300$ K, $T_c = 3000$ K, $d = 100$ nm, $H_b = 10^{15}$ cm^{-3}, $N_v = 10^{19}$ cm^{-3}, and $p(0) = 10^{17}$ cm^{-3}. It is very clear from Figure 4.8 that at low voltages the numerically calculated plot agrees well with that of Equation 4.20. As the voltage increased, the numerically calculated plot deviated from Equation 4.20 and approached the V^2 law. V^2 law was followed for a considerable range of voltage. As the voltage increased further, the curve again deviated from the V^2 law and approached the Ohm's law asymptotically. It is a very important result, which reveals that at higher voltages the J–V curves may approach the V^2 law, but for a nonzero Schottky barrier the curves will ultimately reach Ohm's law. The dashed curve was practically vertical up to almost 10^{-8} A/cm^2 and represented almost zero current up to 0.019 V, which represents V'_{TFL}. The interpretation of this curve is as follows. Each applied voltage can support only a fix amount of charge in the material. The fix charges H'_b (filled traps) produce a voltage, which opposes the applied voltage. At $V < 0.019$ V, the free carriers cannot spread and do not constitute a current. The voltage generated by the fixed charges is equal to V'_{TFL}. When the applied voltage exceeds V'_{TFL} (0.019 V, in the present case) the current begins to flow and, as mentioned earlier, it follows Mott's V^2 law but within the asymptotic limit as voltage tends to infinity the curve approaches the Ohm's law. It is very important to note that the occurrence of the V^2 law depends on the values of $p(0)$ and H'_b. If $p(0)$ is less than H'_b, the V^2 law will not be observed and the post TFL curve will directly reach Ohm's law. These results can also be reproduced analytically as discussed later.

Beyond V'_{TFL} two cases are important: (1) $p(0) \gg H'_b$ and (2) $p(0) \ll H'_b$. The minimum value of u_a will be $H'_b/p(0)$. Let us first consider the case when $p(0) \gg H'_b$ then $H'_b/p(0)$ will be less than 1. As the applied voltage increases, the value of $p(d)$ also increases, and at a sufficiently high voltage $H'_b/p(d)$ becomes much less than 1 or $u_a\left(H'_b/p(d)\right)$ tends to become $\ll 1$. In this case the expansion of Equation 4.38 gives

$$w_a = \frac{1}{2}\left[u_a^2 - \left(\frac{H'_b}{p(0)}\right)^2\right]. \tag{4.44}$$

Similarly the expansion of Equation 4.39 gives

$$v_a = \frac{1}{3}\left[u_a^3 - \left(\frac{H'_b}{p(0)}\right)^3\right]. \tag{4.45}$$

Now putting the values of w_a and v_a from Equations 4.44 and 4.45 in Equation 4.43, we get

$$J = \frac{3}{2}q\mu H'_b\left[\frac{u_a + (H'_b/p(0))}{u_a{}^2 + (H'_b/p(0))^2 + u_a(H'_b/p(0))}\right]\frac{V}{d}. \tag{4.46}$$

For large values of V, u_a tends to $H'_b/p(0)$, which is less than 1. As u_a tends to $H'_b/p(0)$, we considered the case when u_a is less than 1 but much more than $H'_b/p(0)$. In this case Equation 4.46 gives

$$J = \frac{3}{2} q\mu H_b' \frac{V}{d}\left(\frac{1}{u_a}\right). \tag{4.47}$$

When the values of u_a are less than 1, Equation 4.38 gives $u_a = (2w_a)^{1/2}$ and ultimately Equation 4.46 gives

$$J = \frac{9}{8} \mu\varepsilon\varepsilon_0 \frac{V^2}{d^3}, \tag{4.48}$$

which is the well-known Mott's V^2 law. When u_a became equal to $H_b'/p(0)$, Equation 4.46 gives

$$J = qp(0)\mu \frac{V}{d}. \tag{4.49}$$

This is the well-known Ohm's law. Equations 4.48 and 4.49 reveal that after the traps are filled, the current will first go to Mott's V^2 law and then ultimately it reaches Ohm's law at high voltages. This statement is also supported by Figure 4.9, which shows the electric field and charge carrier profiles for different currents. Filled and open symbols are for the profiles of charge carriers and electric field, respectively, when the current follows the V^2 law. Note that when the current follows the V^2 law, the charge carrier density varies as $1/\sqrt{x}$, while the electric field as $\sqrt{x}$. Values of the parameters used in these calculations were the same as those for Figure 4.8. Solid curves represent the charge carrier profiles (C1, C2, and C3), while dash-dotted curves are for electric field profiles (F1, F2, and F3). The profiles C3 and F3 are for 10^{-6} A/cm^2. When current increases further, say, for 10^{-3} A/cm^2, the charge carrier (C2) and field (F2) profiles agree perfectly with the filled and open symbols, respectively. It is a clear evidence that at 10^{-3} A/cm^2, the current follows the V^2 law. On further increase in the current, say, to 1 A/cm^2, the charge carrier (C1) and electric field (F1)

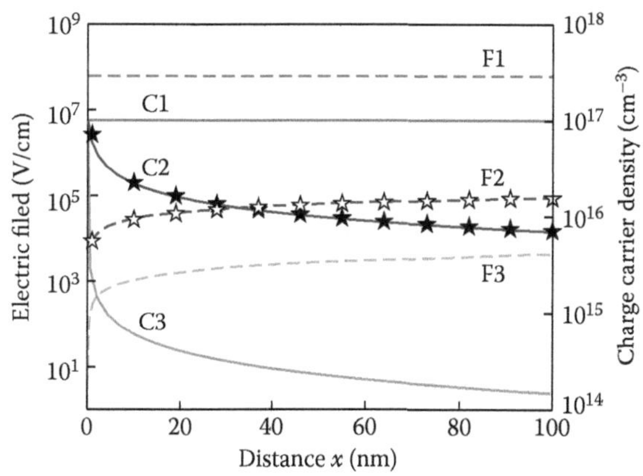

FIGURE 4.9
The electric field (F1, F2, and F3) and charge carrier (C1, C2, and C3) distributions for 1, 10^{-3} and 10^{-6} A/cm^2. The values of parameters are the same as those used in Figure 4.8. (Reproduced from P. Kumar et al., 2009. *Eur. Phys. J. E* 28: 361–368 with kind permission of *The European Physical Journal* (EPJ).)

profiles become perfectly flat, which show the ohmic conduction in the sample. Therefore, ultimately at high voltages the current follows Ohm's law.

Using Equation 4.46 and the condition $1 \gg u_a \gg H_b' / p(0)$, the range of validity of V^2 law was found to be $(4/3)(qH_b'd^2/\varepsilon\varepsilon_0) \ll V \ll (4/3)(qp(0)d^2/\varepsilon\varepsilon_0)$. If $p(0)$ is very large, the range of validity of the V^2 law will also be large. It is obvious that if $p(0) \ll H_b'$ the range of validity of V^2 law will be zero and the V^2 law will not be observed. It can also be shown analytically in a similar way as discussed earlier. For $p(0) \ll H_b'$ we can write u_a in terms of $H_b'/p(0)$ like $u_a = (H_b'/p(0)) + \delta u_a$. As $u_a \to H_b'/p(0)$, $\delta u_a \to 0$. Considering this concept, as $\delta u_a \to 0$, w_a and v_a will be given by

$$w_a = \delta u_a \left(\frac{H_b'}{H_b' + p(0)} \right), \tag{4.50}$$

and

$$v_a = \delta u_a \left(\frac{H_b'^2}{p(0)\left(H_b' + p(0)\right)} \right). \tag{4.51}$$

Putting the value of w_a and v_a from Equations 4.50 and 4.51 in Equation 4.44 we got

$$J = qp(0)\mu \frac{V}{d}. \tag{4.52}$$

This is Ohm's law. Therefore, for higher values of applied voltages the current tends to Ohm's law. Figure 4.10 shows the plots of Equation 4.43 for the two cases (1) when $p(0) \gg H_b'$ and (2) when $p(0) \ll H_b'$. These calculations have been made for $H_b = 10^{16}$ cm^{-3} and for two values of $p(0)$, that is, $p(0) = 10^{17}$ cm^{-3} and $p(0) = 10^{13}$ cm^{-3}. For these two cases the values of

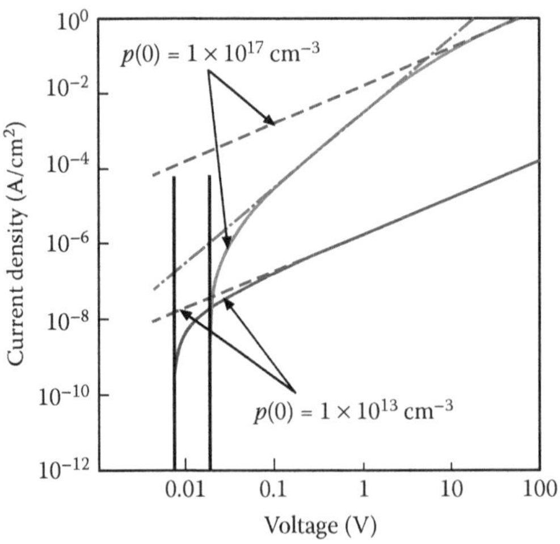

FIGURE 4.10
Post-TFL J–V characteristics for two different values of $p(0)$. The dash-dotted line represents the V^2 law and the dashed lines are for Ohm's law. The dark solid vertical lines are the values of V_{TFL}', calculated using Equation 4.42 for the two respective values of $p(0)$. (Reproduced from P. Kumar et al., 2009. *Eur. Phys. J.* E 28: 361–368 with kind permission of *The European Physical Journal* (EPJ).)

H_b' are 6.31×10^{14} cm^{-3} and 2.51×10^{14} cm^{-3}, respectively. The values of the rest of the parameters are the same as those used for Figure 4.8. The dash-dotted line represents the V^2 law and the dashed lines represent Ohm's law for the two values of $p(0)$. The solid vertical lines are the plots of Equation 4.42 and represent the V'_{TFL} in the two cases. It is clear that when $p(0) > H_b'$, the current first follows the V^2 law and then goes to Ohm's law asymptotically. When $p(0) < H_b'$, no V^2 dependence is observed and the current goes directly to the Ohm's law asymptotically.

A comparison of experimental data on P3HT diode with the preceding simulated results is shown in Figure 4.11. Note that at larger current densities the sheet resistance of indium tin oxide (ITO) plays an important role and some part of the applied voltage will drop across it. The applied voltage had been corrected for the voltage drop across the ITO series resistance. Symbols in Figure 4.11 represent the experimental data measured at different temperatures. The dash-double dotted curves are for Equation 4.20 at the respective temperatures. At low voltages, the characteristics show ohmic behavior that can be attributed to the background doping and thermally excited charge carriers. In the intermediate voltage range the characteristics followed Equation 4.20 (where $J \propto V^{l+1}$ and $l > 1$). As the voltage increased further, the characteristics bent down and deviated from Equation 4.20. The dashed lines are for Ohm's law Equation 4.52 at the respective temperatures. The dotted line representing a slope of 2 is also drawn to guide the eyes. Note that at higher voltages the experimental data had a slope less than 2 and approached Ohm's law. No V^2 dependence of the current had been observed. Therefore, from the absence of V^2 dependence and Ohm's law approach of the experimental data, it could be inferred that the diodes had nonzero injection Schottky barriers and H_b' was much more than $p(0)$, which reduced further at low temperatures. Assuming $N_v = 6 \times 10^{19}$ cm^{-3}, the total trap density in P3HT was calculated to be $H_b = 7.7 \times 10^{18}$ cm^{-3} from the curve fitting. From the slope of the curve in the voltage range where Equation 4.20 was valid, the value of T_c was calculated to be 800 K. The solid curves in Figure 4.11 represent the numerically calculated J–V

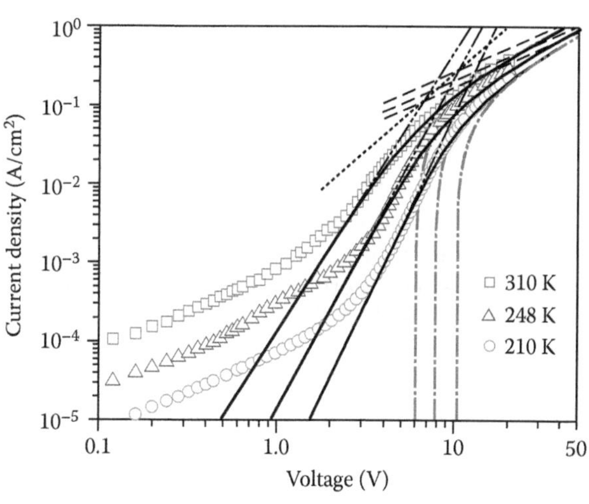

FIGURE 4.11
Comparison of experimental J–V characteristic of the ITO/PEDOT:PSS/P3HT/Au diode with our new model. Symbols represent the experimental data at different temperatures. The dashed lines are for Ohm's law and the solid curves are from numerical calculations. (Reproduced from P. Kumar et al., 2009. *Eur. Phys. J. E* 28: 361–368 with kind permission of *The European Physical Journal* (EPJ).)

characteristic and the dash-dotted curves are for Equation 4.43 at the respective temperatures. It is clear from the figure that at very high voltages the curves approached Ohm's law. The experimental data showed very good agreement with the numerically calculated *J–V* curve in the whole voltage range at different temperatures. The values of different parameters were $\varepsilon = 3$, $T_c = 800$ K, $d = 97$ nm, $H_b = 7.7 \times 10^{18}$ cm^{-3}, $N_v = 6 \times 10^{19}$ cm^{-3}, electron density in metal electrode $N_0 = 10^{19}$ cm^{-3} and $\mu = 2.9 \times 10^{-4}$, 1×10^{-3} and 2.1×10^{-3} cm^2 V^{-1} s^{-1} at 310, 248, and 210 K, respectively. The calculated value of Schottky barrier corresponding to $p(0)$ for P3HT diode was found to be ~0.2 eV, which was in good agreement with the difference of the work function of PEDOT:PSS (5.0 eV) and HOMO of P3HT (5.2 eV) [65]. The values of V'_{TFL} for the P3HT sample were calculated to be 6.0, 7.74, and 10.15 V at 310, 248, and 210 K, respectively. Beyond the V'_{TFL}, the experimental data showed good agreement with Equation 4.43.

4.4.2.4 Hopping Transport

In hopping transport the charge carriers move via hopping from one energy site to other. In disordered organic materials, usually two kinds of disorders are discussed, namely, diagonal (energetic) and off-diagonal (positional) disorders. Diagonal disorder corresponds to fluctuation in energy of HOMO or LUMO levels (energy sites) of the material, whereas off-diagonal disorder corresponds to fluctuations in strength of interactions between molecules [66]. Off-diagonal disorder causes distribution of electronic couplings that can develop conducting pathways as well as dead ends for the charge carriers. The conformational freedom causes distribution in torsional angles between adjacent molecules that leads to conjugation segments of different lengths. A change in conjugation length results in different HOMO and LUMO energies of the molecules, a diagonal disorder. Diagonal disorder is also caused by polarization effects from the surrounding molecules. Usually the energetic disorder or distribution of density of states (DOS) in organic semiconductors is assumed to be in Gaussian shape given by

$$Q(\varepsilon) = \frac{1}{(2\pi)^{1/2}\sigma} \exp\left(-\frac{\varepsilon^2}{2\sigma^2}\right), \tag{4.53}$$

where σ is the standard deviation of distribution and is generally 50–100 meV [67,68]. Gaussian distribution is evident from the absorption bands of disordered organic materials. Early descriptions of hopping transport in disorder materials was presented in the 1970s by Scher et al. [69,70]. Later, more work on hopping transport was done by Bassler and coworkers [71–74 and references therein]. Details on hopping transport can be found in several reviews and textbooks [66,75–78]. I will here just give a brief introduction to this transport phenomenon.

In pure diffusive charge transport the mobility of charge carriers is given by the Einstein–Smoluchowski relation $\mu = qD/k_BT$, where D is the diffusion coefficient. In the presence of an electric field the mobility is defined by the ratio of drift velocity (v_d) and applied electric field (F), that is, $\mu = v_d/F$. In the hopping mechanism the diffusion coefficient D is given by $D = a^2k_{ET}$, where a is the spacing between hopping sites and k_{ET} is the rate of electron transfer or hopping rate [66]. Both the disorder and polaronic effects contribute to hopping transport in organic semiconductors [73]. Holstein [79,80] and Friedman [81] suggested polaron transport models based on the perturbation theory. Though these models are limited, they are used as benchmarks for development of new theoretical models. According to the polaron theory of Holstein, the hopping rate between two neighboring sites is given by [79]

$$k_{ET} = \frac{t^2}{\hbar^2\omega_0}\left[\frac{\pi}{g^2 \csc h(\hbar\omega_0/2k_BT)}\right]^{1/2} \exp\left[-2g^2 \tanh\left(\frac{\hbar\omega_0}{4k_BT}\right)\right], \tag{4.54}$$

where $\hbar\omega_0$ is the phonon energy, t is the transfer integral, and g is the coupling constant of electron–phonon interaction. The polaron binding energy in this case is given by $E_{pol} = \hbar\omega_0 g^2$.

The hopping transport in disordered organic materials is described by the Gaussian disorder model (GDM) [82]. In disordered materials, the Miller–Abrahams [83] and Marcus [84] models are the most frequently used models for hopping rates. According to the Miller–Abrahams model the hopping rate is related to energetic disorder of density of states. The hopping rate from site i to site j is given by

$$k_{ij} = v\exp(-2\gamma R_{ij}) \times \begin{cases} \exp\left(-\dfrac{\varepsilon_j - \varepsilon_i}{k_BT}\right) & \varepsilon_j > \varepsilon_i \\ 1 & \varepsilon_j < \varepsilon_i \end{cases} \tag{4.55}$$

where v is the attempt hopping frequency, R_{ij} is the separation between sites i and j, γ is the overlap factor that is related to electronic coupling matrix, and ε_i and ε_j are the site energies. For the upward jump in energy ($\varepsilon_j > \varepsilon_i$) the last exponential term is a Boltzmann factor which becomes 1 for downward jump ($\varepsilon_j < \varepsilon_i$). The charge carriers proceed randomly and due to energetic disorder the charge carriers tend to relax into the tail of DOS distribution. In the presence of an electric field (F) the Boltzmann factor for upward jump incorporates an additional term ($-qrF$), where r is the vector connecting two sites. The upward jumps are accelerated by electric field but the downward jumps are not. However, the downward jumps always happen irrespective of difference in the site energies. The upward jump is characterized by absorption of a single phonon, whereas the downward jump is characterized by emission of phonon that compensates ($\varepsilon_j - \varepsilon_i$). The Miller–Abrahams model is valid for energies ($\varepsilon_j - \varepsilon_i$) not exceeding the maximum energy of acoustical and optical phonons coupled to the electron transfer reaction. The upward hopping is a thermally activated process, whereas the downward hopping is temperature independent. The hopping continues until the charge carriers reach the exit electrode. The Miller–Abrahams model is applicable at low temperatures and to weak electron–phonon couplings. According to the Marcus model the hopping rate between two sites i and j is given by

$$k_{ij} = \frac{t^2}{\hbar}\left[\frac{\pi}{k_BT\lambda_{reorg}}\right]^{1/2} \exp\left[-\frac{(\lambda_{reorg} + \varepsilon_j - \varepsilon_i)^2}{4\lambda_{reorg}k_BT}\right] \tag{4.56}$$

where $\lambda_{reorg} = 2E_{pol}$. The hopping rate is a function of reorganization energy or the polaron energy (E_{pol}). This rate is similar to that proposed by Holstein in the classical limit [79]. The Marcus model is applicable at high temperatures and for strong electron–phonon coupling.

Bassler and coworkers have beautifully explained the hopping transport in disorder materials via Monte Carlo simulations, where hopping rates and the density of hopping states are two main inputs [85]. Temperature and the electric field are two very important parameters that help in hopping of charge carriers. Both the temperature and electric field help the charge carriers to overcome energy barriers introduced by the diagonal disorder. The electric field reduces the energy barriers for upward hopping and the otherwise trapped charge carriers in tail states of DOS also hop further. In the presence of only the diagonal disorder, Monte Carlo simulation gives Poole–Frenkel-type dependence of mobility on the electric field [86]. Electric field dependence of mobility becomes more prominent for increased diagonal disorder. Increase in the electric field also increases the diffusion coefficient. Monte Carlo simulation suggests violation of the Einstein–Smoluchowski relation for very high electric fields and high energetic disorder. In the presence of both the diagonal and off-diagonal disorders, Monte Carlo simulation gives the following dependence of charge carrier mobility on temperature and electric field [66]:

$$\mu(\sigma,\Sigma,F)=\mu_0\exp\left(-\left(\frac{2\sigma}{3k_BT}\right)^2\right)\times\begin{cases}\exp\left[C\left(\left(\frac{\sigma}{k_BT}\right)^2-\Sigma^2\right)F^{1/2}\right] & \Sigma\geq 1.5\\ \exp\left[C\left(\left(\frac{\sigma}{k_BT}\right)^2-2.25\right)F^{1/2}\right] & \Sigma<1.5\end{cases} \quad (4.57)$$

where Σ is standard deviation associated with off-diagonal (positional) disorder and C is a constant. Mobility shows $F^{1/2}$ dependence on the electric field, and in the limit of zero electric field Equation 4.57 reduces to

$$\mu(\sigma)=\mu_0\exp\left(-\left(\frac{2\sigma}{3k_BT}\right)^2\right), \quad (4.58)$$

which shows $1/T^2$ dependence of $\ln(\mu)$. If the positional disorder (Σ) become more than the energetic disorder (σ), then the field dependence of mobility can become negative, that is, mobility would decrease with an increase in electric field. For example, for the sites that are weakly electronically coupled, the jump in the direction of electric field is blocked. The charge carriers will try to circumvent this jump and may jump against the electric field. But the jump against electric field will be blocked for high fields; now the transport becomes completely drift controlled and the mobility of charge carriers varies as $\mu \propto 1/F$.

As the hopping charge transport in disordered organic materials is controlled by both the disorder and polaronic effects, an adequate model incorporating both facts was lacking. In 2003, Fishchuk et al. developed a hopping transport model considering the superposition of disorder and polaron effects with the Marcus hopping rate [87]. The mobility obeyed following temperature and electric field dependence:

$$\mu=\mu_0\exp\left(-\frac{E_a}{k_BT}-\frac{1}{8Q^2}\left(\frac{\sigma}{k_BT}\right)^2\right)\exp\left(\left(\frac{\sigma}{2k_BT}\right)^{3/2}\frac{1}{Q^2}\sqrt{\frac{qaF}{\sigma}}\right)\exp\left(-\frac{qaF}{4k_BTQ^2}\right) \quad (4.59)$$

where $E_a = E_{pol}/2$, $Q=\sqrt{1-(\sigma^2/8qak_BT)}$, and a is the distance between neighboring hopping sites. Later many improvements were made to the GDM to describe the electric field dependence of charge carrier mobility like the effects of charge carrier concentration and partial filling of DOS at large carrier concentrations [74,88–91]. The modified model is called extended Gaussian disorder (EGD) model and is used frequently to describe charge carrier mobility in organic devices. The model introduced by Torricelli et al. [92] accounted for temperature, carrier density, and electric field effects on the charge carrier mobility. The model is based on the concept of percolation in a variable range hopping (VRH) system with exponential density of states and the mobility is give by

$$\mu[T, p(x), F(x)] = \frac{\sigma_0}{q}\left[\frac{T_0^4 \sin(\pi(T/T_0))}{B_c(2\alpha)^3 T_f^2(2T_f - T)T}\right]^{T_0/T_f} p^{(T_0/T_f)-1}, \tag{4.60}$$

where σ_0 is a parameter for conductivity, α^{-1} is the effective overlap parameter between localized states, T_0 is a parameter that depends in the material disorder, B_c is the critical number for the onset of percolation, and $T_f = \{T^\beta + (\gamma q F/(\alpha k_B))^\beta\}^{1/\beta}$, where β and γ are fitting parameters that depend on the materials. For calculation of *J–V* characteristics from hopping models, the current density and Poisson's equation are solved numerically using necessary boundary conditions. In the current density equation, the charge carrier mobility is taken as a function for temperature, electric field, and charge carrier density as given earlier.

4.5 Determination of Charge Carrier Mobility in Organic Semiconductors

The charge carrier mobility in organic semiconductors could be determined experimentally in several ways. Some of the most frequently used techniques are space charge limited current (SCLC) conduction, time of flight (TOF), and field-effect transistors (FETs). In the SCLC conduction method, the organic materials are sandwiched between two electrodes to formulate a diode, and its *J–V* characteristics are measured. The diodes are single-carrier devices that allow only one type of carrier to pass through when a voltage is applied. The energy barriers at the contacts should not be very high so that the current is bulk limited not injection limited. SCLC depends on film thickness where it decreases with an increase in thickness. Comparing the experimental data with the bulk transport models presented earlier, the charge carrier mobility can be calculated.

In the TOF technique as well a film of organic semiconductor is sandwiched between two electrodes, where one of the electrodes is optically transparent. The film is illuminated in the proximity of the transparent electrode with a laser pulse in the presence of a small voltage applied across the electrodes. Laser illumination generates a sheet of excitons, which are dissociated by a suitable quencher at the interface where the exciton sheet is generated. One type of charge carriers is transferred ultrafast to the quencher and the other type is left in the semiconductor. Depending on the kind of quencher (electron transporter or hole transporter) the applied voltage is biased in a way that the charge sheet in the semiconductor migrates through the semiconductor toward another electrode and the current flow is measured with time. Depending on the material properties, the obtained transient behavior may be sharp or dispersed. For ordered materials the transient behavior is sharp, whereas for disordered materials a broadening in the signal is observed. From transient behavior, the time taken by the charge sheet to travel the semiconductor film, usually called time of flight (TOF) is obtained. If t is the time taken by the charge sheet to travel the semiconductor film, the charge carrier mobility will be given by

$$\mu = \frac{v}{F} = \frac{d}{tF}, \tag{4.61}$$

where d is the film thickness and F is the applied electric field. For zero background doping and no injection of charge carriers by applied voltage, the diode can be considered a parallel plate capacitor with a uniform electric field between the plates and Equation 4.61 would modify to

$$\mu = \frac{d^2}{tV} \tag{4.62}$$

Because the charge carrier mobility in organic semiconductors varies with temperature and electric field, the mobility determined by the TOF technique follow the empirical law [93–95]

$$\mu(T,F) = \mu_0 \exp\left[-\frac{\Delta}{k_B T} + \beta\left(\frac{1}{k_B T} - \frac{1}{k_B T_0}\right)\sqrt{F}\right], \tag{4.63}$$

where μ_0 is a constant, β is a Poole–Frenkel-type coefficient, Δ is the zero field activation energy, and T_0 is generally more than T. This equation has been widely used in the literature to calculate charge carrier mobility in organic semiconductors. Sometime T_0 may be less than T and the mobility slightly decreases in the electric field [96]. This is like a situation when off-diagonal disorder in the Bassler's Monte Carlo simulation [66] dominates the diagonal disorder.

Measurement of *I–V* characteristics of a semiconductor in an FET structure also leads to determination of its charge carrier mobilities. As in linear and saturated regions, organic FETs follow the *I–V* expressions derived for inorganic FETs [97], and the mobility can be found from curve fitting. The expressions for linear and saturated regions of the source–drain *I–V* characteristics of FETs are written as [41]

$$I_{SD} = \frac{W}{L}\mu C(V_G - V_T)V_{DS}, \tag{4.64}$$

and

$$I_{SD} = \frac{W}{2L}\mu C(V_G - V_T)^2. \tag{4.65}$$

Here V_{DS} is the voltage applied across the drain and source, W is the channel width, L is the channel length, C is the gate dielectric capacitance, V_G is the gate voltage, and V_T is the threshold voltage. Charge carriers transport in the semiconductor channel between the source and drain in the vicinity of the oxide-semiconductor interface. The traps in the organic semiconductor and at the dielectric interface affect the charge transport. Due to traps and carrier generation, the electric field distribution does not remain constant; and due to different electric field distributions in high and low voltage regions, the mobility calculated from the saturated region is generally found to be more than that calculated from the linear region. Charge carrier mobility calculated from an FET structure has been observed to depend on trap density, dielectric constant of the gate insulator, gate voltage, and charge carrier density [98–102]. Charge carrier mobilities in organic semiconductors extracted out from transistor and diode structures are observed to be quite

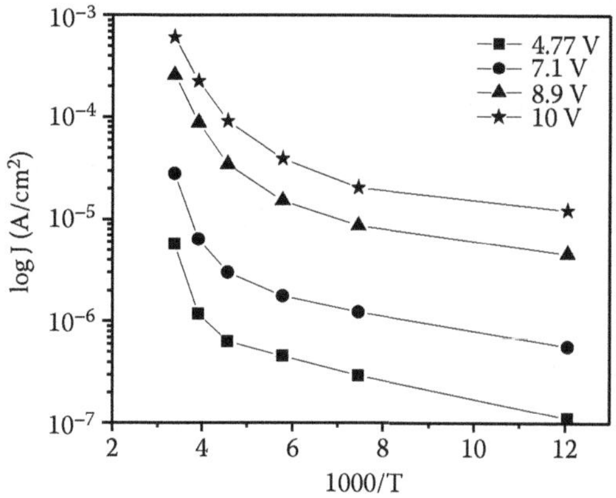

FIGURE 4.12
Log J versus $1000/T$ plots of the Al/Alq_3/Al electron-only diode at different voltages with Alq_3 thickness of 160 nm. (Reprinted with permission from P. Kumar et al., *Jpn. J. Appl. Phys.* 45, 7621–7624, copyright 2006 The Japan Society of Applied Physics.)

different [88]. This difference in mobility is attributed to the difference in charge carrier densities in the two configurations. The charge carriers in the transistor transport at the surface of the semiconductor and the charge carrier density is much more than that injected in the diode.

My colleagues and I studied the J–V characteristics of an electron-only diode of molecular tri(8-hydroxyquinoline) aluminum (Alq_3) complex at different temperatures and observed that the characteristics at higher temperatures could be fitted with the band transport model with exponential distribution of traps in energy space [103]. The characteristics temperature of trap distribution was found to be 2300 K, with a trap density of 2×10^{17} cm^{-3}. However at low temperatures (below 219 K), the observed characteristics could be fitted with the hopping transport model only where mobility is given by Equation 4.63. The values of μ_0 and T_0 were found to be 1×10^{-15} cm^2 V^{-1} s^{-1} and 600 K, respectively. Figure 4.12 shows log J versus $1000/T$ plots of the diode at different voltages. The plots showed two linear regions for each voltage: one at high temperature and other at low temperature. The slope of high temperature regime was larger than that of the low temperature regime and suggests two different conduction mechanisms in the two regions. The activation energies at low temperatures were measured from the straight lines in the low temperature region shown in Figure 4.12, and the zero field activation energy was calculated to be 0.05 eV from extrapolation of activation energy versus $\sqrt{V/d}$ to zero fields [103]. Such behavior was also observed by other authors in different materials [59,104] and could be understood from general physical consideration. At room temperature, a sufficient number of charge carriers (electrons in the present case) is ionized from traps into the LUMO and band transport becomes appropriate. At lower temperatures, however, the number of electrons ionized in the LUMO becomes small. Now the mobility model, which is predominantly controlled by hopping of charge carriers, becomes more important. Note that transport by the band model and mobility model occurs at all temperatures, and it is their relative magnitudes that change in going from high temperatures to low temperatures. If the characteristics energy of traps is small, then the mobility model might not be observed even at low temperatures as a sufficient number of traps might still be ionized [105].

4.6 Factors That Affect Charge Carrier Mobility in Organic Semiconductors

Charge carrier mobility in organic semiconductors has been observed to be affected by a number of parameters, including disorder, traps, impurities, temperature, electric field, charge carrier density, molecular weight, and molecular packing. There are several reports in the literature on mobility variation with the variation in these parameters, but I will briefly discuss here only a few of them. Ordered materials possess high charge carrier mobilities compared to their disordered phases. The degree of disorder of an organic film can be altered and as a result the mobility can be varied up to many orders of magnitude by changing its deposition conditions [106,107]. The presence of traps hinders the motion of charge carriers via trapping and detrapping processes. The traps below the Fermi level are filled, whereas those above are empty and control the charge carrier mobility. The trapped charge carriers are released again into extended states via thermal excitation. The effect of trapping on mobility is demonstrated by continuous time random walk (CTRW) and multiple trapping and release (MTR) models [69,108,109], which give the following expression of mobility for homogenous distribution of traps:

$$\mu = \mu_0 \alpha \exp\left(-\frac{E_t}{k_B T}\right) \tag{4.66}$$

where α is the ratio of density of transport energy levels and trap density, and E_t is the energy of traps. If the trap distribution is not uniform, then the mobility will behave differently with temperature. The thermally stimulated current (TSC) technique is used to determine the trap distribution, in which a sample is cooled to low temperatures and illuminated so that all the traps are filled via photogenerated charge carriers and when heated a current is obtained due to the release of charge carriers. The impurities of some other materials present in the organic semiconductors may also work as trapping sites and affect the charge carrier mobilities [110].

In ordered materials the charge carrier mobility generally decreases with increase in temperature. This corresponds to the band transport where charge carriers are scattered by the lattice phonon. However, the presence of traps can affect mobility adversely where reduction of temperature will keep the charge carriers trapped and the mobility would decrease. In highly disordered materials where transport takes place via hopping, the mobility is thermally activated. An increase in thermal energy helps the charge carriers to overcome energy barriers between hopping sites and the charge carrier mobility often follows Arrhenius-type temperature dependence given by

$$\mu = \mu_0 \exp\left(-\frac{\Delta}{k_B T}\right). \tag{4.67}$$

The activation energy Δ increases with an increase in disorder. The theoretical justification of Arrhenius-type behavior of mobility in organic semiconductors is not available in the literature. However, the Monte Carlo simulation, performed by Bassler and coworkers, gave a different expression, which could also fit the experimental data within a temperature range [111,112]. Generally the charge carrier mobility in organic semiconductors is observed to increase with high electric fields and follows the Poole–Frenkel-type behavior given by Equation 4.63. Due to the Poole–Frenkel effect, the electric field reduces the energy barrier for hopping to the next site and assists the charge carriers in transportation.

Charge carrier mobility in organic semiconductors is also affected by their concentrations [99,102]. At low charge carrier densities, trapping and detrapping effects become important, whereas at high carrier concentrations the traps are filled and charge carriers experience trap-free transport. Also at high concentrations the charge carriers experience an increased scattering resulting in reduced mobility. Figure 4.13 shows the variation in hole mobility as a function of hole density measured by Tanase et al. in the diode and FET structures of P3HT and MDMO-PPV [99]. High mobilities at high carrier concentrations correspond to those from FET structure whereas low mobilities at low carrier concentrations correspond to those from diode structure. Mobility of charge carriers in organic semiconductors has also been observed to depend on their molecular weight and crystallinity. Different charge carrier mobilities in different materials are quite expected, but it can also be different even for the same materials of different molecular weights and different conjugation lengths. For example, Goh et al. investigated the hole mobility in regioregular P3HT of different molecular weights in hole-only diode configuration [113]. The characteristics of materials with high and medium molecular weights could be fitted to the trap-free SCLC model where hole mobility was independent of the electric field. But for low molecular weight P3HT, the hole mobility was observed to be field dependent. The hole mobility calculated via curve fitting increased from 1.33×10^{-5} to 3.3×10^{-4} cm^2 V^{-1} s^{-1} as the molecular weight increased from 2.9 to 31.1 kg/mol (or kDa). Temperature dependence of characteristics suggested the hole mobility to be temperature dependent throughout the investigated temperature range. The mobility showed Arrhenius-type temperature dependence and is shown in Figure 4.14 for P3HT of different molecular weights.

The effect of molecular weight on charge carrier mobility has also been investigated in transistor configuration [114,115]. The field-effect mobility calculated from transistor characteristics also increased with an increase in molecular weight of the semiconductor. Increases in charge carrier mobility with increases in molecular weight of the organic semiconductor have been correlated with traps concentration, regioregularity, polydispersity, film morphology, conjugation length, and crystallinity of the film. The lower molecular weight chains may have higher concentration of traps as they exhibit early termination

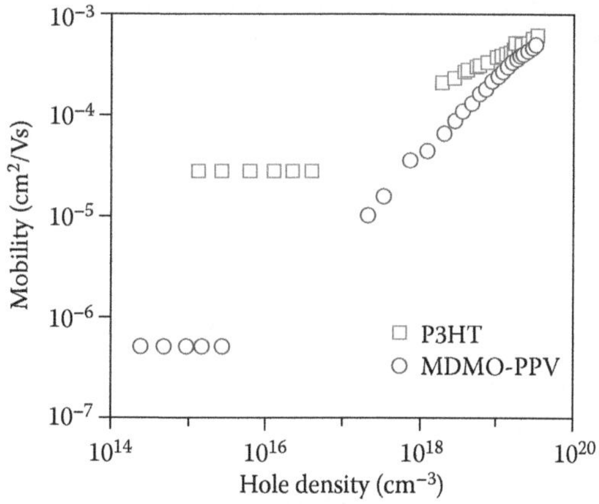

FIGURE 4.13
Variation in mobility as a function of hole density measured in diode and FET structures of P3HT and MDMO-PPV. (Modified from C. Tanase et al., 2003. *Phys. Rev. Lett.* 91: 216601.)

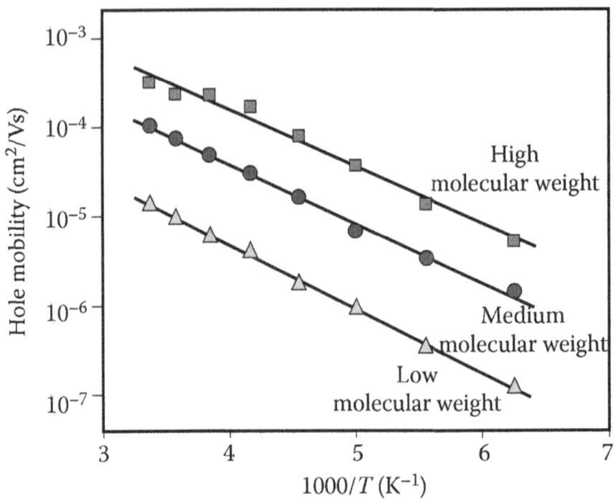

FIGURE 4.14
Mobility versus 1000/*T* plots of P3HT with different molecular weights. Squares are for high molecular weight (31.1 kg/mol), circles are for medium molecular weight (9.72 kg/mol), and triangles are for low molecular weight (2.9 kg/mol) P3HT. Solid lines are fitting of the empirical relation of mobility with temperature. (Reprinted with permission from C. Goh et al., Molecular-weight-dependent mobilities in regioregular poly(3-hexyl-thiophene) diodes, *Appl. Phys. Lett.* 86, 122110. Copyright 2005, American Institute of Physics.)

of polymerization. Regioregularity has a large impact on packing of the polymer chain and has a significant effect on the charge carrier mobility [116]. Higher polydispersity might cause lower mobility because longer chains with longer conjugation length have low band gap, which might act as a trapping site for shorter conjugation length molecules [117]. Variation in molecular weight causes different film morphology with different size of crystalline grains [114]. Kline et al. observed that the low molecular weight (3.2 kg/mol) P3HT film exhibits rodlike structure, whereas that of high molecular weight (31 kg/mol) P3HT has an isotropic nodule structure [114]. The lower molecular weight films were more crystalline, which was further supported by x-ray diffraction (XRD) studies. It was notable that high crystallinity was giving lower mobility. This had been attributed to well-defined grain boundaries in crystalline films that limited the charge carrier mobility. In high molecular weight P3HT film the longer chains connected the ordered regions and provided pathways for transportation of charge carriers from one region to another giving higher mobilities.

4.7 Modeling of *J–V* Characteristics of Organic Solar Cells (OSCs)

Device modeling and understanding of the physics behind device operation help us in mapping the parameters that affect the device performance. Modeling is a powerful tool and a correct model will help us in achieving highly efficient and stable OSCs. The nature of dark and illuminated *J–V* characteristics of OSCs is observed to be different from that of conventional solar cells [118]. However, due to unavailability of suitable models the characteristics of OSCs were compared with the conventional model. My colleagues and I developed a model for OSCs, considering the facts of organic semiconductors that were

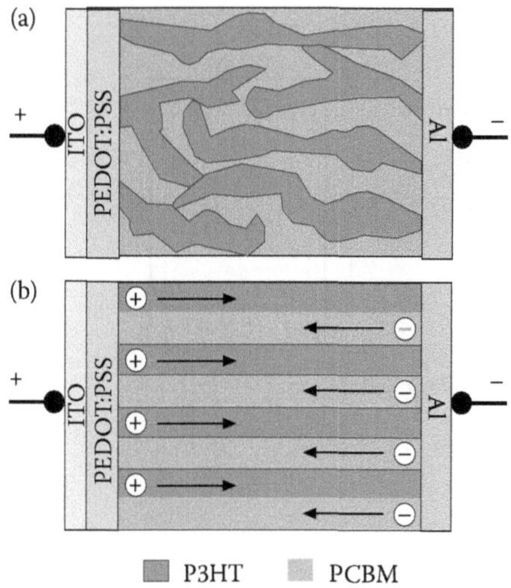

FIGURE 4.15
Schematic representation of conducting paths in the ITO/PEDOT:PSS/P3HT:PCBM/Al solar cell. (a) The interpenetrating P3HT and PCBM paths distributed in a random zigzag way and (b) the distribution of P3HT and PCBM used in the modeling. (Reprinted with permission from P. Kumar et al., A model for the *J–V* characteristics of P3HT:PCBM solar cells, *J. Appl. Phys.* 105, 104507. Copyright 2009, American Institute of Physics.)

not accounted for in the conventional model [118]. The model was developed for BHJ OSCs, considering P3HT:PCBM as the active organic layer sandwiched between ITO/PEDOT:PSS and Al electrodes (see Figure 4.15). In the blend film P3HT and PCBM form the percolated path ways, as shown in Figure 4.15a, and the charge carriers transport through these paths. For the purpose of modeling we replace the structure shown in Figure 4.15a with that shown in Figure 4.15b. The holes were considered to flow through P3HT channels and electrons through PCBM channels. The electron and hole currents were modeled separately and the resultant terminal current was obtained by adding the two currents. Due to the fact that electrons and holes flow in different channels and are separated, the electron–hole recombination and ambipolar effects are suppressed. For modeling of *J–V* characteristics the P3HT:PCBM blend was assumed to be a single virtual semiconductor having electron and hole mobilities equal to electron mobility of PCBM and hole mobility of P3HT, respectively. The HOMO and LUMO levels of the virtual semiconductor corresponded to the HOMO of P3HT and LUMO of PCBM, respectively. The schematic energy level diagram of the solar cell is shown in Figure 4.16, where Figure 4.16a shows the energy levels of respective materials before intimate contact, Figure 4.16b represents the energy level diagram before intimate contact with the virtual semiconductor, and Figure 4.16c shows the situation when the virtual semiconductor and contacting electrodes are brought into intimate contact to formulate the solar cell and a forward bias V is applied. This design is like a metal–insulator–metal (MIM) structure and the MIM diode physics would be applicable.

According to MIM model, two electrodes of different work functions develop an internal built-in electric field and the corresponding voltage is known as built-in voltage (V_{bi}). For undoped organic semiconductors with large nonzero Schottky barriers at the electrodes, the built-in electric field becomes constant throughout the semiconductors [118–120] and the potential distribution $\psi(x)$ in forward bias is given by

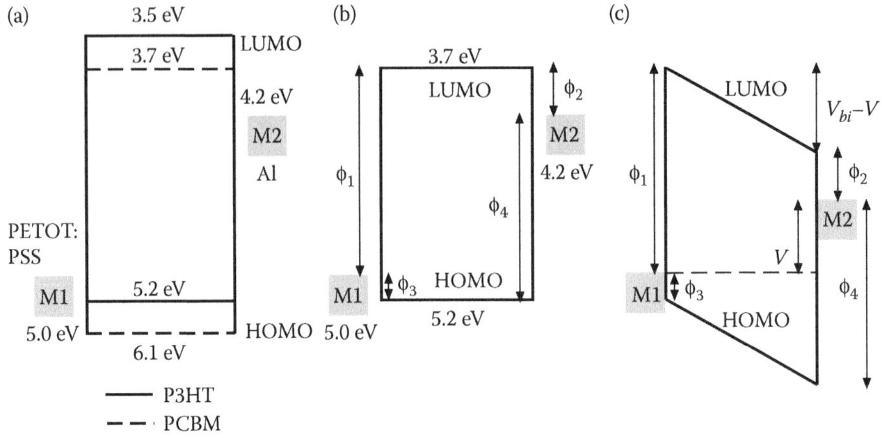

FIGURE 4.16
Schematic energy level diagrams of (a) PEDOT:PSS/P3HT:PCBM/Al solar cell before the materials are brought into intimate contact; (b) virtual semiconductor (the blend of P3HT:PCBM is assumed to be a single virtual semiconductor) before it is brought into intimate contact with the electrodes; and (c) the virtual semiconductor is brought into intimate contact with the electrode materials under forward bias. (Reprinted with permission from P. Kumar et al., A model for the *J–V* characteristics of P3HT:PCBM solar cells, *J. Appl. Phys.* 105, 104507. Copyright 2009, American Institute of Physics.)

$$q\psi(x) = \left(\frac{\phi_1 - \phi_2 - qV}{d}\right)x - \phi_1, \tag{4.68}$$

where d is the semiconductor thickness; and ϕ_1 and ϕ_2 are electron injection barriers from the anode and cathode, respectively. The electron current in forward bias in dark will be given by

$$J_n = q\mu_n n(x)F(x) + qD_n \frac{\partial n(x)}{\partial x}, \tag{4.69}$$

where $n(x)$ is electron density, $F(x)$ is the electric field, and D_n is the diffusion coefficient for electrons. Since $F(x) = (-\partial\psi(x)/\partial x))$, Equation 4.69 could be written as

$$J_n = qD_n\left(-\frac{q}{k_BT}n(x)\frac{\partial\psi(x)}{\partial x} + \frac{\partial n(x)}{\partial x}\right). \tag{4.70}$$

Multiplying with integrating factor $\exp(-q\psi(x)/k_BT)$ on both sides and integrating w.r.t. x over the entire thickness of semiconductor, Equation 4.70 gives

$$J_n = \frac{qD_nN_c(\exp(qV/k_BT)-1)}{\int_0^d \exp((-q\psi(x))/k_BT)dx}. \tag{4.71}$$

Using the boundary conditions,

$$n(0) = N_c \exp\left(\frac{-\phi_1}{k_BT}\right), \tag{4.72}$$

$$n(d) = N_c \exp\left(\frac{-\phi_2}{k_B T}\right), \tag{4.73}$$

$$q\psi(0) = -\phi_1, \tag{4.74}$$

$$q\psi(d) = -(\phi_2 + qV), \tag{4.75}$$

where N_c is the effective density of states in conduction band, Equation 4.71 gives

$$J_n = \frac{qD_n N_c(\exp(qV/kT) - 1)}{\int_0^d \exp(-q\psi(x)/kT)dx} \tag{4.76}$$

Assuming $(\phi_1 - \phi_2) = qV_{bi}$ and using Equation 4.68, the denominator of Equation 4.76 can be written as

$$\int_0^d \exp\left(\frac{-q\psi(x)}{k_B T}\right)dx = \frac{dk_B T}{q(V_{bi} - V)}\exp\left(\frac{\phi_1}{k_B T}\right)\left(1 - \exp\left(\frac{-q(V_{bi} - V)}{k_B T}\right)\right). \tag{4.77}$$

Now Equation 4.76 can be written as

$$J_n = \frac{q^2 D_n N_c (V_{bi} - V)\exp(-\phi_1/k_B T)(\exp(qV/k_B T) - 1)}{dk_B T\left(1 - \exp(-q(V_{bi} - V)/k_B T)\right)}. \tag{4.78}$$

A similar expression exists for the hole current with corresponding energy barriers. The diffusion and drift components of the current can also be calculated independently. Figure 4.17a shows the plot of diffusion and drift components of the electron current as a function of x. The currents had been calculated for 0.5 V and the values of rest of the parameters are given in the figure caption. The two currents flow in opposite direction. Figure 4.17b shows the terminal current (J_{term}) (which is the difference of diffusion and drift components) as a function of x, at different voltages. For the given set of parameters, the diffusion current is observed more than the drift current at low voltages, and as expected the terminal current is observed to be independent of x and increases with the applied voltage. This was the analysis of dark current only. To get the illuminated current, the photocurrent was added to the dark current.

4.7.1 Equivalent Circuit Model

The equivalent circuit model of a solar cell is important to understand the nature of characteristics and its qualitative and quantitative losses. Although the basic principle is the same, OSCs differ in many ways from conventional c-Si solar cells, namely, materials, structural design, active layer possessing, large excitons binding energy, low charge carrier mobilities, and rapid degradation. To a reasonable limit, the conventional circuit model (one-diode model) has also been used to interpret the behavior of OSCs [121,122]. Since diode and photovoltaic characteristics of OSCs happen to be quite different from

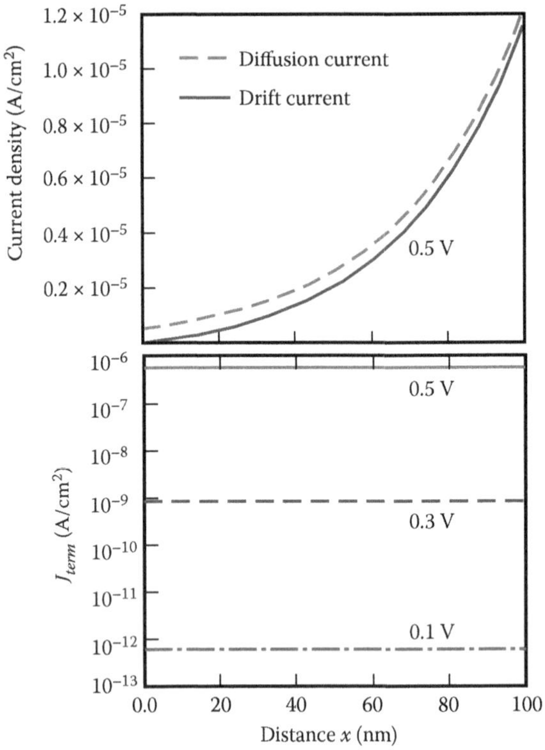

FIGURE 4.17
(a) Diffusion and drift currents at 300 K in the double Schottky device at 0.5 V and (b) terminal currents (J_{term}) at different applied voltages as a function of x. The values of the parameters used in calculation are $\phi_1 = 0.88$ V, $\phi_2 = 0.30$ V, $d = 100$ nm, $N_c = 1 \times 10^{19}$ cm^{-3}, and $\mu = 1 \times 10^{-4}$ cm^2 V^{-1} s^{-1}. (Reprinted with permission from P. Kumar et al., A model for the *J–V* characteristics of P3HT:PCBM solar cells, *J. Appl. Phys.* 105, 104507. Copyright 2009, American Institute of Physics.)

those of inorganic counterparts, different equivalent circuit models were developed to explain the characteristics of OSCs [39,123–126]. The single-diode model [120,121], two-diode model [127], and three-diode model [128] are the most important models and are shown in Figure 4.18. The light contribution in these models is represented by a constant current source, generating a photocurrent density J_{ph}. Single-diode and two-diode models are extensively used for inorganic solar cells [41]. The two-diode model consists of an additional diode connected in parallel to the original diode in the same polarity. For the second diode, the reverse saturation current density corresponds to the deep-level recombination in the space charge region and is generally 3–7 orders of magnitude larger than that for the original diode. Mazhari presented an improved three-diode model for OSCs under illumination [128]. My colleagues and I analyzed the validity of these circuit models and found that all of these models had their limitations and did not explain some of the important features of OSCs [129]. With detailed investigations and deep analysis of *J–V* characteristics of the solar cells based on the P3HT:PCBM blend, an improved equivalent circuit model was developed. The improved circuit model proved quite useful in interpreting many important features of OSCs.

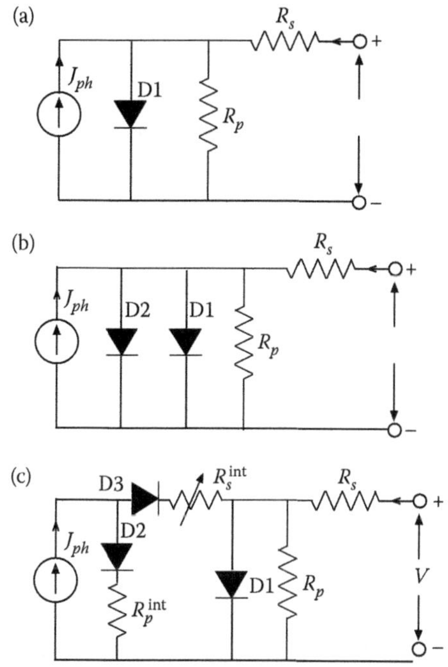

FIGURE 4.18
Equivalent circuits of (a) one-diode, (b) two-diode, and (c) three-diode models for solar cells. (Reproduced with permission from A. Gaur and P. Kumar, 2014. *Prog. Photovolt.: Res. Appl.*, 22: 937.)

4.7.2 Conventional One-Diode Circuit Model

Before understanding the behavior of OSCs, let us have a look at the characteristics and circuit model of conventional inorganic solar cells. For a conventional Si p-n junction solar cell, the J–V relation in the dark is given by [41]

$$J = J_s\left(\exp\left(\frac{q(V - JAR_s)}{nk_BT}\right) - 1\right) + \frac{V - JAR_s}{AR_p}, \tag{4.79}$$

where J_s is the reverse saturation current density, n is the ideality factor, R_s and R_p are the parasitic series and shunt resistances, respectively, and A is the active area of the cell. Under illumination J–V relation modifies to [41],

$$J = J_s\left(\exp\left(\frac{q(V - JAR_s)}{nkT}\right) - 1\right) + \frac{V - JAR_s}{AR_p} - J_{ph}, \tag{4.80}$$

where J_{ph} is the photogenerated current density in the cell. In the dark $J_{ph} = 0$. It is assumed here that for a given illumination intensity, the photocurrent remains constant in magnitude and is independent of applied voltage. This equation is based on the superposition principle of current densities, and in forward bias the direction of flow of the photocurrent is opposite that of the dark current. For a given set of parameters, Figure 4.19 shows the

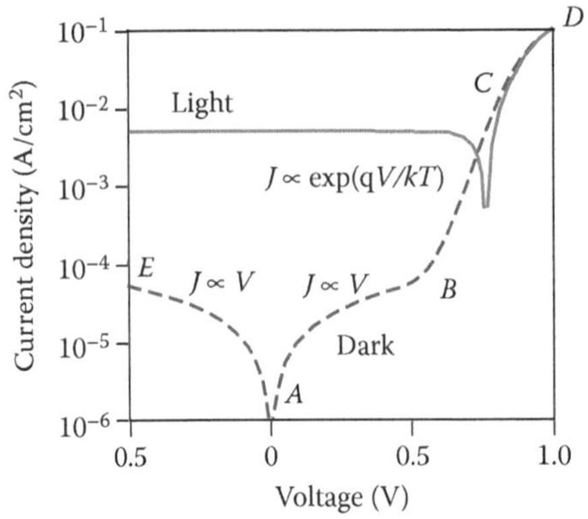

FIGURE 4.19
Calculated dark (dashed) and illuminated (solid) J–V characteristics of a Si solar cell with $J_s = 3.38 \times 10^{-11}$ A/cm², $n = 1.6$, $R_s = 10\ \Omega$, $R_p = 1 \times 10^5\ \Omega$, $A = 0.1$ cm², $J_{ph} = 5$ mA/cm², and T = 295 K. (Reproduced with permission from A. Gaur and P. Kumar, 2004. *Prog. Photovolt.: Res. Appl.* 22: 937.)

plots of Equation 4.80 in the dark and under illumination. The corresponding equivalent circuit is shown in Figure 4.18a. The forward bias characteristic in the dark can be divided into three regions, mainly *A–B*, *B–C*, and *C–D*. Region *A–B* shows ohmic conduction and is because of leakage current through R_p until the current through diode D1 is sufficiently large. Region *B–C* shows exponential variation of current with voltage and is controlled by diode D1, whereas in region *C–D* the current becomes very large and the potential drop across R_s becomes important, therefore the current no longer remains exponential and is controlled by R_s along with diode D1. In reverse bias the characteristic shows ohmic nature in the entire voltage range *A–E* and is because of leakage current through R_p only, as D1 is OFF and the leakage current is more than the reverse saturation current density of D1.

In conventional solar cells J_{sc} is supposed to be equal to the photocurrent of the cell (J_{ph}), however, it is not true as some part of photocurrent would recombine through R_p. If R_p is very large and R_s is very low, J_{sc} may be approximately equal to J_{ph}. V_{oc} in conventional solar cells can be found from

$$J_s\left(\exp\left(\frac{qV_{oc}}{nkT}\right)-1\right)+\frac{V_{oc}}{AR_p}-J_{ph}=0. \tag{4.81}$$

Calculations show that for low values of R_p, V_{oc} becomes small as the large part of the photocurrent would pass through R_p, and photocurrent contribution through R_s would decrease. Therefore, to make the total current zero at output (condition for V_{oc}), a small applied voltage will be required. For smaller values of R_p, V_{oc} increases with R_p, however, for large values of R_p, V_{oc} gets saturated. R_s is found to have no effect on V_{oc} but on the photocurrent extracted out of the cell. Based on the superposition principle, the photocurrent contribution extracted out of the cell (J'_{ph}) can be obtained by subtracting the dark current from the illuminated current. Figure 4.20 shows the variation in J'_{ph} as a function of applied voltage, obtained from the difference of two currents from Equations 4.80 and 4.79. It is seen that J'_{ph} remains equal to J_{ph} in reverse bias, whereas in forward bias it remains almost equal to J_{ph} for low voltages and decreases as the applied voltage increases. It is because when in reverse bias D1 is OFF and the current through R_p is small, J'_{ph} may be equal to J_{ph}, but when in forward bias D1 is ON, a large part of the photocurrent would recombine through D1

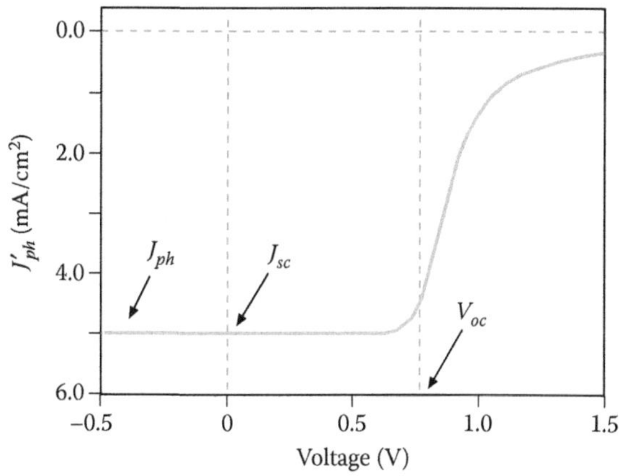

FIGURE 4.20
Photocurrent contribution (J'_{ph}) in a diode under illumination, obtained by subtracting the dark current from the light current of Figure 4.19. (Reproduced with permission from A. Gaur and P. Kumar, 2014. *Prog. Photovolt.: Res. Appl.* 22: 937.)

and R_p giving a reduced J'_{ph}. Or in other words, under illumination some additional current would flow through R_p and D1, which will increase the voltage drop across R_p, therefore the voltage drop across and the current through R_s would decrease. At V_{oc}, J'_{ph} is equal and opposite to the dark current through R_s due to applied voltage, which results in a net zero current through R_s. When the applied voltage increases beyond V_{oc}, J'_{ph} decreases first rapidly and then very slowly with voltage, and at very large voltage it becomes negligible. It is worth noting that for $V > 0$, the photocurrent generated in the cell (J_{ph}) is not actually changing but it is constant, and only its contribution at the output terminals (J'_{ph}) changes with the applied voltage due to some recombination through D1 and R_p.

4.7.3 Improved Circuit Model for OSCs

Charge transport in organic materials is expected to be space charge limited [40], however, at low voltages the effect of V_{bi} due to asymmetric electrodes and diffusion of charge carriers cannot be ignored. Calculations show that if the contacts are not ohmic in the MIM diode, then the internal electric field remains constant for low applied voltages; however, at large applied voltages the effect of diffusion may become negligible and the field distribution will no longer remain constant. For ohmic contacts the field will not be constant in the vicinity of the contact, however, it will be constant far away from the contacts [118,130]. The solution of current density and Poisson's equations for electrons in organic semiconductors, in steady state and in the low voltage region gives Equation 4.78. In a semiconductor if recombination/generation is considered it can be either band to band or trap assisted. Since we have considered electron and hole transport through different channels, the probability of band-to-band recombination/generation is reduced and trap-assisted recombination/generation becomes more probable. The current due to trap-assisted recombination can be given by integration of trap-assisted Shockley–Read–Hall recombination rate [41], and now Equation 4.78 would modify to

$$J_n = J_{n0}(V)\left(\exp\left(\frac{qV}{nk_BT}\right)-1\right), \tag{4.82}$$

where

$$J_{n0}(V) = \frac{q^2 D_n N_c (V_{bi} - V)\exp(-\phi_1 / nk_B T)}{dnk_B T\left(1 - \exp\left((-q(V_{bi} - V))/nk_B T\right)\right)}$$

and n is the ideality factor. Hole current density can also be given by a similar expression and the total current through the diode would be given by

$$J = \left(J_{n0}(V) + J_{p0}(V)\right)\left(\exp\left(\frac{qV}{nk_B T}\right) - 1\right), \quad (4.83)$$

or

$$J = J_s(V)\left(\exp\left(\frac{qV}{nk_B T}\right) - 1\right) \quad (4.84)$$

where $J_s(V) = J_{n0}(V) + J_{p0}(V)$. For a practical case where parasitic resistances R_s and R_p become important, the current through the diode would be given by

$$J = J_s(V - JAR_s)\left(\exp\left(\frac{q(V - JAR_s)}{nk_B T}\right) - 1\right) + \frac{V - JAR_s}{AR_p}. \quad (4.85)$$

R_s and R_p in OSCs come from the resistivity of organic materials, contact metal electrodes, and metal–organic interfaces. Equation 4.85 is similar to Equation 4.79, except the reverse bias current density is not constant but a function of applied voltage. Equation 4.85 explains well the nature of dark characteristics of OSCs in forward bias, as that explained by Equation 4.79 for Si solar cells. However at very high voltages, the effect of diffusion will become negligible and the charge carrier transport would be controlled by SCLC conduction [131], and Equation 4.85 will no longer be valid. Efforts were made to fit the experimentally measured data of P3HT:PCBM solar cell with Equation 4.85. Even for the best fitting of Equation 4.85 with experimental data in forward bias, the plot of Equation 4.85 did not follow the experimental data in reverse bias. However, Equation 4.85 showed a reasonable fit with experimental data in reverse bias for low voltages only. The low voltage region in reverse bias corresponded to ohmic conduction, which was in good agreement with Equation 4.85, but at higher voltages the current varied exponentially with applied voltage (Schockley kind). Investigations on different cells, prepared via different processing conditions, revealed that the nature of characteristics depends on the quality of cell. In some cases this exponential variation may not happen even for very large voltages and in some cases the variation may not be exponential but space charge limited. This different nature of characteristics in reverse bias forced us to rethink about the issue and to develop a suitable model.

Such a nature of characteristics could be explained by introduction of another diode D2 in the equivalent circuit, connected in parallel but in opposite polarity to the existing diode D1 as shown in Figure 4.21. Due to the nonlinear nature of characteristics in reverse bias, the use of a diode is justified. The current through diode D2 may be exponential in voltage or space charge limited depending upon the cell structure and its properties. Diode D2 will have different values of parameters than those for D1. For low reverse bias, the leakage current through R_p would be more than the current through D2, however, for

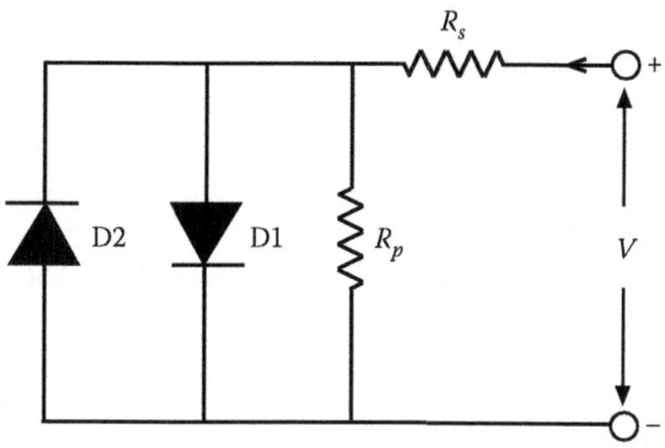

FIGURE 4.21
Improved equivalent circuit for OSCs in the dark.

high reverse bias the current through D2 would dominate. It is worth mentioning that even with pure ohmic conduction in reverse bias, the characteristics can be modeled by introduction of D2, though the current through it would be negligible in that case. As in the present case, characteristics in reverse bias show exponential variation with voltage, the Equation 4.85 modifies to

$$J = J_{s1}(V - JAR_s)\left(\exp\left(\frac{q(V - JAR_s)}{n_1 k_B T}\right) - 1\right) - J_{s2}\left(\exp\left(\frac{-q(V - JAR_s)}{n_2 k_B T}\right) - 1\right) + \frac{V - JAR_s}{AR_p}, \quad (4.86)$$

where $J_{s1}(V - JAR_s)$ and J_{s2} are, respectively, the reverse bias current densities in D1 and D2, and n_1 and n_2 are the ideality factors of the respective diodes. Figure 4.22 shows the

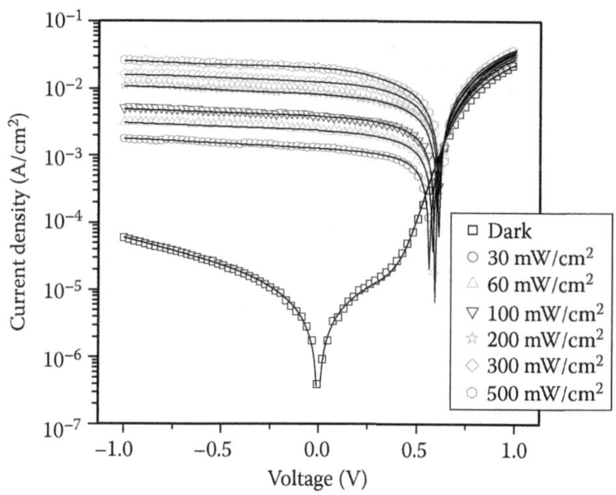

FIGURE 4.22
Dark and illuminated *J–V* characteristics (symbols) of the ITO/PEDOT:PSS/P3HT:PCBM/Al solar cell measure at room temperature in ambient environment. The solid curves are the plot of Equation 4.87 for different illumination conditions where J_{ph} (*V*) is given by Equation 4.88. (Reproduced with permission from A. Gaur and P. Kumar, 2014. *Prog. Photovolt.: Res. Appl.* 22: 937.)

experimental dark and illuminated *J*–*V* characteristics of P3HT:PCBM solar cell at different illumination intensities (symbols). The lowest solid curve is the fit of Equation 4.86 to dark characteristics. An excellent agreement is observed with dark *J*–*V* characteristics in both the forward and reverse biases. The corresponding values of parameters are $N_c = N_v = 10^{19}$ cm^{-3}, $d = 104$ nm, $\mu = 2.6 \times 10^{-4}$ cm^2 V^{-1} s^{-1}, $\phi_1 = 1.3$ eV, $\phi_2 = 0.87$ eV, $V_{bi} = 0.66$ eV, $n_1 = 1.35$, $R_s = 1\ \Omega$, $R_p = 1.23 \times 10^6\ \Omega$, $A = 0.04$ cm^2, $T = 295$ K, $J_{s2} = 4 \times 10^{-7}$ A/cm^2, and $n_2 = 12$. Note that for different nature of characteristics in reverse bias, the second term in Equation 4.86 (corresponding to D2) would have to modify accordingly and the values of parameters will be different.

To model the illuminated *J*–*V* characteristics, it is of utmost important to calculate the photogenerated current first in OSCs. Photocurrent in an OSC can be obtained by calculating the EQE, where optical losses, actual absorption profile, charge carrier generation, and losses during transportation play important roles [132–134]. Once the EQE is known, the photocurrent generated in the cell can be calculated by integrating EQE over the absorption spectral range. The central assumption of conventional model that J_{ph} is constant for a given illumination and independent of applied voltage does not hold for OSCs. The photocurrent in OSCs is constant for constant illumination, but it varies with applied voltage. In order to examine the validity of this assertion, we did calculate the J'_{ph} for different illumination intensities and the same is shown in Figure 4.23 as a function of applied voltage. J'_{ph} was calculated by subtracting the dark current from the current under illumination. It was found that for all illumination intensities, J'_{ph} remained negative in direction for $V < 0$ and varied slightly linearly, whereas for $V > 0$ it decreased first slowly and then rapidly with applied voltage. For all illumination intensities J'_{ph} became zero at 0.66 V and for $V > 0.66$ V, it became positive in direction and started increasing with applied voltage. These results were quite contrary to those observed in Figure 4.20. Also according to the model proposed by Mazhari, J'_{ph} can be zero but the direction will not change [128]. Therefore, due to the vanishing character for $0 < V < 0.66$ V and change in direction for $V > 0.66$ V, J_{ph} could not be considered constant with applied voltage, as assumed by many authors [121,122,135].

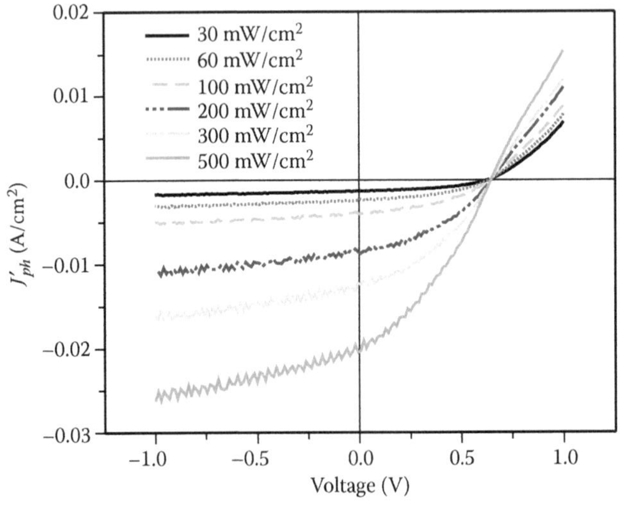

FIGURE 4.23
Variation of extracted photocurrent contribution (J'_{ph}) with applied voltage at different illumination intensities. (Reproduced with permission from A. Gaur and P. Kumar, 2014. *Prog. Photovolt.: Res. Appl.* 22: 937.)

These results suggested J_{ph} to be a function of applied voltage. Therefore, under illumination the J–V relation for OSCs could be written as

$$J = J_{s1}(V - JAR_s)\left(\exp\left(\frac{q(V - JAR_s)}{n_1 k_B T}\right) - 1\right) - J_{s2}\left(\exp\left(\frac{-q(V - JAR_s)}{n_2 k_B T}\right) - 1\right) + \frac{V - JAR_s}{AR_p} - J_{ph}(V), \tag{4.87}$$

where $J_{ph}(V)$ is the photocurrent generated in the cell. The function that defines the relation between J_{ph} and V is important to understand. The relation between J_{ph} and V was established by critically analyzing the J–V characteristics under illumination.

It is worth mentioning that in BHJ solar cells the exciton dissociation efficiency is ~100%, but after exciton dissociation, the charge collection efficiency depends on the drift length of the charge carriers, which is given by $\mu\tau E_{eff}$, where μ is the carrier mobility, τ is the carrier lifetime, and E_{eff} is the effective electric field in the cell, which can be written as $(V_{bi} - V)/d$. Along with diffusion, the internal electric field provides necessary force to collect the charge carriers at respective electrodes. Charge carriers do not recombine within the drift length, but beyond the drift length charge carriers do not survive and recombine through traps. If drift length of charge carriers is more than the thickness of the active layer, then only all photogenerated charge carriers are expected to be collected, otherwise only the charge carriers within the drift length will be collected and the rest will recombine. For $V = V_{bi}$, E_{eff} will vanish and there will be no extraction of photogenerated charge carriers (if diffusion is ignored) and all of them will recombine. This is exactly the situation seen in Figure 4.23, where the photocurrent became zero at 0.66 V. For $V > 0.66$ V, the direction of the photocurrent is reversed and it is consistent with the situation that for $V > V_{bi}$ the direction of E_{eff} is reversed and because of that the direction of charge collection is reversed. These observations suggest J_{ph} to be a function of E_{eff}. Based on the drift length concept, Brabec et al. proposed J_{ph} to be a linear function of E_{eff} given by $J''_{ph}(\mu\tau E_{eff}/d^2)$, where J''_{ph} is the maximum possible photocurrent generated in the cell and is constant for a given illumination intensity and independent of applied voltage [120]. If $\mu\tau E_{eff} > d$, J_{ph} will be equal to J''_{ph}, though the direction will depend on the direction of E_{eff}, which will depend on the magnitude of applied voltage V. For $V < 0$, the photocurrent will flow in the same direction as that of the dark current. Though for $0 < V < V_{bi}$, the magnitude of the photocurrent would decrease but the direction will still be the same as that of the dark current. For $V > V_{bi}$ the direction of the photocurrent will change and the magnitude will increase with applied voltage until it saturates. It is important to note that the photocurrent usually becomes zero at $V = V_{bi}$, but it is not a necessary condition for all organic solar cells. For example, at $V = V_{bi}$ the drift component of the photocurrent becomes zero but even then we may have photocurrent due to diffusion of charge carriers. This is an important situation for bilayer solar cells. In that case V_{oc} will be more than V_{bi} and the intersection voltage of dark and illuminated characteristics, where the photocurrent becomes zero, will be more than V_{bi}. If diffusion is not important, then the photocurrent is determined by the drift and the intersection point will be equal to V_{bi} and in that case V_{oc} will be less than V_{bi}. In BHJ solar cells diffusion becomes negligible if the intensity distribution is uniform. As we are modeling the characteristics for BHJ solar cells, where charges are generated throughout the cell, we can ignore the effect of diffusion of charge carriers, and the photocurrent becoming zero at $V = V_{bi}$ would be a good approximation.

According to the aforementioned model, J_{ph} is expected to be equal to J''_{ph} for high internal electric fields. Calculations show that even at $V = 0$ V, the internal electric field in the present sample due to V_{bi} was sufficient enough for $\mu\tau E_{eff} > d$ and the photocurrent was expected to be saturated at 0 V, but rather it varies slightly linearly in reverse bias. This was because of some polaron–pair recombination at the donor–acceptor interface due to coulombic attraction between electrons and holes. Now, the application of additional reverse bias would provide sufficient force for collection of all photogenerated charge carries and give a saturated photocurrent. This statement was proved with the measurements performed at a relatively higher voltage range. Figure 4.24 shows dark and illuminated *J–V* characteristics of the cell measured in the voltage range of ±3.5 V. Variation in J'_{ph} with applied voltage in this case is shown in the inset of the figure. In reverse bias, initially for up to ~−1.2 V, J'_{ph} varies slightly linearly with applied voltage and then interestingly it saturates at ~−1.5 V. Two solid lines have been plotted in the inset just to show the linear variation and saturation in J_{ph}'. But more interesting, for higher voltages beyond −2.7 V, J'_{ph} started decreasing. Do not think that this reduction is because of the reduction in J_{ph} after saturation; it is because J_{ph} is saturated at J''_{ph} but the dark current is still increasing with applied voltage, which results in reduction in the difference of two currents and J'_{ph} tends to zero. In forward bias J'_{ph} varied nonlinearly in the fourth quadrant, where it decreased first slowly and then rapidly with applied voltage and became zero at 0.66 V. As the voltage increased further, J'_{ph} got reversed in direction and started increasing rapidly. These results are consistent with those observed for the lower voltage range (see Figure 4.23). It was interesting that after ~1.7 V, J'_{ph} started decreasing and could be attributed to the same effect observed in reverse bias beyond −2.7 V, where dark current increased with applied voltage and photocurrent saturated therefore J'_{ph} decreased and tended to zero.

We tried to fit the characteristics measured under illumination with Equation 4.87 with the same values of parameters used in the dark and considering J_{ph} a linear function of E_{eff}. The characteristics did not show agreement in forward bias ($V > 0$) but a reasonably good fit was observed in reverse bias ($V < 0$). However, a rough fitting was observed for $0 < V < V_{bi}$ with a slight reduction in shunt resistance. As we can see that for $0 < V < V_{bi}$, J'_{ph} varies nonlinearly with applied voltage, therefore the assumption that J_{ph} varies linearly

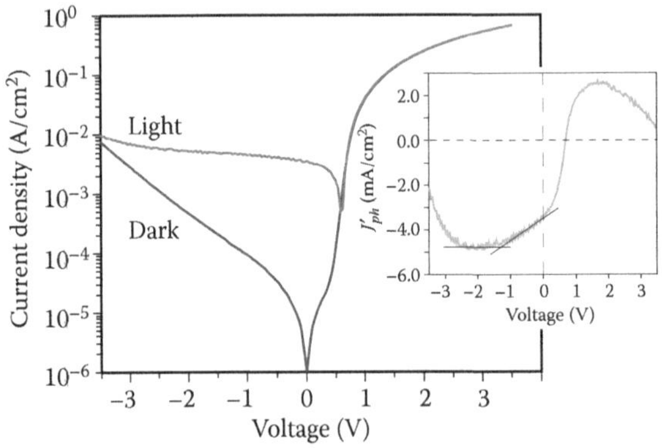

FIGURE 4.24
Dark and illuminated (100 mW/cm²) *J–V* characteristics of the ITO/PEDOT:PSS/P3HT:PCBM/Al solar cell with higher voltage range. Inset shows the variation of extracted photocurrent contribution (J'_{ph}) with applied voltage. Solid lines in the inset were plotted to guide the eyes about trend in variation. (Reproduced with permission from A. Gaur and P. Kumar, 2014. *Prog. Photovolt.: Res. Appl.* 22: 937.)

with applied voltage gave a rough fitting. An excellent fitting was observed for $V < 0$ and $V < V_{bi}$ by considering J_{ph}, an exponential function of $(V_{bi} - V)$, of the form

$$J''_{ph}\left(1-\exp\left(\frac{V-V_{bi}-J_{ph}(V)AR_{s2}}{n_3k_BT}\right)\right),$$

where n_3 is the ideality factor and R_{s2} is the effective series resistance under illumination. The fitting was excellent for all illumination intensities and with the same values of the diode parameter as those in the dark, except reduction in shunt resistance and increase in series resistance with illumination intensity. As illumination intensity increases, shunt resistance reduces further and series resistance increases. Reduction in shunt resistance under illumination was accepted, as it had also been observed by many authors [118,120]. Reduction in shunt resistance could be attributed to photoinduced doping and could be represented by adding a parallel resistance R_{p2} where photoconductivity is proportional to incident illumination intensity. In order to estimate the contribution of this effect to the existing shunt resistance the photoconductivity had to be weighted by the probability of the charge carriers of one type to penetrate the barrier presented by the selective electrodes. Waldauf et al. called this weighing factor the contact permeability (cp) [136]. The cp could be estimated by curve fitting the J–V characteristics of an illuminated solar cell under reverse bias. On the other hand, increase in series resistance could be attributed to difficulty in extraction of charge carriers due to space charge effect, which increases with illumination intensity and could be represented by adding a series resistance R_{s2} to the photocurrent extraction. For $V > V_{bi}$, the characteristics did show agreement with none of the above functions. Since for $V > V_{bi}$ the direction of photocurrent extraction was changed, the photogenerated electrons and holes collected at opposite electrodes. Though in principle there should have not be any change in photovoltaic behavior, but as the interfaces for charge extraction were changed the photophysical properties of the cell changed and even the magnitude of the extracted photocurrent changed. However, excellent agreement could be achieved with the experimental data by considering J_{ph} an exponential function of the form

$$-J''_{ph}\left(1-\exp\left(\frac{V-V_{bi}-J_{ph}(V)AR_{s2}}{n_4k_BT}\right)\right)$$

until it saturated to some multiple of J''_{ph}. Here n_4 is the ideality factor. It is the same function used for $V < V_{bi}$, except the ideality factor was different. With these assumptions we were able to have excellent fit for $V > V_{bi}$ for all illuminated intensities. As a whole, $J_{ph}(V)$ could be written as

$$J_{ph}(V) = J''_{ph} \quad \text{for } V < 0,$$

$$J_{ph}(V) = J''_{ph}\left(1-\exp\left(\frac{V-V_{bi}-J_{ph}(V)AR_{s2}}{n_3k_BT}\right)\right) \quad \text{for } 0 < V < V_{bi}$$

and

$$J_{ph}(V) = -J''_{ph}\left(1-\exp\left(\frac{V-V_{bi}-J_{ph}(V)AR_{s2}}{n_4k_BT}\right)\right) \quad \text{for } V > V_{bi}. \tag{4.88}$$

TABLE 4.1

Variation in Fitted Values of R_s and R_p with Illumination Intensity

Illumination Intensity (mW/cm²)	R_s (Ω)	R_p (Ω)
30	1	1.0×10^5
60	10	7.8×10^4
100	60	4.9×10^4
200	120	2.5×10^4
300	160	2.2×10^4
500	180	1.6×10^4

Source: Reproduced with permission from A. Gaur and Kumar, P., *Prog. Photovolt.: Res. Appl.* 22 (2014): 937.

Dark solid curves in Figure 4.22 are the plots of Equation 4.87 where J_{ph} (V) is given by Equation 4.88. The modeled characteristics show excellent agreement with experimental data at different illumination intensities. The values of R_s and R_p for different illumination intensities are given in Table 4.1. Values for the rest of the parameters were the same as those in the dark with $n_3 = 6.7$ and $n_4 = 8.0$. While calculating the photocurrent, we considered equal contribution from both photogenerated electrons and holes.

Based on the aforementioned experimental observations the dark equivalent circuit model (Figure 4.21) was extended to that for under illumination and is shown in Figure 4.25. Under illumination we may consider the addition of a current source to the dark circuit with recombination and collection provisions. Since photophysical properties of studied cells were different for $V < V_{bi}$ and $V > V_{bi}$, the circuit could be modeled by considering two independent light sources, one of which contributes for $V < V_{bi}$ and other for $V > V_{bi}$. The two sources were connected in such a way that one was active for $V < V_{bi}$ and the other for $V > V_{bi}$. In that case the two sources were considered connected through an ideal Zener diode with Zener voltage equal to V_{bi}, which was connected in reverse bias such that it was short circuited for $V > V_{bi}$ and open circuited for $0 < V < V_{bi}$. As discussed

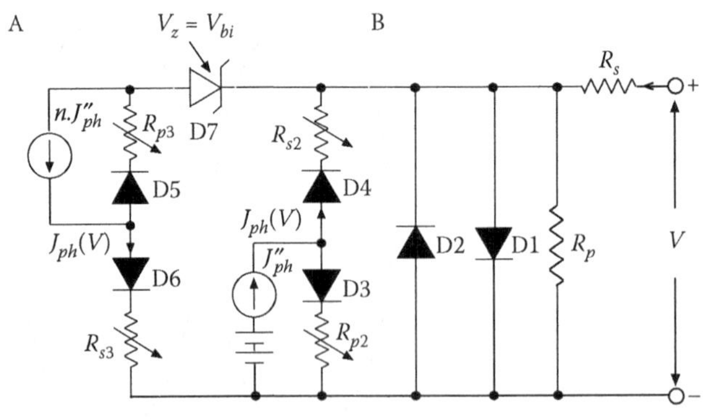

FIGURE 4.25
Improved equivalent circuit for OCs under illumination. (Reproduced with permission from A. Gaur and P. Kumar, 2014. *Prog. Photovolt.: Res. Appl.* 22: 937.)

earlier recombination and collection happens in both cases, therefore such behavior could be described by introduction of two additional diodes along with series and shunt resistances such that one diode was for recombination losses and the other was for charge extraction [128].

To understand in detail we take the two cases separately; the first case is $V < V_{bi}$. As discussed earlier, the illumination part cannot be represented by a constant current source, but it should be a source whose output changes with applied voltage until it reaches its maximum value, which is constant for a given illumination intensity. The maximum photocurrent corresponds to the net generation rate of excitons in the cell. Therefore, the current source (encircled portion B, Figure 4.25) comprised a voltage source with potential nearly equal to V_{bi} and a reservoir of charge carriers, which could allow the charge carriers to flow, giving a maximum current of J''_{ph}. For low internal electric fields, all the charge carriers will not be extracted out, and recombination will happen at the dissociating interface and in the bulk of the active layer. Therefore, recombination and extraction of charge carriers were accounted through photoinduced R_{p2} and R_{s2} along with diodes D3 and D4, in such a way that it represented characteristics of PSCs. Use of photoinduced R_{p2} and R_{s2} was essential to justify the reduction in shunt resistance under illumination. The value of R_{p2} changed with illumination intensity. Exponential variation of photocurrent could be justified by the use of diodes. The diodes D3 and D4 were connected in such a way that they became important under illumination only. In the dark, these diodes were reverse biased and had no impact on J–V characteristics of the cell. The current through D3 along with shunt resistance R_{p2} corresponded to the recombination current, whereas the current through D4 along with series resistance R_{s2} corresponded to $J_{ph}(V)$. In the short-circuit condition, the voltage drop across D3 will be ~V_{bi}, and it will result into current flow in the entire circuit through D4, and J_{sc} will be given by

$$J_{sc} = J_{ph}(0) - J_{RD1}(0) - J_{RR_p}(0), \tag{4.89}$$

where $J_{RD1}(0)$ corresponds to the recombination current through D1 and $J_{RR_p}(0)$ represents the recombination current through R_p at zero applied voltage. Note that $J_{ph}(0)$ will not be equal to J''_{ph} because of some of the recombination losses through D3 and R_{p2}. If R_{p2} is small the internal recombination will be large, and $J_{ph}(0)$ will be significantly less than J''_{ph}, which will result into reduced J_{sc}. Therefore to get maximum J_{sc}, the internal recombination losses should be minimized. The recombination losses can be minimized by using the materials of high charge carrier mobilities with optimum nanoscale morphology. However for a given case in reverse bias ($V < 0$), as the applied voltage would increase, the voltage across D3 would decrease, and eventually when D3 is OFF the recombination through D3 would become zero and we would get the maximum extractable photocurrent J''_{ph}. This is consistent with the results observed in Figure 4.24. If the current near the short-circuit condition shows good saturation, J_{sc} will be equal to J''_{ph}. If the current does not show saturation at the short-circuit condition, it represents great recombination losses, which should be minimized to get high efficiencies. For $0 < V < V_{bi}$, the voltage drop across D4 would decrease, which will result in a reduction in output photocurrent contribution (J'_{ph}) and an increase in recombination losses. However, the nonlinear reduction in J'_{ph} in this region (see Figure 4.23) can be understood in terms of the recombination of charge carriers through D3, R_{p2}, D1, and R_p, and exponential variation of J_{ph} through D4. At a particular +ve voltage, the voltage drop across D4 would become zero and there will be no extraction of photocurrent and all the photogenerated charge carriers will recombine through D3 and R_{p2}. In this situation $J_{ph}(V)$ would become zero resulting in $J'_{ph} = 0$. This is the situation that will happen at

$V = V_{bi}$, which is 0.66 V in the present case (see Figure 4.23). In the open-circuit condition, J'_{ph} will be equal and opposite to the current due to applied voltage, which will result in a net zero current through R_s. Now V_{oc} of the cell can be determined from

$$J_{s1}(V_{oc})\left(\exp\left(\frac{qV_{oc}}{n_1 k_B T}\right)-1\right)+\frac{V_{oc}}{AR_p}+J_{ph}(V_{oc})=0. \tag{4.90}$$

For the open-circuit condition, D2 will be OFF and the current through it will be negligible. Therefore at any given illumination intensity, the maximum V_{oc} can be obtained by extracting the maximum photocurrent via minimizing the recombination losses. The recombination losses can be minimized by increasing R_p and R_{p2} through improved nanoscale morphology or by reducing R_{s2} through improved charge carrier mobilities and reduced trap density. For no recombination losses, the photocurrent would be maximum and we will get maximum V_{oc}.

For $V > V_{bi}$, D4 will be reverse biased and no photocurrent would be extracted out through this circuit and it becomes passive. Now in that case the second circuit component (encircled portion A, Figure 4.25) would become active, as the Zener diode D7 would now be short circuited. Now the direction of flow of photocurrent will get reversed and it will add to the dark current. In this case the current source comprises a reservoir of charge carriers but not the voltage source. The charge carriers are extracted out by the applied electric field. The maximum current through this reservoir can be some multiple of J''_{ph}. Though for $V > V_{bi}$ the direction of photocurrent is changed, but recombination and collection will happen in this case as well. Similar to the previous case, this situation could also be described by introduction of two more diodes such that one diode (D5) was for recombination losses and the other (D6) was for charge extraction. D5 and D6 were connected in such a way that they had importance under illumination only. In the dark they were reverse biased and did not contribute. The current through D5 along with R_{p3} corresponded to recombination losses and current through D6 along with R_{s3} corresponded to the extracted current J_{ph} (V). In principle R_{s3} and R_{p3} are equal to R_{s2} and R_{p2}, respectively. Due to exponential variation of the photocurrent with applied voltage, the use of diodes is justified. For $V > V_{bi}$, D5 will be in reverse biased and photogenerated charge carriers will be extracted out through D6. The extracted photocurrent increases exponentially with applied voltage until it saturates to some multiple of J''_{ph}. The change in series resistance under illumination could be explained by variation in photoinduced R_{p3} and R_{s3}. Variation in photoinduced series and shunt resistances cannot be ignored with variation in temperature, as it is a crucial parameter for determination of J_{sc}, V_{oc}, and *FF*.

Based on the aforementioned model we could explain various other important features observed for polymer solar cells (PSCs). Reduction in J'_{ph} at very high reverse bias (Figure 4.24) was because the photocurrent through D4 had become saturated and the current due to applied voltage through D2 increased rapidly, therefore the difference between the illuminated and dark current (J'_{ph}) decreased. In case of forward bias when photocurrent had saturated through D6, the current through D1 due to applied voltage increased very rapidly and the difference between light and dark current (J'_{ph}) decreased. J'_{ph} increased with slightly in illumination intensity and could be attributed to the increase in J''_{ph} due to the increased exciton generation rate. The characteristics were measured, even for larger voltage ranges (±5 V), but it could not explain the observed results. This can be attributed to the change in device physics at very high voltages, as current may become space charge limited and the aforementioned model will not be applicable. Figure 4.26 shows the

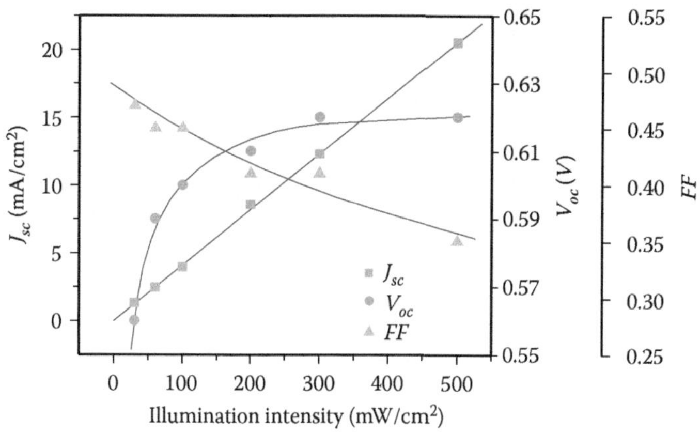

FIGURE 4.26
Variation in J_{sc}, V_{oc}, and FF with illumination intensity for the ITO/PEDOT:PSS/P3HT:PCBM/Al solar cell. (Reproduced with permission from A. Gaur and P. Kumar, 2014. *Prog. Photovolt.: Res. Appl.* 22: 937.)

measured J_{sc}, V_{oc}, and FF for different illumination intensities. J_{sc} increases linearly, and FF decreases almost linearly, whereas V_{oc} increases first rapidly and then saturates with illumination intensity. The saturated V_{oc} was less than V_{bi}. Increases in J_{sc} with increases in illumination intensity could be attributed to increases in $J_{ph}(0)$ due to increases in J''_{ph}. Calculations showed that the current under illumination exhibited poor saturation near the short-circuit condition for low charge carrier mobilities, however, saturation could be observed at large reverse bias. High charge carrier mobilities would result in saturation very near to the short-circuit condition. Increases in V_{oc} with illumination intensity could be attributed to increases in $J_{ph}(V_{oc})$, resulting in a large voltage drop across R_p in the open-circuit condition (see Equation 4.89). For very large illumination intensity V_{oc} would saturate but will remain less than V_{bi} as some voltage would drop across R_{s2} as well. Similarly, reduction in FF with illumination intensity could be attributed to reduction in shunt resistance and increase in series resistance. It was also observed that J_{sc} and V_{oc} changed with temperature, where J_{sc} decreased and V_{oc} increased with reduction in temperature [137]. Reduction in J_{sc} with temperature can be attributed to reduction in R_{p2} due to reduction in charge carrier mobilities. Reduction in R_{p2} would result in large recombination losses resulting reduction in $J_{ph}(0)$ and J_{sc}, whereas increase in V_{oc} can be attributed to increase in V_{bi} at low temperatures [138]. It is seen from the earlier model that J''_{ph} can be obtained in reverse bias with large applied voltages. Once J''_{ph} is known, R_{p2} can be calculated from the condition when J'_{ph} becomes zero at $V = V_{bi}$, as in that case the current through R_{p2} would be J''_{ph} and R_{p2} can be obtained from $(V_{bi} - JAR_s) / J''_{ph}$, where J is the output current at $V = V_{bi}$. Once R_{p2} is known R_{s2} can also be determined from Kirchhoff's law. Optimization of R_{s2} and R_{p2} would minimize the internal losses and would lead to efficient devices. Depending upon the device properties, the extrinsic series and shunt resistances would also have significant impact on the performance even with optimized internal resistances.

The aforementioned model could also explain degradation in OSCs and a good agreement was observed with the characteristics of degraded solar cells [139]. A comparison of Equations 4.86 and 4.87 with initial (0 h) dark and illuminated characteristics of the ITO/PEDOT:PSS/P3HT:PCBM/Al solar cell is shown in Figure 4.27. Symbols represent experimental data, whereas the dashed and solid curves are the plots of Equations 4.86 and

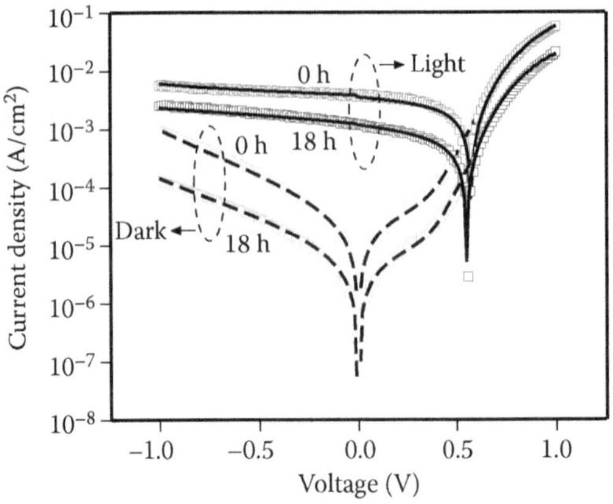

FIGURE 4.27
Comparison of initial (0 h) and after 18 h dark and illuminated *J–V* characteristics of the ITO/PEDOT:PSS/P3HT:PCBM/Al solar cell with Equations 4.86 and 4.87, where $J_{ph}(V)$ is given by Equation 4.88. Symbols represent experimental data, whereas the dashed and solid curves, respectively, correspond to the plots of Equations 4.86 and 4.87 for the respective cases. (Reprinted with permission from P. Kumar and A. Gaur, Model for the *J–V* characteristics of degraded polymer solar cells, *J. Appl. Phys.* 113, 94505. Copyright 2013, American Institute of Physics.)

4.87, respectively, where $J_{ph}(V)$ was given by Equation 4.88. The model exhibited an excellent agreement with experimental data for $N_c = 10^{19}$ cm^{-3}, $J_{s2} = 9.7 \times 10^{-6}$ A/cm^2, $R_s = 2\ \Omega$, $R_p = 3.0 \times 10^5\ \Omega$, $V_{bi} = 0.76$ V, $\phi_1 = 1.3$ eV, $\phi_2 = 0.92$ eV, $\mu = 8.9 \times 10^{-5}$ cm^2 V^{-1} s^{-1}, $n_1 = 1.9$, $n_2 = 10$, $n_3 = 11$, $n_4 = 11$, and $J''_{ph} = 3.8\,\text{mA/cm}^2$. Under illumination R_s and R_p changed to 5 and $1.9 \times 10^4\ \Omega$, respectively. The values of V_{bi}, ϕ_1, and ϕ_2 were calculated from the energy levels of organic and electrodes materials, values of n_1 and n_2 were calculated from the slope of dark characteristics in forward and reverse biases, respectively, and the value of R_p was calculated from the slope of the *J–V* characteristics in reverse bias near $V = 0$ V. While calculating the photocurrent, we assumed equal contribution from photogenerated electrons and holes with same charge carrier mobilities. Even after 18 h the characteristics showed excellent agreement with Equations 4.86 and 4.87 for $N_c = 10^{19}$ cm^{-3}, $J_{s2} = 1.9 \times 10^{-6}$ A/cm^2, $R_s = 4\ \Omega$, $R_p = 1.2 \times 10^6\ \Omega$, $V_{bi} = 0.76$ V, $\phi_1 = 1.3$ eV, $\phi_2 = 0.92$ eV, $\mu = 3.25 \times 10^{-5}$ cm^2 V^{-1} s^{-1}, $n_1 = 1.8$, $n_2 = 11$, $n_3 = 9$, $n_4 = 15$, and $J''_{ph} = 1.23\,\text{mA/cm}^2$. Under illumination R_s and R_p changed to 8 and $2.7 \times 10^4\ \Omega$, respectively. We observed reduction in J_{s2}, μ, and J_{sc}, whereas an increase in R_s and R_p, which was expected because of inferior electronic properties of the active layer due to degradation. The increase in R_p could be understood in terms of change in the nanoscale morphology, phase separation, and formation of dead zones in the active layer, which minimized shorts and reduced leakage current.

After ~41 h, *FF* reduced substantially and the illuminated characteristics exhibited S-shaped curves on the linear scale. Detailed analysis of dark and illuminated characteristics showed that the dark characteristics still follow Equation 4.86 with increased values of R_s, R_p, n_1, and n_2. Under illumination the *J–V* characteristics could be written as Equation 4.87, but the photocurrent contribution did not follow Equation 4.88. We also tried to fit the experimental data by considering J_{ph} to be a linear function of the effective electric field (E_{eff}), but no agreement was observed. Now it became very important to find out the function that defines relation between J_{ph} and *V*. In BHJ solar cells the exciton dissociation efficiency is ~100%, but due to chemical and physical degradation of the active layer, reduction in dissociation efficiency cannot be ignored. Because of the introduction of traps due to degradation, the mobility and lifetime of the charge carrier will decrease; also due to reduction in V_{bi}, the E_{eff} will decrease, which will result in reduction in drift length ($\mu\tau E_{eff}$) of the charge carriers. Due to reduction in drift and diffusion lengths, the extraction of charge carriers reduces and recombination losses increase. Most of the polaron–pair recombinations of charge carriers is expected at the donor–acceptor interface. Only the charge carriers in the vicinity of the electrodes will be collected and we would not get maximum photocurrent out of the cell; therefore, we observed rapid reduction in photocurrent extraction. However, maximum charge extraction was expected at high electric fields because it would provide sufficient force for dissociation of CSTs and collection of charge carries.

For *S*-shaped characteristics, the photocurrent for $V < 0$ was observed to vary almost linearly because of recombination losses, which could be represented by recombination through photoinduced shunt resistance. We analyzed the illuminated characteristics and J'_{ph} in detail for $V < V_{bi}$ and $V > V_{bi}$ and found that they could be explained only when $J_{ph}(V)$ varied as

$$J_{ph}(V) = J_{sc}\left(\frac{V_{bi} - V - V_{ind} + JAR_{s2}}{V_{bi}}\right)^{m} \quad \text{for } V < V_{bi},$$

where $m > 1$. Here, V_{ind} corresponded to the induced voltage due to the polarization effect [140]. Since for a degraded solar cell, we did not have maximum extraction of photogenerated charge carriers at zero voltage; we considered J_{sc} instead of J''_{ph}. For $V > V_{bi}$, the direction of photocurrent got reversed because the direction of E_{eff} was changed and flow of the photocurrent became in the same direction as that of the current due to applied voltage. The illuminated characteristics for $V > V_{bi}$ could be fitted only when $J_{ph}(V)$ varied as

$$J_{ph}(V) = J_{sc}\left(1 - \left(\frac{V_{bi} + V_{ind} - JAR_{s2}}{V_{bi}}\right)^{w}\right) \quad \text{for } V > V_{bi},$$

where $0 < w < 1$. This change in nature of $J_{ph}(V)$ was because of the change in charge transport properties of the active layer due to degradation, and the transport was expected to be SCLC conduction, due to trapping and different charge carrier mobilities. The charge carriers with low mobility will be extracted out of the device slowly and will form a space charge in the device [141]. The nature of space charge limited conduction changes with time due to changes in trap density and charge carrier mobilities.

Depending on the active layer thickness, doping, and difference in charge carrier mobilities, the space charge effect may occur even in fresh devices. In thicker devices there are the regions with zero electric field, which cause inefficient collection of photogenerated charge carriers. If charge carriers are not extracted out, they will recombine at the donor–acceptor interface or recombine through the traps. Therefore under illumination, the linear nature of the current in third quadrant is because of recombination of charge carriers at the donor–acceptor interfaces and through traps. During extraction, the charge carries recombine at several places resulting in a reduced extracted photo current. The nature of the current will depend upon recombination. The space charge will also inhibit the extraction to opposite charge carriers via attraction force. However, when the electric field increases, the extraction increases and the space charge reduces, which results in enhanced photocurrent. The photocurrent becomes zero for $V = V_{bi}$. At $V = V_{oc}$, the photocurrent is not zero but the output contribution of the photocurrent is equal and opposite to that in the dark, and the net current becomes zero. The photocurrent contribution shows almost the same behavior as that of the illuminated current. For reverse bias if photocurrent is not saturated, it is because of recombination through shunts. When applied voltage increases ($0 < V < V_{bi}$) or the effective voltage in the solar cell decreases, the extracted photocurrent would decrease. For $V > V_{bi}$, the device was not able to extract out the generated charge carriers due to space charge and trapping effects. Therefore, for the present degraded solar cell $J_{ph}(V)$ could be written as

$$J_{ph}(V) = J_{sc}\left(\frac{V_{bi} - V - V_{ind} + JAR_{s2}}{V_{bi}}\right)^{m} \quad \text{for } V < V_{bi},$$

and

$$J_{ph}(V) = J_{sc}\left(1 - \left(\frac{V_{bi} + V_{ind} - JAR_{s2}}{V_{bi}}\right)^{w}\right) \quad \text{for } V > V_{bi}. \tag{4.91}$$

Figure 4.28 compares the experimental data beyond 41 h, with Equation 4.87, where $J_{ph}(V)$ is given by Equation 4.91. Symbols represent the experimental data, whereas dashed and solid curves correspond to the plots of Equations 4.86 and 4.87. The values of parameters, corresponding to the characteristics measured at different time intervals are given in Table 4.2. It was observed that as device degradation increases, the values of m and w increased, which corresponded to difficulty in charge carrier extraction and was because of increases in R_s and R_p. Even for the degraded solar cells, we observed that the equivalent circuit model in the dark and under illumination would be the same as that for a fresh sample. The voltage dependence of the extracted photocurrent in degraded solar cells is changed and becomes space charge limited.

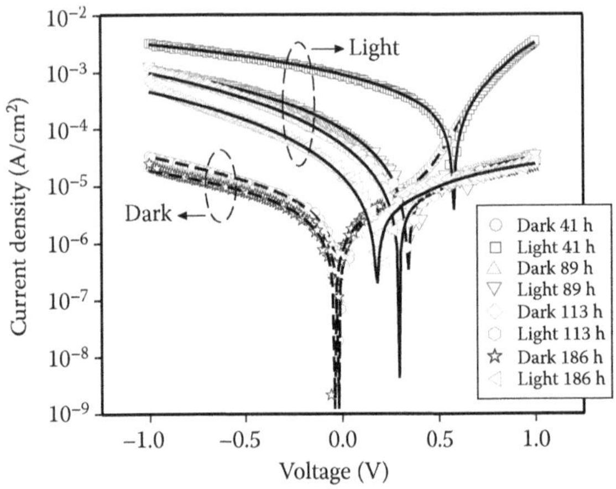

FIGURE 4.28
Comparison of dark and illuminated J–V characteristics of the ITO/ PEDOT:PSS/P3HT:PCBM/Al solar cell with Equations 4.86 and 4.87, where $J_{ph}(V)$ is given by Equation 4.91. Symbols represent experimental data, whereas the dashed and solid curves, respectively, correspond to the plots of Equations 4.86 and 4.87 for the respective cases. (Reprinted with permission from P. Kumar and A. Gaur, Model for the J–V characteristics of degraded polymer solar cells, *J. Appl. Phys.* 113, 94505. Copyright 2013, American Institute of Physics.)

TABLE 4.2

Degradation in Photovoltaic Parameters of ITO/PEDOT:PSS/P3HT:PCBM/Al Solar Cell with Time

S.No.	Time (h)	J_{sc} (mA/cm²)	R_s (Ω) (in dark)	R_p (Ω) (in dark)	ϕ_1 (eV)	ϕ_2 (eV)	V_{bi} (eV)	n_1	n_2	μ (cm²/Vs)	J_{s2} (mA/cm²)	R_s (Ω) (in light)	R_p (Ω) (in light)	V_{ind}
1.	41	0.93	10	1.3×10^6	1.3	0.9	0.75	2	13	3.0×10^{-5}	3.5×10^{-7}	12	4.5×10^5	0.01
2.	89	0.11	100	1.6×10^6	1.0	0.9	0.45	3.4	14	8.5×10^{-7}	1.2×10^{-7}	110	2.8×10^6	0.02
3.	113	0.06	150	1.7×10^6	1.0	0.9	0.45	3.9	14	3.0×10^{-8}	1.2×10^{-7}	200	2.9×10^6	0.02
4.	186	0.019	200	1.7×10^6	0.95	0.75	0.38	4.4	14	6.0×10^{-9}	1.2×10^{-7}	250	2.9×10^6	0.02

Source: Reprinted with permission from P. Kumar and A. Gaur, Model for the *J–V* characteristics of degraded polymer solar cells, *J. Appl. Phys.* 113, 94505. Copyright 2013, American Institute of Physics.

4.8 Effect of Temperature and Illumination Intensity on Cell Performance

Solar illumination intensity and ambient temperature are very important parameters that affect the performance of OSCs. However, there are only a few reports in the literature on this issue. The fundamental knowledge of such effects and the basic physics of OSCs will help further to improve their performance. I discuss here some of the important investigations carried out by my groups in this direction.

My colleagues and I studied the effect of temperature on the performance of CuPc/C_{60} planer heterojunction solar cells [142]. The *J*–*V* characteristics were measured in the temperature range 125–300 K in the dark and under 80 mW/cm^2 irradiance of a W-halogen lamp. Temperature variation strongly influenced the characteristics and photovoltaic parameters of the cells. Initially the current density in the dark decreased with a reduction in temperature but at very low temperatures the revere bias current became independent of temperature. This behavior of current suggested a different charge transport phenomenon at low temperatures. At higher temperatures the characteristics could be explained by bulk limited conduction, but at low temperature when charges do not have sufficient energy to overcome the injection barrier, the current is decided by the tunneling phenomenon. Reduction in the reverse bias current with temperature was relatively slow compared to that in forward bias. At 300 K the device showed typical diode characteristics with a rectification ratio of 2.98×10^2 at ±1 V, which reduced to 5.5 at 125 K. Under illumination the J_{sc} decreased with reduction in temperature, whereas V_{oc} increased almost linearly. The temperature dependence of *FF* was nearly similar to that of J_{sc} and was understood in terms of temperature dependent series and shunt resistances that originate from bulk resistivity and temperature-dependent charge transport properties of organics. As regards the efficiency, it increased first and then decreased with reduction in temperature. At higher temperatures the reduction in $J_{sc} \times FF$ with temperature was less compared to the corresponding increase in V_{oc}; therefore, the efficiency had a negative temperature coefficient and increased with reduction in temperature. However, at low temperatures the strong temperature dependence of $J_{sc} \times FF$ overwhelmed the weak temperature dependence of V_{oc} and η had a positive temperature coefficient, therefore efficiency decreased with further reduction in temperature. In a conventional Si solar cell the temperature dependence of V_{oc} is given by [41]

$$V_{oc} = \frac{nk_BT}{q}\ln\left(\frac{J_{sc}}{J_s}+1\right). \tag{4.92}$$

This equation has been found not valid for OSCs [142], therefore more detailed investigations were performed to understand the temperature dependence of V_{oc} in OSCs [138]. The solar cells were first subjected to different illumination intensity and the effect on V_{oc} was measured. V_{oc} increased with illumination intensity first rapidly and then slowly and got saturated at high illumination intensities. J_{sc} increased linearly with illumination intensity. Figure 4.29 shows the *J*–*V* characteristics of the ITO/CuPc(20 nm)/C_{60} (40 nm)/BPhen (8 nm)/Al (150 nm) solar cell measured at different illumination intensities.

All the illuminated and dark *J*–*V* characteristics were observed to intersect at a common point. This point corresponds to the voltage where the photocurrent becomes zero. If the

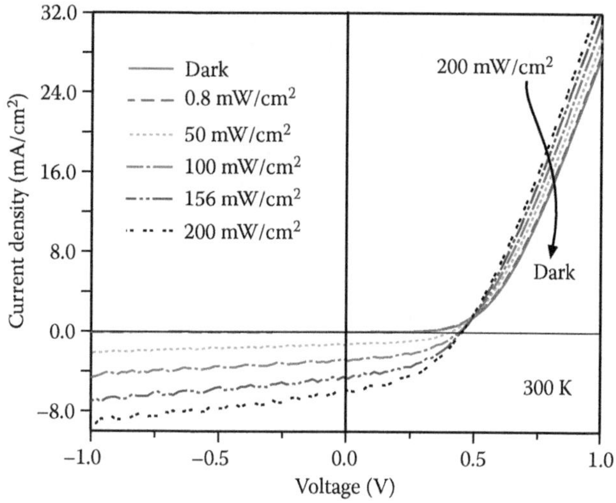

FIGURE 4.29
Dark and illuminated *J–V* characteristics of the ITO/CuPc (20 nm)/C_{60} (40 nm)/BPhen (8 nm)/Al (150 nm) solar cell measured at 300 K. (Reprinted with permission from P. Kumar et al., Effect of illumination intensity and temperature on open circuit voltage in organic solar cells, *Appl. Phys. Lett.* 94, 183505. Copyright 2009, American Institute of Physics.)

diffusion component of the photocurrent is neglected, this intersection point corresponds to V_{bi} to a good approximation. At room temperature this intersection point was 0.49 V. Now to investigate the effect of temperature on V_{oc} the illumination intensity was kept at 200 mW/cm² where V_{oc} was saturated at 0.44 V and the *J–V* characteristics were measured at different temperatures. The reduction in temperature had significant effect on saturated V_{oc}. The saturated V_{oc} increased monotonically with reduction in temperature. The increase in V_{oc} at low temperatures went beyond V_{bi} at room temperature. This was an important finding and to understand the reason behind the illumination intensity was varied keeping the sample at low temperatures. At a given temperature the magnitude of intersection voltage for dark and illuminated *J–V* characteristics was more than that at higher temperatures. These observations suggested a shift in V_{bi} toward higher values with reduction in temperature. Calculations were made for the effect of temperature on V_{bi} and charge carrier distribution. In thermal equilibrium the current density and Poisson's equation for elections are written as

$$J = q\mu_n n(x)F(x) + qD_n \frac{\partial n(x)}{\partial x} = 0, \tag{4.93}$$

and

$$\frac{\partial F(x)}{\partial x} = -\frac{qn(x)}{\varepsilon\varepsilon_0}, \tag{4.94}$$

where x is the distance from the anode ($x = 0$ at anode and $x = d$ at cathode); ε and ε_0 are the dielectric constant of semiconductor and permittivity of free space, respectively; $n(x)$

is the electron density distribution; $F(x)$ is the electric field distribution; and D_n is the diffusion coefficient for electrons. Using the Einstein relation $D_n = \mu_n k_B T/q$, Equations 4.93 and 4.94 give

$$\frac{\partial^2 F(x)}{\partial x^2} + \frac{q}{k_B T} F(x) \frac{\partial F(x)}{\partial x} = 0. \tag{4.95}$$

Since $F(x) = -\partial\psi(x)/\partial x$, integration of Equation 4.95 gives

$$-\psi(x) = \frac{2k_B T}{q} \ln\left\{\cosh\left[\sqrt{\frac{qC}{2k_B T}}(x-D)\right]\right\} + E, \tag{4.96}$$

where C, D, and E are the first, second, and third constants of integration, and their values were obtained solving Equations 4.94 and 4.96 numerically using boundary conditions [138]:

$$n(0) = N_c \exp\left(-\frac{\phi_1}{k_B T}\right), \tag{4.97}$$

$$n(d) = N_c \exp\left(-\frac{\phi_2}{k_B T}\right), \tag{4.98}$$

$$q\psi(0) = -\phi_1, \tag{4.99}$$

$$q\psi(d) = -\phi_2. \tag{4.100}$$

Here N_c is the effective density of states in conduction band, and ϕ_1 and ϕ_2 are the electron injection barriers at the anode and cathode, respectively. The charge density distribution was obtained from Equation 4.94 as

$$n(x) = -\frac{\varepsilon\varepsilon_0 C}{q}\left\{1 - \tanh^2\left[\sqrt{\frac{qC}{2k_B T}}(x-D)\right]\right\}. \tag{4.101}$$

Equation 4.96 has been plotted along with Equation 4.101 in Figure 4.30 for different temperatures. Solid curves are for Equation 4.96 and the dashed curves represent Equation 4.101. The values of parameters used in these calculations are given in the caption of the figure. Reduction in temperature results in reduction in diffusion coefficient D_n, which led to reduction in band bending. The change in temperature led to change in diffusion current, which led to change in the drift current or in other terms the band bending. Therefore, charge carrier density decreased with a reduction in temperature. Assuming V_{bi} as the voltage required for the flat band condition to occur, it can be seen from Figure 4.30 that V_{bi} increased with reduction in temperature. As V_{oc} has direct dependence on V_{bi}, the variation in saturated V_{oc} with temperature could be attributed to the variation of V_{bi} due to temperature variation. Also the incremental variation of V_{oc} beyond the V_{bi} value at room temperature could be attributed to an increase in V_{bi} itself with reduction

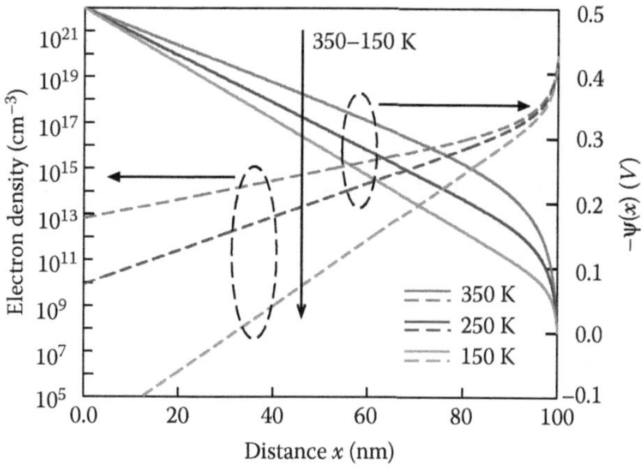

FIGURE 4.30
Distribution of potential $\psi(x)$ and electron density in a semiconductor at different temperatures with an ohmic cathode. The values of parameters used in the calculations are $\phi_1 = 0.5$ eV, $\phi_2 = 0$, $d = 100$ nm, $\varepsilon = 3$, and $N_c = 10^{20}$ cm^{-3}. (Reprinted with permission from P. Kumar et al., Effect of illumination intensity and temperature on open circuit voltage in organic solar cells, *Appl. Phys. Lett.* 94, 183505. Copyright 2009, American Institute of Physics.)

in temperature. It was observed that for large injection Schottky barriers and/or low temperatures, V_{bi} was equal to the work function difference of the cathode and anode (ΔW). For very low Schottky barriers or an ohmic contact, V_{bi} was observed to be less than ΔW. In the vicinity of ohmic contacts, an accumulation of charge carriers takes place, which results in a significant band bending. This band bending leads to reduction in the voltage (δV) at which the flat band condition is reached in the bulk of the device.

4.9 Origin of V_{oc} in OSCs

It was been pointed out earlier that the conventional relationship used for determination of V_{oc} is not valid for OSCs. The exact origin on V_{oc} in OSCs has been very complicated to tell, as it depends on a number of parameters. To understand the origin of V_{oc} in OSCs, various groups performed a number of experiments and presented several models to explain the experimental observations. Some of the very important results and models proposed from different groups are discussed here. Koster et al. [143] studied the effect of illumination intensity on the V_{oc} of a polymer solar cell and found that the light intensity dependence of V_{oc} was in contradiction to the conventional model. They presented an alternate model, which was based on the notation that at zero current the quasi-Fermi levels are constant throughout the device, and it incorporates both the drift and diffusion of charge carriers. According to their model, the V_{oc} in OSCs should be given by

$$V_{oc} = \frac{E_{gap}}{q} - \frac{kT}{q} \ln\left(\frac{(1-P)\gamma N_c^{\,2}}{PG}\right), \tag{4.102}$$

where E_{gap} is the difference between the HOMO of the donor and the LUMO of the acceptor, P is the dissociation probability of excitons, G is the generation rate of excitons, γ is the Langevin recombination constant, and N_c is the effective density of states in conduction band. The model showed good agreement with the experimental observation at different temperatures and illumination intensities and suggested V_{oc} to be a bulk property. But it was difficult to predict the effect of active layer thickness on V_{oc} using this model.

Brabec et al. [144] synthesized and used a series of fullerene derivatives, methanofullerene $PC_{61}BM$, azafulleroid 5, and ketolactam 6, as the electron acceptors to investigate the effect of acceptor strength on V_{oc}. For these studies, MDMO-PPV was used as the light absorber and electron donor. ITO/PEDOT:PSS-coated substrates were used as the anode and four different metal electrodes—Au, Ag, Al, and Ca—were used as the cathode. The variation in V_{oc} with variation in acceptor strength and the work function of the cathode as measured by Brabec et al. is shown in Figure 4.31. The V_{oc} in these systems has been found to follow

$$V_{oc} = \left(A_{ox} - S_1 E_{red(A)}\right) - S_2\left(\phi_M - E_{red(A)}\right) + C \quad (4.103)$$

where ϕ_M is the work function of cathode; C is a constant describing the interface potential for the ideal ohmic contact; $E_{red(A)}$ is the reduction potential of the acceptor; A_{ox} is a constant

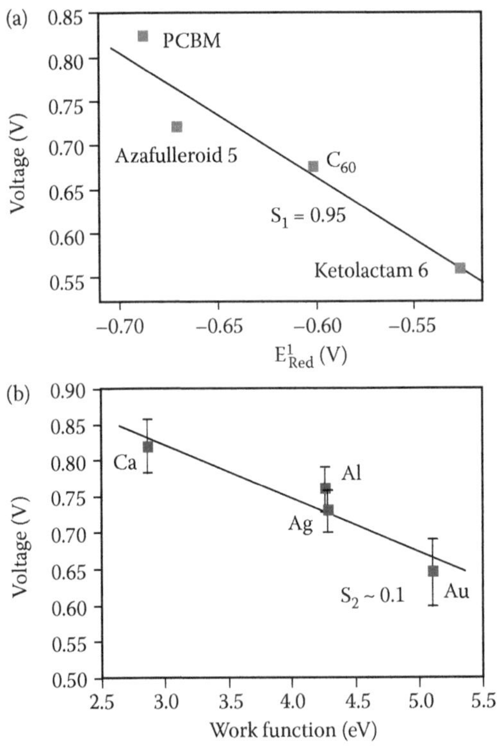

FIGURE 4.31
Variation in V_{oc} with (a) acceptor strength and (b) cathode work function in BHJ polymer solar cells based on the MDMO-PPV light absorber. (Redrawn with permission from C. J. Brabec et al., 2001. *Adv. Funct. Mater.* 11: 374–380.)

representative of all the contributions to V_{oc} from the positive electrode and is expected to be properly described by the oxidation potential of the conjugated polymer; and S_1 and S_2 are the quality factors for ohmicity of the contact and the index of interface behavior, respectively. A linear fit to the experimental data showed a slope of ~1 (S_1) for the variation of V_{oc} with the acceptor strength, whereas a slope of ~0.1 (S_2) was observed for the linear fit to the variation of V_{oc} with work function of the electrodes. These results show that the variation in work function of the electrode has a less significant effect compared to acceptor strength. This electrode insensitive V_{oc} behavior was attributed to the Fermi level pinning between the metal cathode and fullerene derivative via charged interfacial states. The linear variation of V_{oc} in the range 25–60°C was correlated with the temperature dependence of quasi-Fermi levels of the donor and acceptor materials [145]. To explain the illumination intensity dependence of V_{oc}, Brabec et al. presented a model that was directly based on the MIM model and related to the electrode materials used. These two observations suggested the dependence of V_{oc} on both the bulk and electrode materials. The thickness dependence of V_{oc} was still not clear from these models.

Frohne et al. [146] investigated the effect of the anodic work function on the performance of MDMO-PPV-based solar cells, where they altered electrochemically the work function of PEDOT. With an increase in equilibrium potential of PEDOT, the V_{oc} was observed to increase. This increase in V_{oc} had been attributed to increase in V_{bi}. The V_{bi} changed with the change in equilibrium potential of PEDOT, and the Fermi level of PEDOT was not pinned to any of the energy levels of the active layer. When the work function of PEDOT was equal to the reduction potential of PCBM, it did not show any photovoltaic response, as the cathode was pinned to PCBM, and it resulted in zero V_{bi} in the sample. Ramsdale et al. [147] investigated the effect of cathode work function on the polymer bilayer solar cells based on electron-donating poly(9,9′,dioctylfluorene-co-bis-*N*,*N*′-(4-butylphenyl)bis-*N*,*N*′-phenlyenediamine) and electron-accepting poly(9,9′-dioctylfluorene-co-benzothiazole). Au, Al, Cr, Cu, and Ca were used as cathode materials. They found a scaling of V_{oc} with difference of work function of anode and cathode, with an additional contribution of ~1 V, which was attributed to the accumulation of charge carriers at the donor–acceptor interface. This accumulation of charge carriers at the donor–acceptor interface gave rise to the diffusion current, which was compensated by a drift current at open circuit. Liu et al. modified the CuPc/C_{60} interface by insertion of a very thin layer (0.5 nm) of molybdenum trioxide (MoO_3) and observed an increase of 0.4 V in V_{oc}. This increase in V_{oc} had been attributed to the increase in the energy level offset between the HOMO of CuPc and the LUMO of C_{60}. Insertion of a 4 nm thin layer of tungsten oxide (WO_3) between ITO and CuPc by Chen et al. increased the V_{oc} remarkably and this increase in V_{oc} had been attributed to the improved anodic work function [148].

Mihailetchi et al. [119] investigated the effect of variation of cathode work function on V_{oc} in MDMOPPV:PCBM-based solar cells. ITO/PEDOT:PSS was used as the anode, whereas LiF/Al, Ag, Au, and Pd were used as the cathode, and interestingly a total variation of more than 0.5 V was observed in V_{oc}. However, as discussed earlier in the experiments of Brabec et al. [144], only a variation less than 200 mV was observed in V_{oc} with the variation of cathode work function by more than 2.2 eV in the same photoactive medium. To explain this discrepancy, Mihailetchi et al. performed detailed investigations and presented a model, according to which OSCs follow the MIM model. According to the MIM model, V_{oc} is determined by V_{bi} and scales with work function difference of electrodes. They demonstrated that for nonohmic contacts, V_{oc} is determined by the difference of work functions of the electrodes, whereas for ohmic contacts, V_{oc} is governed by the LUMO and HOMO levels of acceptor and donor materials, respectively. In both cases, the magnitude

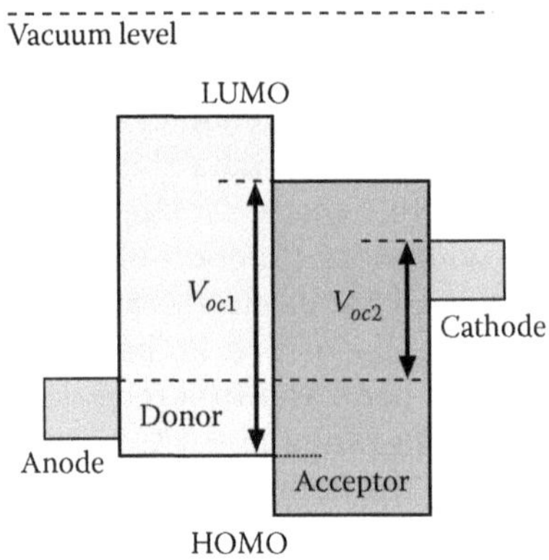

FIGURE 4.32
Schematic representation of V_{oc} in OSCs for ohmic (V_{oc1}) and nonohmic (V_{oc2}) contacts.

of V_{oc} is shown schematically in Figure 4.32. In the case of ohmic contacts, the work functions of the anode and cathode match with the HOMO of the donor and the LUMO of the acceptor, respectively, and the maximum V_{oc} would be V_{oc1}. In case of nonohmic contact, the MIM model is followed, and the maximum V_{oc} will be V_{oc2}. The deviation of the MIM model in the results of Brabec et al. [144] has been explained by the pinning of the Fermi level of electrodes to the LUMO level of fullerene. It must be noted that according to the MIM model, if the contact is ohmic, an accumulation of charge carriers will take place in the vicinity of the contact that will result in a band bending. This band bending causes a reduction in the internal electric field. For example, in the experiments performed by Mihailetchi et al. [119], ITO/PEDOT:PSS made ohmic contact with the HOMO of MDMO-PPV, and LiF/Al made ohmic contact with the LUMO of PCBM; therefore, there was a band bending at both electrodes. On the other hand, with Au and Ag as the cathode, there was band bending only at the anode not at the cathode. For ohmic contacts, V_{oc} not equal to the HOMO, and LUMO difference of donor and acceptor materials can be attributed to the band bending at ohmic contacts and other bulk losses. It has been demonstrated that only the Fermi level of the cathode with a work function lower than the LUMO of PCBM will be pinned, whereas other metals will follow the MIM model. The only metal that might show deviation from the MIM model is Au, and this can be attributed to the presence of interface dipoles. V_{oc} varied linearly with the energy level of the cathode and pinned when the LUMO level of PCBM (~3.7 eV) is reached.

To understand the origin of V_{oc} in OSCs my colleagues and I studied the effect of illumination intensity, temperature, and active layer thickness [130,138]. The effects of illumination intensity and temperature on V_{oc} studied by us have already been discussed in the previous sections of this chapter. However, the V_{oc} decreased nonlinearly with an increase in active layer thickness as shown in Figure 4.33 [130]. We developed a model to explain the observed behavior of V_{oc} and found that the variation in V_{bi} affects the V_{oc} in OSCs. Uhrich et al. studied both the bilayer and BHJ solar cells based on some small molecules and concluded that V_{oc} depends on the cell architecture [149]. It can either be determined primarily by the V_{bi} or by the difference of energy levels of donor and acceptor materials. They concluded that in the studied BHJ solar cells, V_{oc} is primarily determined by V_{bi}, whereas in the case of bilayer solar cells, it equals the quasi-Fermi levels at the donor–acceptor

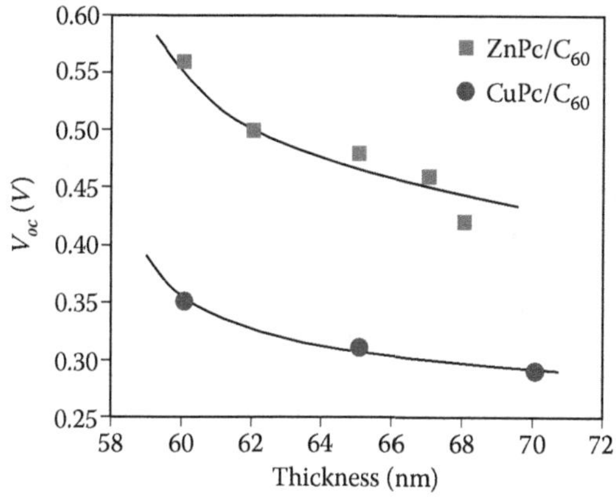

FIGURE 4.33
V_{oc} as a function of active layer thickness in small molecular OSCs. (From P. Kumar et al., *Jpn. J. Appl. Phys.* 48, 121501. Copyright 2009, The Japan Society of Applied Physics.)

interface, which is controlled by the LUMO and HOMO energy levels of the acceptor and donor, materials respectively. CTSs also play an important role in controlling V_{oc} and the maximum value of V_{oc} is determined by the energy of CTSs.

The aforementioned investigations from different groups show that both the V_{bi} and the HOMO and LUMO energy levels of the donor and acceptor materials have direct control on V_{oc} in OSCs, but it is their relative effect that dominates at a time. Now it will be very important to investigate the precise conditions when one of these factors will dominate so that the properties of the cells could be tailored accordingly. Therefore, more detailed investigations have to be carried out for full control on V_{oc} in OSCs.

References

1. T. Forster, 1959. Transfer mechanisms of electronic excitation, 10th Spiers Memorial Lecture, The Royal Society of Chemistry.
2. D. L. Dexter, 1953. A theory of sensitized luminescence in solids, *J. Chem. Phys.* 21: 836–850.
3. R. C. Powell and Z. G. Soos, 1975. Singlet exciton energy transfer in organic solids, *J. Lumin.* 11: 1–45.
4. S. C. J. Meskers, J. Habner, M. Oestreich, and H. Bassler, 2001. Dispersive relaxation dynamics of photoexcitations in a polyfluorene film involving energy transfer: Experiment and Monte Carlo simulations, *J. Phys. Chem. B* 105: 9139–9149.
5. C. Madigan and V. Bulovic, 2006. Modeling of exciton diffusion in amorphous organic thin films, *Phys. Rev. Lett.* 96: 046404.
6. T. Ahn, N. Wright, and C. J. Bardeen, 2007. The effects of orientational and energetic disorder on Forster energy migration along a one-dimensional, *Chem. Phys. Lett.* 446: 43–48.
7. M. E. Kose, P. Graf, N. Kopidakis, S. E. Shaheen, K. Kim, and G. Rumbles, 2009. Exciton migration in conjugated dendrimers: A joint experimental and theoretical study, *Chem. Phys. Chem.* 10: 3285–3294.
8. S. M. Menke and R. J. Holmes, 2014. Exciton diffusion in organic photovoltaic cells, *Energy Environ. Sci.* 7: 499–512.

9. P. E. Shaw, A. Ruseckas, and I. D. W. Samuel, 2008. Exciton diffusion measurements in poly(3-hexylthiophene), *Adv. Mater.* 20: 3516.
10. Y. Tamai, Y. Matsuura, H. Ohkita, H. Benten, and S. Ito, 2014. One-dimensional singlet exciton diffusion in poly(3-hexylthiophene) crystalline domains, *J. Phys. Chem. Lett.* 5: 399–403.
11. A. L. Ayzner, S. C. Doan, B. T. deVillers, and B. J. Schwartz, 2012. Ultrafast studies of exciton migration and polaron formation in sequentially solution-processed conjugated polymer/fullerene quasi-bilayer photovoltaics, *J. Phys. Chem. Lett.* 3: 2281–2287.
12. D. E. Markov, E. Amsterdam, P. W. M. Blom, A. B. Sieval, and J. C. Hummelen, 2005. Accurate measurement of the exciton diffusion length in a conjugated polymer using a heterostructure with a side-chain cross-linked fullerene layer, *J. Phys. Chem. A* 109: 5266–5274.
13. S. R. Scully and M. D. McGehee, 2006. Effects of optical interference and energy transfer on exciton diffusion length measurements in organic semiconductors, *J. Appl. Phys.* 100: 034907.
14. C. Goh, S. R. Scully, and M. D. McGehee, 2007. Effects of molecular interface modification in hybrid organic–inorganic photovoltaic cells, *J. Appl. Phys.* 101: 114503.
15. L. Luer, H. J. Egelhaaf, D. Oelkrug, G. Cerullo, G. Lanzani, B. H. Huisman, and D. de Leeuw, 2004. Oxygen-induced quenching of photoexcited states in polythiophene films, *Org. Electron.* 5: 83.
16. D. E. Markov and P. W. M. Blom, 2006. Anisotropy of exciton migration in poly(p-phenylene vinylene), *Phys. Rev. B* 74: 085206.
17. M. A. Stevens, C. Silva, D. M. Russell, and R. H. Friend, 2001. Exciton dissociation mechanisms in the polymeric semiconductors poly(9,9-dioctylfluorene) and poly(9,9-dioctylfluorene-co-benzothiadiazole), *Phys. Rev. B* 63: 165213.
18. A. J. Lewis, A. Ruseckas, O. P.M. Gaudin, G. R. Webster, P. L. Burn, and I. D. W. Samuel, 2006. Singlet exciton diffusion in MEH-PPV films studied by exciton–exciton annihilation, *Org. Electron.* 7: 452–456.
19. S. Cook, H. Liyuan, A. Furube, and R. Katoh, 2010. Singlet annihilation in films of regioregular poly(3-hexylthiophene): Estimates for singlet diffusion lengths and the correlation between singlet annihilation rates and spectral relaxation, *J. Phys. Chem. C* 114: 10962–10968.
20. J. E. Kroeze, T. J. Savenije, M. J. W. Vermeulen, and J. M. Warman, 2003. Contactless determination of the photoconductivity action spectrum, exciton diffusion length, and charge separation efficiency in polythiophene-sensitized TiO_2 bilayers, *J. Phys. Chem. B* 107: 7696–7705.
21. H. Najafov, B. Lee, Q. Zhou, L. C. Feldman, and V. Podzorov, 2010. Observation of long-range exciton diffusion in highly ordered organic semiconductors, *Nat. Mater.* 9: 938–943.
22. E. Engel, K. Leo, and M. Hoffmann, 2006. Ultrafast relaxation and exciton–exciton annihilation in PTCDA thin films at high excitation densities, *Chem. Phys.* 325: 170–177.
23. F. Fennel and S. Lochbrunner, 2012. Förster-mediated spectral diffusion in disordered organic materials, *Phys. Rev. B* 85: 094203.
24. V. Dyakonov and E. Frankevich, 1998. On the role played by polaron pairs in photophysical processes in semiconducting polymers, *Chem. Phys.* 227: 203–217.
25. A. C. Morteani, P. Sreearunothai, L. M. Herz, R. H. Friend, and C. Silva, 2004. Exciton regeneration at polymeric semiconductor heterojunctions, *Phys. Rev. Lett.* 92: 247402.
26. M. C. Scharber, N. A. Schultz, N. S. Sariciftci, and C. J. Brabec, 2003. Optical and photocurrent detected magnetic resonance studies on conjugated polymer/fullerene composites, *Phys. Rev. B* 67: 85202.
27. D. Veldman, Ö. Ipek, S. C. J. Meskers, J. Sweelssen, M. M. Koetse, S. C. Veenstra, J. M. Kron, S. S. van Bavel, J. Loos, and R. A. J. Janssen, 2008. Compositional and electric field dependence of the dissociation of charge transfer excitons in alternating polyfluorene copolymer/fullerene blends, *J. Am. Chem. Soc.* 130: 7721–7735.
28. L. Goris, A. Poruba, L. Hod'akova, M. Vanecek, K. Haenen, M. Nesladek, P. Wagner, D. Vanderzande, L. De Schepper, and J. V. Manca, 2006. Observation of the subgap optical absorption in polymer–fullerene blend solar cells, *Appl. Phys. Lett.* 88: 52113.
29. A. C. Morteani, R. H. Friend, and C. Silva, 2005. Exciton trapping at heterojunctions in polymer blends, *J. Chem. Phys.* 122: 244906.

30. K. Vandewal, A. Gadisa, W. D. Oosterbaan, S. Bertho, F. Banishoeib, I. V. Severen, L. Lutsen, T. J. Cleij, D. Vanderzande, and J. V. Manca, 2008. The relation between open-circuit voltage and the onset of photocurrent generation by charge-transfer absorption in polymer: Fullerene bulk heterojunction solar cells, *Adv. Funct. Mater.* 18: 2064–2070.
31. K. Tvingstedt, K. Vandewal, A. Gadisa, F. Zhang, J. Manca, and O. Inganäs, 2009. Electroluminescence from charge transfer states in polymer solar cells, *J. Am. Chem. Soc.* 131: 11819–11824.
32. M. Hallermann, S. Haneder, and E. Da Como, 2008. Charge-transfer states in conjugated polymer/fullerene blends: Below-gap weakly bound excitons for polymer photovoltaics, *Appl. Phys. Lett.* 93: 053307.
33. C. Yin, 2007. Photovoltaic properties and exciplex emission of polyphenylenevinylene-based blend solar cells, *Appl. Phys. Lett.* 90: 092117.
34. M. Hallermann, I. Kriegel, E. Da Como, J. M. Berger, E. von Hauff, and J. Feldmann, 2009. Charge transfer excitons in polymer/fullerene blends: The role of morphology and polymer chain conformation, *Adv. Funct. Mater.* 19: 3662–3668.
35. C. Deibel, T. Strobel, and V. Dyakonov, 2010. Role of the charge transfer state in organic donor–acceptor solar cells, *Adv. Mater.* 22: 4097–4111.
36. K. Vandewal, K. Tvingstedt, J. V. Manca, and O. Inganäs, 2010. Charge-transfer states and upper limit of the open circuit voltage in polymer:fullerene organic solar cells, *IEEE J. Select. Top. Quant. Electron.* 16: 1676–1684.
37. R. A. Street and M. Schoendorf, 2010. Interface state recombination in organic solar cells, *Phys. Rev. B* 81: 205307.
38. J. Lee, K. Vandewal, S. R. Yost, M. E. Bahlke, L. Goris, M. A. Baldo, J. V. Manca, and T V. Voorhis, 2010. Charge transfer state versus hot exciton dissociation in polymer–fullerene blended solar cells, *J. Am. Chem. Soc.* 132: 11878–11880.
39. Y. Yuan, T. J. Reece, P. Sharma, S. Poddar, S. Ducharme, A. Gruverman, Y. Yang, and J. Huang, 2011. Efficiency enhancement in organic solar cells with ferroelectric polymers, *Nat. Mater.* 10: 296.
40. S. C. Jain, M. Willander, and V. Kumar, 2007. *Conducting Organic Materials and Devices*, San Diego: Academic Press.
41. S. M. Sze, 1981. *Physics of Semiconductor Devices*, New York: Wiley.
42. I. D. Parker, 1994. Carrier tunneling and device characteristics in polymer light-emitting diodes, *J. Appl. Phys.* 75: 1656.
43. Y. Kawabe, M. M. Morrell, G. E. Jabbour, S. E. Shaheen, B. Kippelen, and N. Peyghambarian, 1998. A numerical study of operational characteristics of organic light-emitting diodes, *J. Appl. Phys.* 84: 5306.
44. J. M. Lupton and I. D. W. Samuel, 1999. Temperature-dependent single carrier device model for polymeric light emitting diodes, *J. Phys. D: Appl. Phys.* 32: 2973.
45. M. A. Abkowitz, H. A. Mizes, and J. S. Facci, 1995. Emission limited injection by thermally assisted tunneling into a trap-free transport polymer, *Appl. Phys. Lett.* 66: 1288.
46. A. J. Campbell, D. D. C. Bradley, and D. G. Lidzey, 1997. Space-charge limited conduction with traps in poly(phenylene vinylene) light emitting diodes, *J. Appl. Phys.* 82: 6326.
47. A. J. Campbell, M. S. Weaver, D. G. Lidzey, and D. D. C. Bradley, 1998. Bulk limited conduction in electroluminescent polymer devices, *J. Appl. Phys.* 84: 6737.
48. A. B. Walker, A. Kambili, and S. J. Martin, 2002. Electrical transport modelling in organic electroluminescent devices, *J. Phys.: Cond. Matter* 14: 9825.
49. S. J. Martin, J. M. Lupton, I. D. W. Samuel, and A. B. Walker, 2002. Modelling temperature-dependent current–voltage characteristics of an MEH-PPV organic light, *J. Phys.: Cond. Matter* 14: 9925.
50. P. E. Burrows, Z. Shen, V. Bulovic, D. M. McCarty, S. R. Forrest, J. A. Cronin, and M. E. Thompson, 1996. Relationship between electroluminescence and current transport in organic heterojunction light-emitting devices, *J. Appl. Phys.* 79: 7991.
51. B. K. Crone, P. S. Davids, I. H. Campbell, and D. L. Smith, 2000. Device model investigation of bilayer organic light emitting diodes, *J. Appl. Phys.* 87: 1974.

52. N. F. Mott and R. W. Gurney, 1940. *Electronic Process in Ionic Crystal*, New York: Dover Publications (reprinted 1964).
53. W. Helfrich and P. Mark, 1962. Raumladungsbeschränkte Ströme in Anthrazen als Mittel zur Bestimmung der Beweglichkeit von Defektelektronen, *Z. Phys.* 166: 370; W. Helfrich and P. Mark, 1962. Ein Nachweis der Raumladung bei raumladungsbeschränkten Defektelektronenströmen in Anthrazen, *Z. Phys.* 168: 495.
54. K. C. Kao and W. Hwang, 1981. *Electrical Transport in Solids*, Oxford: Pergamon Press.
55. M. A. Lampert and P. Mark, 1970. *Current Injection in Solids*, New York: Academic Press.
56. P. Kumar, S. C. Jain, A. Misra, M. N. Kamalasanan, and V. Kumar, 2006. Characteristics of a conducting organic diode with finite (non-zero) Schottky barrier, *J. Appl. Phys.* 100: 114506.
57. M. Pope and C. E. Swenberg, 1982. *Electronic Processes in Organic Crystals*, Oxford: Oxford University Press.
58. M. Stobel, J. Staudigel, F. Steuber, J. Blassing, J. Simmerer, and A. Winnacker, 2000. Space-charge-limited electron currents in 8-hydroxyquinoline aluminum, *Appl. Phys. Lett.* 76: 115.
59. A. K. Kapoor, S. C. Jain, J. Poortmans, V. Kumar, and R. Mertens, 2002. Temperature dependence of carrier transport in conducting polymers: Similarity to amorphous inorganic semiconductors, *J. Appl. Phys.* 92: 3835.
60. P. Kumar, S. C. Jain, V. Kumar, A. Misra, S. Chand, and M. N. Kamalasanan, 2007. Current–voltage characteristics of an organic diode: Revisited, *Synth. Met.* 157: 905.
61. P. Kumar, S. C. Jain, V. Kumar, S. Chand, and R. P. Tandon, 2008. Trap filled limit and high current–voltage characteristics of organic diodes with non-zero Schottky barrier, *J. Phys. D: Appl. Phys.* 41: 155108.
62. P. Kumar, S. C. Jain, V. Kumar, S. Chand, and R. P. Tandon, 2009. Effect of non-zero Schottky barrier on the *J-V* characteristics of organic diodes, *Eur. Phys. J. E Soft Matter* 28: 361–368.
63. V. Kumar, S. C. Jain, A. K. Kapoor, W. Geens, T. Aernauts, J. Poortmans, and R. Mertens, 2002. Carrier transport in conducting polymers with field dependent trap occupancy, *J. Appl. Phys.* 92: 7325.
64. S. C. Jain, A. K. Kapoor, W. Geens, J. Poortmans, R. Mertens, and M. Willander, 2002. Trap filled limit of conducting organic materials, *J. Appl. Phys.* 92: 3752–3754.
65. J. Y. Kim, S. H. Kim, H. H. Lee, K. Lee, W. Ma, X. Gong, and A. J. Heeger, 2006. New architecture for high-efficiency polymer photovoltaic cells using solution-based titanium oxide as an optical spacer, *Adv. Mater.* 18: 572–576.
66. V. Coropceanu, J. Cornil, D. A. da Silva Filho, Y. Olivier, R. Silbey, and J. L. Bredas, 2007. Charge transport in organic semiconductors, *Chem. Rev.* 107: 926–952.
67. P. W. M. Blom, M. J. M deJong, and M. G. van Munster, 1997. Electric-field and temperature dependence of the hole mobility in poly(p-phenylene vinylene), *Phys. Rev. B* 55: R656.
68. H. Sirringhaus, 2005. Device physics of solution-processed organic field-effect transistors, *Adv. Mater.* 17: 2411.
69. H. Scher and M. Lax, 1973. Stochastic transport in a disordered solid. I. *Theory, Phys. Rev. B* 7: 4491.
70. H. Scher and E. W. Montroll, 1975. Anomalous transit-time dispersion in amorphous solids, *Phys. Rev. B* 12: 2455.
71. H. Bassler, G. Schonherr, M. Abkowitz, and D. M. Pai, 1982. Hopping transport in prototypical organic glasses, *Phys. Rev. B* 26: 3105.
72. M. Silver, G. Schoenherr, and H. Bassler, 1982. Dispersive hopping transport from an exponential energy distribution of sites, *Phys. Rev. Lett.* 48: 352.
73. I. I. Fishchuk, A. Kadashchuk, S. T. Hoffmann, S. Athanasopoulos, J. Genoe, H. Bassler, and A. Kohler, 2013. Unified description for hopping transport in organic semiconductors including both energetic disorder and polaronic contributions, *Phys. Rev. B* 88: 125202.
74. I. I. Fishchuk, V. I. Arkhipov, A. Kadashchuk, P. Heremans, and H. Bässler, 2007. Analytic model of hopping mobility at large charge carrier concentrations in disordered organic semiconductors: Polarons versus bare charge carriers, *Phys. Rev. B* 76: 045210.

75. M. Pope, C. E. Swenberg, and M. Pope, 1999. *Electronic Processes in Organic Crystals and Polymers*, 2nd ed., New York: Oxford University Press.
76. E. Silinsh and V. Capek, 1994. *Organic Molecular Crystals: Interaction, Localization, and Transport Phenomena*, New York: American Institute of Physics.
77. H. Bottger and V. V. Bryksin, 1985. *Hopping Conduction in Solids*, Deerfield Beach, FL: VCH.
78. B. Movaghar, 1987. Theory of conduction in molecular solids, *J. Mol. Electron.* 3: 183.
79. T. Holstein, 1959. Studies of polaron motion: Part I. The molecular-crystal model, *Ann. Phys.* 8: 325–342.
80. T. Holstein, 1959. Studies of polaron motion: Part II. The "small" polaron, *Ann. Phys.* 8: 343.
81. L. Friedman, 1964. Density matrix formulation of small-polaron motion, *Phys. Rev.* 135: A233.
82. V. I. Arkhipov, I. I. Fishchuk, A. Kadashchuk, and H. Bassler, 2007. Charge transport in neat and doped random organic semiconductors, in *Semiconducting Polymers: Chemistry, Physics and Engineering*, vol. 1, G. Hadziioannou and G. G. Malliaras (eds.), p. 275, Weinheim, Germany: Wiley.
83. A. Miller and E. Abrahams, 1960. Impurity conduction at low concentrations, *Phys. Rev.* 120: 745.
84. R. A. Marcus, 1993. Electron transfer reactions in chemistry: Theory and experiment, *Rev. Mod. Phys.* 65: 599.
85. H. Bassler, 1993. Charge transport in disordered organic photoconductors a Monte Carlo simulation study, *Phys. Status Solidi B* 175: 15–56.
86. V. I. Arkhipov, I. I. Fishchuk, A. Kadashchuk, and H. Bassler, 2006. Charge transport in disordered organic semiconductors, in *Photophysics of Molecular Materials*, G. Lanzani (ed.), pp. 261–366, Weinheim, Germany: Wiley-VCH.
87. I. I. Fishchuk, A. Kadashchuk, H. Bassler, and S. Nespurek, 2003. Nondispersive polaron transport in disordered organic solids, *Phys. Rev. B* 67: 224303.
88. W. F. Pasveer, J. Cottaar, C. Tanase, R. Coehoorn, P. A. Bobbert, P. W. M. Blom, D. M. de Leeuw, and M. A. J. Michels, 2005. Unified description of charge-carrier mobilities in disordered semiconducting polymers, *Phys. Rev. Lett.* 94: 206601.
89. R. Coehoorn, W. F. Pasveer, P. A. Bobbert, and M. A. J. Michels, 2005. Charge-carrier concentration dependence of the hopping mobility in organic materials with Gaussian disorder, *Phys. Rev. B* 72: 155206.
90. C. Tanase, P. W. M. Blom, and D. M. de Leeuw, 2004. Origin of the enhanced space-charge-limited current in poly(p-phenylene vinylene), *Phys. Rev. B* 70: 193202.
91. F. Torricelli, D. Zappa, and L. Colalongo, 2010. Space-charge-limited current in organic light emitting diodes, *Appl. Phys. Lett.* 96: 113304.
92. F. Torricelli and L. Colalongo, 2009. Unified mobility model for disordered organic semiconductors, *IEEE Electron. Device Lett.* 30: 1048–1050.
93. G. G. Malliaras, Y. Shen, D. H. Dunlap, H. Murata, and Z. H. Kafafi, 2001. Nondispersive electron transport in Alq3, *Appl. Phys. Lett.* 79: 2582.
94. S. C. Tse, H. H. Fong, and S. K. So, 2003. Electron transit time and reliable mobility measurements from thick film hydroxyquinoline-based organic light-emitting diode, *J. Appl. Phys.* 94: 2033.
95. H. Fong, K. Lun, and S. So, 2002. Extrinsic electron traps in tris(8-hydroxyquinoline) aluminium, *Jpn. J. Appl. Phys.* 41: L1122.
96. A. J. Mozer and N. S. Sariciftci, 2004. Negative electric field dependence of charge carrier drift mobility in conjugated, semiconducting polymers, *Chem. Phys. Lett.* 389: 438–442.
97. G. Horowitz, 1998. Organic field-effect transistors, *Adv. Mater.* 10: 365–377.
98. C. D. Dimitrakopoulos, S. Purushothaman, J. Kymissis, A. Callegari, and J. M. Shaw, 1999. Low-voltage organic transistors on plastic comprising high-dielectric constant gate insulators, *Science* 283: 822–824.
99. C. Tanase, E. J. Meijer, P. W. M. Blom, and D. M. de Leeuw, 2003. Unification of the hole transport in polymeric field-effect transistors and light-emitting diodes, *Phys. Rev. Lett.* 91: 216601.

100. A. F. Stassen, R. W. I. de Boer, N. N. Iosad, and A. F. Morpurgo, 2004. Influence of the gate dielectric on the mobility of rubrene single-crystal field-effect transistors, *Appl. Phys. Lett.* 85: 3899.
101. N. Kirova and M. N. Bussac, 2003. Self-trapping of electrons at the field-effect junction of a molecular crystal, *Phys. Rev. B* 68: 235312.
102. H. Houili, J. D. Picon, L. Zuppiroli, and M. N. Bussac, 2006. Polarization effects in the channel of an organic field-effect transistor, *J. Appl. Phys.* 100: 023702.
103. P. Kumar, A. Misra, M. N. Kamalasanan, S. C. Jain, R. Srivastava, and V. Kumar, 2006. Temperature effect on current–voltage characteristics of molecular organic tris(8-hydroxyquinoline) aluminium complex, *Jpn. J. Appl. Phys.* 45: 7621–7624.
104. R. Kiebooms, R. Menon, and K. Lee, 2001. Synthesis, electrical, and optical properties of conjugated polymers, in *Handbook of Advanced Electronic and Photonic Materials and Devices*, vol. 8, H. S. Nalwa (ed.), chap. 1, New York: Academic.
105. P. Kumar, A. Misra, M. N. Kamalasanan, S. C. Jain, and V. Kumar, 2007. Charge transport through conducting organic poly(2-methoxy-5-(2-ethylhexyloxy)-1,4-phenylenevinylene), *J. Phys. D: Appl. Phys.* 40: 561.
106. C. D. Dimitrakopoulos and D. J. Mascaro, 2001. Organic thin-film transistors: A review of recent advances, *IBM J. Res. Develop.* 45: 11.
107. N. Karl, K. H. Kraft, J. Marktanner, M. Munch, F. Schatz, R. Stehle, and H. M. Uhde, 1999. Fast electronic transport in organic molecular solids? *J. Vac. Sci. Technol. A* 17: 2318.
108. P. G. Lecomber and W. E. Spear, 1970. Electronic transport in amorphous silicon films, *Phys. Rev. Lett.* 25: 509.
109. J. Noolandi, 1977. Equivalence of multiple-trapping model and time-dependent random walk, *Phys. Rev. B* 16: 4474.
110. O. D. Jurchescu, J. Baas, and T. T. M. Palstra, 2004. Effect of impurities on the mobility of single crystal pentacene, *Appl. Phys. Lett.* 84: 3061.
111. J. Veres, S. D. Ogier, S. W. Leeming, D. C. Cupertino, and S. M. Khaffaf, 2003. Low-*k* insulators as the choice of dielectrics in organic field-effect transistors, *Adv. Funct. Mater.* 13: 199–204.
112. W. Zhu, S. J. Singer, Z. Zheng, and A. T. Conlisk, 2005. Electro-osmotic flow of a model electrolyte, *Phys. Rev. E* 71: 041501.
113. C. Goh, R. J. Kline, M. D. McGehee, E. N. Kadnikova, and J. M. J. Fréchet, 2005. Molecular-weight-dependent mobilities in regioregular poly(3-hexyl-thiophene) diodes, *Appl. Phys. Lett.* 86: 122110.
114. R. J. Kline, M. D. McGehee, E. N. Kadnikova, J. Liu, and J. M. J. Fréchet, 2003. Controlling the field effect mobility of regioregular polythiophene by changing the molecular weight, *Adv. Mater.* 15: 1519.
115. A. Zen, J. Pflaum, S. Hirschmann, W. Zhuang, F. Jaiser, U. Asawapirom, J. P. Rabe, U. Scherf, and D. Neher, 2004. Effect of molecular weight and annealing of poly(3-hexylthiophene)s on the performance of organic field effect transistors, *Adv. Funct. Mater.* 14: 757.
116. H. Sirringhaus, P. J. Brown, R. H. Friend, M. M. Neilson, K. Bechgaard, B. M. W. LangeveldVoss, A. J. H. Spiering, R. A. J. Janssen, and E. W. Meijer, 1999. Two-dimensional charge transport in self-organized, high-mobility conjugated polymers, *Nature* 401: 685.
117. A. Menon, H. P. Dong, Z. I. Niazimbetova, L. J. Rothberg, and M. E. Galvin, 2002. Polydispersity effects on conjugated polymer light-emitting diodes, *Chem. Mater.* 14: 3668–3675.
118. P. Kumar, S. C. Jain, V. Kumar, S. Chand, and R. P. Tandon, 2009. A model for the *J–V* characteristics of P3HT:PCBM solar cells, *J. Appl. Phys.* 105: 104507.
119. V. D. Mihailetchi, P. W. M. Blom, J. C. Hummelen, and M. T. Rispens, 2003. Cathode dependence of the open-circuit voltage of polymer:fullerene bulk heterojunction solar cells, *J. Appl. Phys.* 94: 6849–6854.
120. P. Schilinsky, C. Waldauf, J. Hauch, and C. J. Brabec, 2004. Simulation of light intensity dependent current characteristics of polymer solar cells, *J. Appl. Phys.* 95: 2816.
121. A. Jain and A. Kapoor, 2005. A new approach to study organic solar cell using Lambert W-function, *Sol. Energy Mater. Sol. Cells* 86: 197.

122. W. J. Potscavage, Jr., A. Sharma, and B. Kippelen, 2009. Critical interfaces in organic solar cells and their influence on the open circuit voltage, *Acc. Chem. Res.* 42: 1758.
123. J. Peet, J. Y. Kim, N. E. Coates, W. L. Ma, D. Moses, A. J. Heeger, and G. C. Bazan, 2007. Efficiency enhancement in low-bandgap polymer solar cells by processing with alkane dithiols, *Nat. Mater.* 16: 497.
124. J. K. J. van Duren, X. Yang, J. Loos, C. W. T. Bulle-Lieuwma, A. B. Sieval, J. C. Hummelen, and R. A. H. Janssen, 2004. Relating the morphology of poly(phenylenevinylene)/methanofullerene blends to solar cell performance, *Adv. Funct. Mater.* 14: 425.
125. G. Li, V. Shrotriya, J. Huang, Y. Yao, T. Moriarty, K. Emery, and Y. Yang, 2005. High efficiency solution processable polymer photovoltaic cells by self-organization of polymer blends. *Nat. Mater.* 4: 864.
126. W. Ma, C. Yang, X. Gong, K. Lee, and A. J. Heeger, 2005. Thermally stable, efficient polymer solar cells with nanoscale control of the interpenetrating network morphology, *Adv. Funct. Mater.* 15: 1617.
127. G. L. Araujo, E. Sanchez, and M. Marti, 1982. Determination of the two exponential solar cell equation parameters from empirical data, *Sol. Cells* 5: 199.
128. B. Mazhari, 2006. An improved solar cell circuit model for organic solar cells, *Sol. Energy Mater. Sol. Cells* 90: 1021.
129. A. Gaur and P. Kumar, 2014. An equivalent circuit model for polymer solar cells, *Prog. Photovolt.: Res. Appl.* 22: 937.
130. P. Kumar, H. Kumar, S. C. Jain, P. Venkatesu, S. Chand, and V. Kumar, 2009. Effect of active layer thickness on open circuit voltage in organic photovoltaic devices, *Jpn. J. Appl. Phys.* 48: 121501.
131. N. I. Cracium, J. J. Brondijk, and P. W. M. Blom, 2008. Diffusion-enhanced hole transport in thin polymer light-emitting diodes, *Phys. Rev. B* 77: 35206.
132. G. Dennler, M. C. Scharber, and C. J. Brabec, 2009. Polymer–fullerene bulk-heterojunction solar cells, *Adv. Mater.* 21: 1323.
133. H. Hoppe, N. Arnold, N. S. Sariciftci, and D. Meissner, 2003. Modeling the optical absorption within conjugated polymer/fullerene-based bulk-heterojunction organic solar, *Sol. Energy Mater. Sol. Cells* 80: 105–113.
134. L. J. A. Koster, E. C. P. Smits, V. D. Mihailetchi, and P. W. M. Blom, 2005. Device model for the operation of polymer/fullerene bulk heterojunction solar cells, *Phys. Rev. B* 72: 85205.
135. M. K. Lee, J. C. Wang, S. F. Horng, and H. F. Meng, 2010. Extraction of solar cell series resistance without presumed current–voltage functional form, *Sol. Energy Mater. Sol. Cells* 94: 578–582.
136. C. Waldauf, P. Schilinsky, J. Hauch, and C. J. Brabec, 2004. Material and device concepts for organic photovoltaics: Towards competitive efficiencies, *Thin Solid Films* 451–452: 503–507.
137. E. A. Katz, D. Faiman, S. M. Tuladhar, J. M. Kroon, M. M. Wienk, T. Fromherz, F. Padinger, C. J. Brabec, and N. S. Sariciftci, 2001. Temperature dependence for the photovoltaic device parameters of polymer–fullerene solar cells under operating conditions, *J. Appl. Phys.* 90: 5343.
138. P. Kumar, S. C. Jain, H. Kumar, S. Chand, and V. Kumar, 2009. Effect of illumination intensity and temperature on open circuit voltage in organic solar cells, *Appl. Phys. Lett.* 94: 183505.
139. P. Kumar and A. Gaur, 2013. Model for the *J–V* characteristics of degraded polymer solar cells, *J. Appl. Phys.* 113: 94505.
140. P. Kumar, A. Sharma, and D. P. Singh, 2013. Effect of voltage sweep direction on the performance evaluation of P3HT:PCBM solar cells, *Prog. Photovolt.: Res. Appl.* 21: 950.
141. T. Kirchartz, T. Agostinelli, M. C. Quiles, W. Gong, and J. Nelson, 2012. Understanding the thickness-dependent performance of organic bulk heterojunction solar cells: The influence of mobility, lifetime, and space charge, *J. Phys. Chem. Lett.* 3: 3470–3475.
142. H. Kumar, P. Kumar, N. Chaudhary, R. Bhardwaj, S. Chand, S. C. Jain, and V. Kumar, 2009. Effect of temperature on the performance of CuPc/C_{60} photovoltaic device, *J. Phys. D.: Appl. Phys.* 42: 15102.
143. L. J. A. Koster, V. D. Mihailetchi, R. Ramaker, and P. W. M. Blom, 2005. Light intensity dependence of open-circuit voltage of polymer:fullerene solar cells, *Appl. Phys. Lett.* 86: 123509.

144. C. J. Brabec, A. Cravino, D. Meissner, N. S. Saricoftci, T. Fromherz, M. T. Rispens, L. Sanchez, and J. C. Hummelen, 2001. Origin of the open circuit voltage of plastic solar cells, *Adv. Funct. Mater.* 11: 374–380.
145. E. A. Katz, D. Faiman, S. M. Tuladhar, J. M. Kroon, M. M. Wienk, T. Fromherz, F. Padinger, C. J. Brabec, and N. S. Sariciftci, 2001. Temperature dependence for the photovoltaic device parameters of polymer–fullerene solar cells under operating conditions, *J. Appl. Phys.* 90: 5343–5350.
146. H. Frohne, S. E. Shaheen, C. J. Brabec, D. C. Muller, N. S. Sariciftci, and K. Meerholz, 2002. Influence of the anodic work function on the performance of organic solar cells, *Chem. Phys. Chem.* 3: 795–799.
147. C. M. Ramsdale, J. A. Barker, A. C. Arias, J. D. MacKenzie, R. H. Friend, and N. C. Greenham, 2002. The origin of the open circuit voltage in polyfluorene-based photovoltaic devices, *J. Appl. Phys.* 92: 4266–4270.
148. M. Y. Chan, C. S. Lee, S. L. Lai, M. K. Fung, F. L. Wong, H. Y. Sun, K. M. Lau, and S. T. Lee, 2006. Efficient organic photovoltaic devices using a combination of exciton blocking layer and anodic buffer layer, *J. Appl. Phys.* 100: 94506.
149. C. Uhrich, D. Wynands, S. Olthof, M. K. Riede, K. Leo, S. Sonntag, B. Maennig, and M. Pfeiffer, 2008. Origin of open circuit voltage in planar and bulk heterojunction organic thin-film photovoltaics depending on doped transport layers, *J. Appl. Phys.* 104: 43107.

5

Degradation and Its Characterization in Organic Solar Cells

5.1 Introduction

Silicon (Si) solar cells presently dominate the photovoltaic (PV) market, but some of their inherent limitations and high cost make them less attractive for cost-effective energy production. Researchers are trying to find ways for cost-effective and efficient energy production by reducing the cost of Si PVs or using alternate materials for PV applications. Organic solar cells (OSCs) could be an effective alternative to Si PVs, as they could reduce the energy production cost from PV modules. Additionally they are thin, lightweight, semitransparent, and mechanically flexible giving them a freedom of incorporation into different surfaces making new types of solar-energy-collecting shapes and designs. But OSCs suffer from the flaws of lower efficiency and deterioration in performance with time. Like efficiency, cost, and processing, stability is another decisive parameter for success and real-life application of OSC technology. The lifetime of Si PV devices ranges up to 25 years or more, whereas the lifetime of OSCs ranges from months to years. Si PV companies are still trying to improve the lifetime up to 40 years. Therefore to compete with conventional PV technologies, OSCs need to be improved tremendously. OSCs of high efficiency with long operational lifetime are still to be realized and it is a real challenge for this technology. Unlike inorganic semiconductors, organic semiconductors are by nature very susceptible to oxygen- and moisture-induced degradation, and degradation happens throughout the device from the top electrode to the bottom electrode. Oxygen and moisture may get introduced during device preparation or even after preparation and react with device components. Understanding and preventing degradation is highly important for the success of this emerging technology. There have been a number of studies on degradation that show that it is a complicated phenomenon and not yet completely understood; however, there is a rough classification between chemical and physical degradation. Chemical degradation includes formation of organic molecules with inferior electrical and optical properties due to reaction of original organic molecules with oxygen, moisture, and electrode materials [1–4], whereas physical degradation is because of delamination of the top electrode, change in the active layer morphology, diffusion of electrode materials into organic layers, and degradation of different interfaces [5–7]. Photoelectrochemical reactions, photooxidation, and oxidation of low work function electrodes are other important chemical degradation processes [8–10]. Degradation happens in both the dark and illumination; however, under illumination the degradation is expected to accelerate due to heating and ultraviolet (UV) light exposure. It is important to note the intensity and spectrum of incident light, the way of device illumination, device temperature, and surrounding conditions, as the rate of degradation and degradation mechanisms may be different for different test conditions. Therefore, where different materials and different test conditions are employed, direct comparison of stability becomes very difficult. In this chapter, I shall discuss how diversified the degradation in OSCs is. The variety of tools and techniques used for characterization of degradation are also discussed.

5.2 Chemical Degradation

Chemical degradation mainly happens due to reactions of oxygen, moisture, and electrode materials with organic molecules. To map water and oxygen diffusion into OSCs, Norrman et al. developed a technique where solar cells were exposed to isotopically labeled water ($H_2{}^{18}O$) and oxygen ($^{18}O_2$) molecules and subsequently subjected to time-of-flight secondary ion mass spectroscopy (TOF-SIMS) analysis. The detection of reaction products helped in mapping the diffusion of oxygen and moisture molecules. The top electrode was found to be the main entrance for moisture and oxygen diffusion, and the sides of the devices were less important [1,11–13]. Oxygen and moisture molecules were found to diffuse throughout from the top electrode to the bottom electrode regardless of whether the device was kept in the dark or under illumination. Microscopic pinholes and the grains in the evaporated top metal electrode are the main channels for oxygen and moisture entrance. Oxygen and moisture diffusion through pinholes oxidizes the underlying organic materials, which expand in all directions to burst the top electrode surface centered on the microscopic pinholes [11]. Illumination further accelerates the oxidation, as UV light readily activates oxygen. It fails the performance of solar cells; therefore for longer lifetime, elimination of pinholes from the top electrode is highly important.

5.2.1 Degradation in Active Organic Materials

There are many reports in the literature where degradation in OSCs and materials has been studied [2,8,9,14–18]. It is found that the organic molecules react with oxygen and water molecules and form new compounds with inferior optical and electrical properties. Some organic molecules degrade slowly, whereas some degrade faster and it depends strongly on the availability of groups that are more vulnerable to degradation. For example P3HT molecules are comparatively more stable than poly(phenylene-vinylene) (PPV) derivatives due to lack of easily oxidizable vinylene groups. PPV derivatives are very sensitive to photooxidation, and under illumination the excited polymer transfers its energy to the adsorbed oxygen molecule to make it in singlet state. Though for this process to take place the triplet state of polymer is required to be higher in energy than the singlet state of oxygen. The singlet oxygen reacts with the vinyl group of PPVs causing chain scission as shown in Figure 5.1 [14]. Cumpston et al. studied the photooxidation of some OSCs with infrared reflection absorption spectroscopy [17] and Neugebauer et al. studied the photochemical degradation of MDMO-PPV using Fourier transform infrared spectroscopy [8]. Bianchi et al. studied the photooxidation of MDMO-PPV and found that after oxidization the vinyl groups convert into carbonyl groups [16]. The degradation rate depends on the nature of the molecule and its substituents. The molecules having electron-donating substituents degrade faster than those having electron-withdrawing groups [15]. Padinger et al. studied degradation in MDMO-PPV:PCBM solar cells [18]. The devices were kept in the

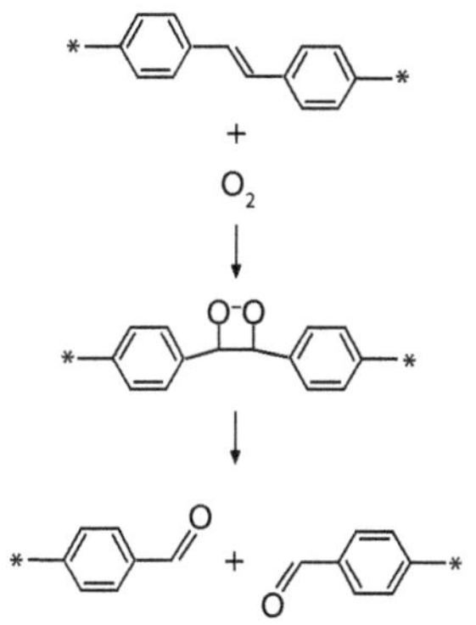

FIGURE 5.1
Chain scission of a PPV derivative by singlet oxygen. (Redrawn with permission from M. Jørgensen, K. Norrman, and F. C. Krebs, 2008. *Sol. Energy Mater. Sol. Cells* 92: 686–714.)

dark and under 80 mW/cm² irradiance of a white light source. Studies showed that the device kept under illumination degraded faster than that kept in the dark. The degradation was accelerated when the device was kept at higher temperatures.

Chambon et al. studied the photo and thermal degradation of thin films of MDMO-PPV, PCBM, and their blends in the presence/absence of oxygen [2]. They found modification in chemical structures of the materials, and the changes were examined via UV–visible absorption and infrared (IR) spectroscopic studies. MDMO-PPV was shown to degrade by a redicalar oxidation, which first oxidized the ether function and then the double bonds, whereas PCBM degradation was mainly photooxidation of C_{60} moiety. In the case of the MDMO-PPV:PCBM blend, both the constituents degraded but the presence of PCBM reduced degradation of MDMO-PPV. Figure 5.2 shows the degradation profile of normalized absorption of IR band of ether function at 1352 cm^{-1} for pristine MDMO-PPV and its blend with PCBM. The ether side chain in MDMO-PPV is very sensitive to radical initiated oxidation and the IR band corresponding to ether bonding vanished in about 100 h under illumination. PCBM avoids not only the photooxidation but also the thermo-oxidation and photolysis of MDMO-PPV. This happens because of quenching of triplet states of

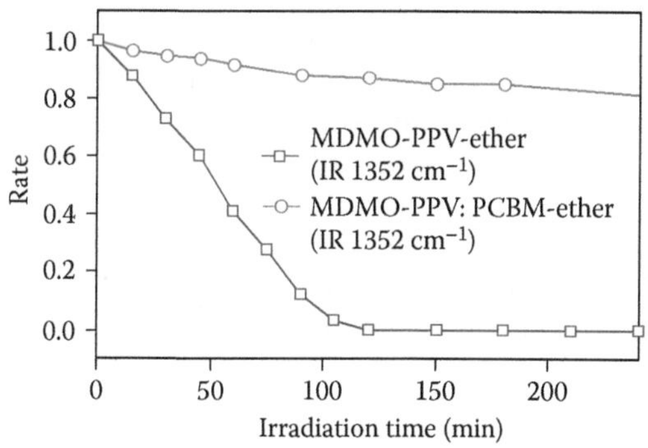

FIGURE 5.2
Degradation in ether bond of MDMO-PPV during photooxidation with and without PCBM doping. (Redrawn with permission from S. Chambon et al., 2007. *Sol. Energy Mater. Sol. Cells* 91: 394–398.)

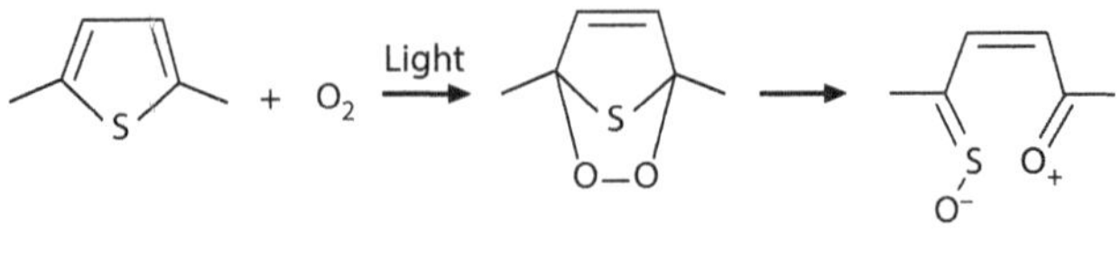

FIGURE 5.3
Oxidation and further decomposition of one of the thiophene derivatives under illumination. (Redrawn with permission from M. Jørgensen, K. Norrman, and F. C. Krebs, 2008. *Sol. Energy Mater. Sol. Cells* 92: 686–714.)

MDMO-PPV, which avoids formation of singlet oxygen and prevents oxidation of vinylene group and bond breaking. More detailed experiments performed for photooxidation analysis of some specific PPV derivatives suggested similar chemical reactions [19,20].

For thiophene derivatives like P3HT, oxygen is supposed to form a reversible charge transfer complex possessing cations and anions which could be detected by electron spin resonance (ESR). Dissociation of the charge transfer complex seems to degrade the polymer. In the process of photooxidation, the adsorbed O_2 takes electrons from the photoexcited polymer and forms singlet O_2, which breaks down the polymer chain to form new entities and new intermediate adducts as shown in Figure 5.3 [2,9,21]. Formation of charged entities represents *p*-type doping in P3HT, and it increases with increase in O_2 concentration [22]. Electron transfer from P3HT to the O_2 molecule results in formation of mobile holes on P3HT and immobile electrons on O_2. O_2-induced degradation of P3HT decreases charge carrier mobility and increases trap densities [22]. Norrman et al. did investigate the O_2 uptake in a P3HT:PCBM solar cell and the fragment ions were studied by TOF-SIMS [4]. It has been found that in P3HT, sulfur reacts with H_2O and/or O_2 molecules and increases $C_xH_ySO_z^-$ and SOx⁻ concentrations. A reduction in $C_xH_yS^-$ ions and increase in $C_xH_yO_z^-$ ions suggest that either the thiophene carbon in P3HT becomes oxygenated or the side chains get oxygenated. O_2 exposure has been observed to increase charge carrier concentration and decrease the carrier mobility with enhanced trap density [23]. In a way, oxygen doping increases conductivity of P3HT by increasing the number of charge carriers [21,24,25]. Luer et al. suggested that metastable oxygen species exhibit slow but irreversible degradation in P3HT [24]. Formation of ionic entities on degradation of P3HT was also confirmed from hysteresis behavior observed in current density–voltage (*J–V*) characteristics of P3HT-based degraded solar cells under illumination [26]. The ionic entities cause polarization effect in the active layer that leads to hysteresis behavior.

For the hybrid solar cells that incorporate inorganic semiconducting oxides such as TiO_2, SnO_2, and ZnO, the oxidation is further accelerated as the oxide semiconductors serves as photocatalysts for oxidation of organic molecules [27–30]. In the presence of UV light, oxide semiconductors like TiO_2 generate a superoxide radical anion $O_2^{\bullet-}$ and cation $TiOH^{\bullet+}$, which can oxidize and degrade many organic molecules [27,28]. The necessity of oxygen for operation in some of the oxide-semiconductor-based hybrid solar cells makes the role of oxygen unclear on the stability of these solar cells [31]. UV light is very harmful for organic molecules, as it leads to irreversible degradation and photobleaching of the materials. However, application of UV filters can prevent UV-light-induced degradation to some extent [32].

5.2.2 Degradation in Electrode Materials

As far as the degradation of electrodes is concerned, electrochemical reactions take place and cause corrosion at both the anode and cathode to degrade their properties. The solar cells based on low work function metal cathodes like Ca, Ba, Mg, Al, and Ag show high performance, but unfortunately these metals have poor stability in ambient air. The diffused oxygen and moisture react and cause oxidation of these metallic electrodes. Degradation phenomena at the anode and cathode interfaces are very complicated. We fabricated OSCs based on Ca/Al and Ca/Ag cathodes and studied chemical thermodynamics of their degradation [33]. The most relevant atmospheric oxidation reactions for these materials can be given by following equations:

$$2Ca + O_2 \rightarrow 2CaO \tag{5.1}$$

$$Ca + H_2O \rightarrow CaO + H_2 \tag{5.2}$$

$$CaO + H_2O \leftrightarrow Ca(OH)_2 \tag{5.3}$$

$$4Ag + O_2 \leftrightarrow 2Ag_2O \tag{5.4}$$

$$2Ag + H_2O \rightarrow Ag_2O + H_2 \tag{5.5}$$

$$4Al + 3O_2 \rightarrow 2Al_2O_3 \tag{5.6}$$

$$2Al + 3H_2O \rightarrow Al_2O_3 + 3H_2 \tag{5.7}$$

Chemical thermodynamics of an oxidation reaction determines the relative ease of the oxidation pathway for a given metal and is commonly presented in the form of an Ellingham diagram [34]. These diagrams plot the Gibbs free energy of formation per mole of oxygen consumption (ΔG_f) as a function of temperature and are typically used in metallurgy to determine preferential oxidation in mixed metal systems (such as alloys) [35]. The Ellingham diagrams for the reaction of the three metal electrode materials with oxygen in the solid state is shown in Figure 5.4a, which shows that the order of oxidation preference at all temperatures is Ca, Al, and Ag. Moreover, from Figure 5.4b it was observed that

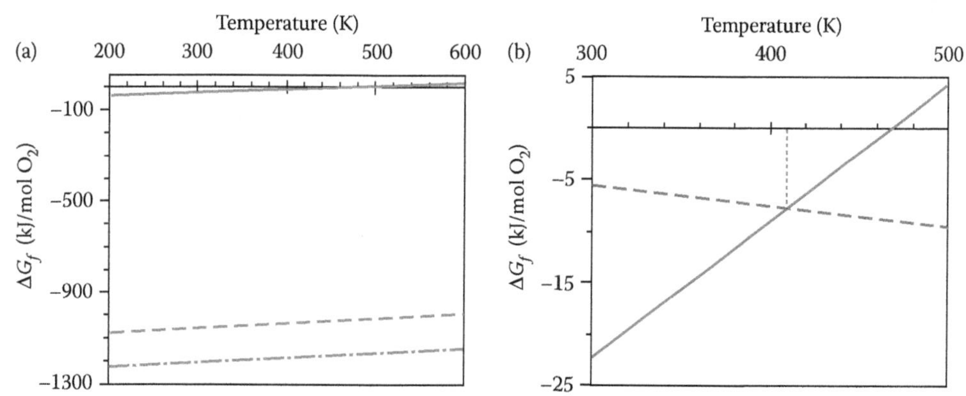

FIGURE 5.4
(a) Ellingham diagram for oxidation of Ca (dash-dotted line), Al (dashed line), and Ag (solid line). (b) Ellingham diagram for the oxidation of Ag (solid line) along with the free energy line for a partial pressure of O_2 of 10^{-4} kPa (dashed line). The intercept of the two lines indicates the equilibrium oxidation temperature for the formation of Ag_2O (dotted line). (Reproduced with permission from P. Kumar et al., 2014. *ACS Appl. Mater. Inter.* 6: 5281.)

at a partial pressure of O_2 of 10^{-4} kPa (~1 ppm) the equilibrium oxidation temperature for Ag_2O is lowered from its value of 462 K (189°C) under ambient conditions to 409 K (136°C). Oxidation of these metal electrodes can also occur through interaction with water. In the case of Ca, this reaction is much more facile than the oxygen reaction and dominates the oxidation process under typical atmospheric conditions. In addition, CaO readily undergoes a further hydrolysis reaction in the presence of water to produce $Ca(OH)_2$, whereas Ag_2O and Al_2O_3 are generally chemically inert to hydrolysis under ambient conditions. Work by Schaube et al. has shown that the equilibrium temperature for Equation 5.3 can be written as [36]

$$\ln\left(\frac{P_{eq}}{10^5}\right) = \frac{-12845}{T_{eq}} + 16.508, \tag{5.8}$$

where P_{eq} is the equilibrium partial pressure of H_2O in Pa, and T_{eq} is the equilibrium temperature in kelvins. Figure 5.5 shows that as the partial pressure of water drops, the equilibrium temperature of CaO hydrolysis reaction also decreases. Moreover, at a partial pressure of H_2O of 10^{-4} kPa (~1 ppm) the equilibrium temperature for CaO hydrolysis is lowered to a value of 424 K (151°C). This analysis reveals that not all of the oxidative processes that occur during degradation of the polymer solar cell (PSC) cathodes are irreversible. Thus, while Al_2O_3 and CaO readily form at all temperatures, Ag_2O will convert back to Ag at ~140–150°C under conditions of low relative O_2 concentration. Likewise, under dry conditions the hydrolysis of $Ca(OH)_2$ is reversed at similar temperatures. OSC fabrication is typically conducted under an inert atmosphere (with O_2 and H_2O concentrations kept below 1 ppm) with processing conditions (such as annealing) typically occurring at 140–150°C. Consequently, these fabrication processes will not just alter the morphology of the active layer but will also change the oxidation state of the metal electrodes. Moreover, it should be possible to reverse certain degradation (oxidation) processes under the correct conditions for particular materials.

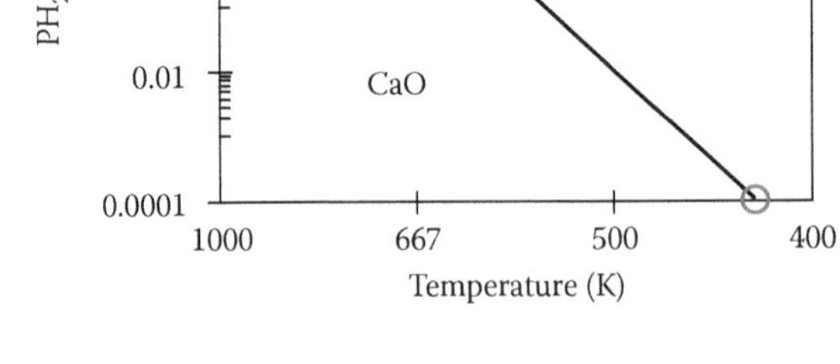

FIGURE 5.5
Variation in the equilibrium temperature of CaO hydrolysis reaction as a function of partial pressure of water. The open circle indicates the equilibrium temperature (424 K) at a partial pressure of H_2O of 10^{-4} kPa. (Reproduced with permission from P. Kumar et al., 2014. *ACS Appl. Mater. Inter.* 6: 5281.)

The metallic electrodes have also been observed to react with organic molecules [33,37,38]. Logdlund et al. proposed formation of some organo-aluminum compounds due to reaction of Al with organic molecules [37] and such species are highly reactive to any proton donor like H_2O or O_2. Due to high electron affinity, fullerene and its derivatives are more likely to react with metal electrodes to form metal-C_{60} bonding [36]. To avoid direct contact of the metal electrode with fullerene, a thin buffer layer or an electron transport layer is introduced between the active layer and the metal cathode. Lithium fluoride (LiF) has been used most frequently as an electron buffer layer whereas tri(8-hydroxy quinoline) aluminum (Alq_3), bathocuproine (BCP) and bathophenanthroline (BPhen) have been used as electron transport materials for OSCs [39–42]. Insertion of a thin electron buffer layer also has many other benefits for OSCs. It retards the diffusion of oxygen and moisture into the active layer and protects the active layer from direct heating by thermally evaporated hot metal electrode. Though BCP improves the cell performance, it readily crystallizes in moist environment and results in a not very long lifetime [39]. Song et al. [40] replaced BCP with Alq_3 in small molecular solar cells and found ~150 times longer life compared to that with BCP. Similarly, replacement of BCP with 1,3,5-tris(2,*N*-phenylbenzimidazolyl) benzene has also shown improvement in the lifetime of the devices [41]. LiF reduces the effective work function of cathode and leads to higher open-circuit voltage (V_{oc}) of the solar cells [42]. There are also few reports that show that insertion of a C_{60} layer between the active organic layer and Al cathode improves the lifetime of solar cells [1,43]. It is expected that a thin layer of Al_2O_3 is formed during preparation or after preparation at the C_{60}–Al interface leading to improved stability.

PEDOT:PPS is an important material and is frequently used as anode buffer layer for efficient solar cells. PEDOT:PSS serves multiple purposes like enhancing the effective work function of the anode, smoothing the anode–organic interface (that is very rough otherwise), and transporting the photogenerated hole from the active organic layer to the anode. But unfortunately it also causes rapid degradation in cell performance. Due to its acidic nature, it etches indium tin oxide (ITO) and hampers its electronic properties. Jong et al. used the Rutherford backscattering (RBS) technique to investigate the effect of PEDOT:PSS on ITO and found decomposition of ITO where indium diffuses from ITO to PEDOT:PSS [44]. In ambient air this effect got accelerated and after 10 days PEDOT:PSS had about 1.2% indium. Indium diffuses into other layers as well up to the counter electrode [1,7,33]. Such changes do modify the work function of ITO and due to creation of more free electrons in ITO, the Fermi level goes up. Also the presence of indium in active layers affects its electronic properties leading to inferior photovoltaic performance. PEDOT:PSS is a hygroscopic material and adsorbs moisture from the surrounding. Though PEDOT:PSS is coated from its aqueous solution, after film preparation, adsorption of moisture results in loss of its electronic properties. Kawano et al. studied degradation in MDMO-PPV:PCBM solar cells with and without PEDOT:PSS and found that PEDOT:PSS accelerates the degradation in humid condition [45]. The hygroscopic PEDOT:PSS before the active layer adsorbs water molecules and results in light-independent degradation. The solar cells having no PEDOT:PSS layer exhibited lower efficiency but better stability. Studies showed that water interaction of PEDOT:PSS appears to formulate insulating patches, which increases the resistivity of the PEDOT:PSS/active layer interface that results in the loss of device current and device efficiency. Norrman et al. proposed that PSS from PEDOT:PSS diffuses and reacts with other components of the devices [11]. They suggested that PSS goes through an oxido-de-silfonato substitution that reacts back with PSS and results in a molecule with two PSS chains linked together via a sulfonic ester group. These molecules form particles in the device and contribute to cell degradation.

5.3 Physical Degradation

Physical degradation includes degradation due to changes in nanophase and interfacial morphologies, delamination of the top electrode, and its diffusion into organic layers. In bulk-heterojunction (BHJ) solar cells, the spatial organization of different constituents and formation of interpenetrating networks is critical for efficient photon harvesting. Also the charge carrier transport, nanophase of active layer, and interfacial morphologies are equally important for efficient photon harvesting. BHJ architecture has proved to be quite efficient, but thermodynamically it is probably not favored, and achievement of an optimum nanophase separation is not easy; it depends on the kinetics of segregation and crystallization of the constituents. Tremendous efforts are being made to get the best morphology out of a donor–acceptor system. In the case of polymer BHJ solar cells, the nanophase morphology depends on a number of parameters such as solvent used, rate of solvent evaporation, ratio of donor–acceptor materials and their compatibility, concentration, temperature, film preparation process, and pre-/postthermal treatments. Along with morphological changes the organic molecules may undergo crystallization with time. Yang et al. extensively investigated the phase separation and effect of morphology on degradation of MDMO-PPV:PCBM solar cells [5]. At large PCBM concentrations, the PCBM formed nanocrystals with random crystallographic orientations, whereas at lower PCBM concentrations, the nanocrystals were very small in size, and the morphology was quite homogeneous. Thermal annealing of the blend films above the glass transition temperature (T_g) of MDMO-PPV accelerated the phase separation because PCBM molecules could diffuse through the MDMO-PPV matrix and formed large single crystals. Figure 5.6 shows the transmission electron microscopy (TEM) images of MDMO-PPV:PCBM blend films for different PCBM concentrations, which were annealed at 130°C for different time durations [5]. Figure 5.6a and b are the TEM images of the films containing 50 wt% PCBM, which were annealed at 130°C for 10 and 60 min, respectively, whereas Figure 5.6c and d are the TEM images of the films with 80 wt% PCBM, which were annealed at 130°C for 10 and 60 min, respectively.

The growth and change in size of PCBM nanocrystals (dark areas in the images) could be clearly seen. Formation of large single crystals increased the dimension and extent of phase segregation and could be seen in the MDMO-PPV:PCBM system at 60°C, which was ~20°C below the T_g of MDMO-PPV. Studies were also carried out on the films prepared using different solvents and techniques. Figure 5.7 shows the TEM images of MDMO-PPV:PCBM (1:4 wt%) thin films prepared by spin coating from toluene and chlorobenzene solutions. Phase segregation could be clearly seen and the size of PCBM-rich domains changed with the choice of solvent. Later, Yang et al. investigated the stability of a P3HT:PCBM device and found that the annealing of the device at 120°C resulted in a stable morphology with device lifetime of more than 1000 h at 70°C under 100 mW/cm^2 [46]. Water-induced degradation studies on normal polymer solar cells based on P3HT:PCBM and MEH-PPV:PCBM showed that water behaves similarly but not identically with molecular oxygen [47]. Norrman et al. did present the studies on the mapping of spatial degradation due to moisture and oxygen in inverted ITO/ZnO/P3HT:PCBM/PEDOT:PSS/Ag polymer solar cells [4]. X-ray photoelectron spectroscopy (XPS) and TOF-SIMS were used to get the information about where and to what extent the degradation took place. Degradation was studied in detail in humid (oxygen free) and oxygen environments in the dark and under illumination. In these inverted devices, the reactions taking place at the active layer–PEDOT:PSS interface were found to be the major candidates of degradation.

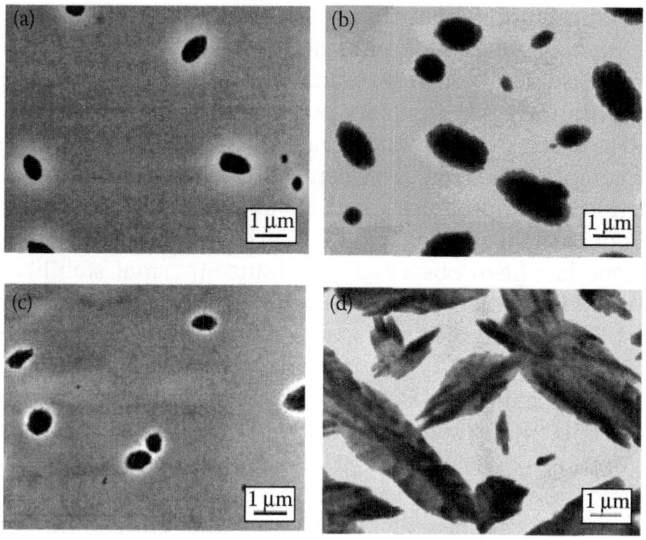

FIGURE 5.6
Formation of PCBM crystals in MDMO-PPV:PCBM blend concentration of 50 wt% (a, b) and 80 wt% (c, d) after annealing at 130°C for 10 min (a, c) and 60 min (b, d). (Reprinted with permission from X. Yang et al., Morphology and thermal stability of the active layer in poly(*p*-phenylenevinylene)/methanofullerene plastic photovoltaic devices, *Macromolecules*, 37, 2151–2158. Copyright 2004 American Chemical Society.)

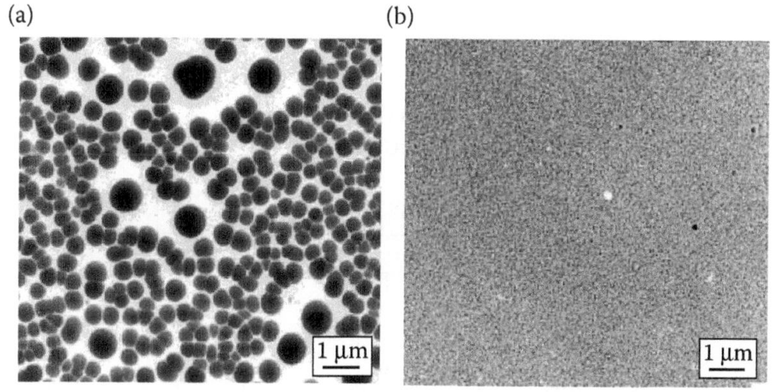

FIGURE 5.7
TEM images of MDMO-PPV:PCBM films (1:4 wt%) prepared by spin coating using (a) toluene and (b) chlorobenzene as the solvent. (Reprinted with permission from X. Yang et al., Morphology and thermal stability of the active layer in poly(*p*-phenylenevinylene)/methanofullerene plastic photovoltaic devices, *Macromolecules*, 37, 2151–2158. Copyright 2004 American Chemical Society.)

PEDOT:PSS exhibited phase separation where the PEDOT phase was responsible for interface degradation in the oxygen environment. In the case of a humid environment, some chemically induced degradation was observed. Contrary to their behavior in normal geometry, the inverted cells exhibited short lifetime in O_2 and long lifetime in saturated H_2O environment. Tromholt et al. did present the studies on the effect of concentrated sunlight (0.2–27 suns) on photovoltaic performance and stability for P3HT:PCBM inverted solar cells [48]. For illumination intensity of <10 sun, J_{sc} (short-circuit current density) was found to increase linearly, whereas V_{oc} increased logarithmically with J_{sc}. The efficiency was observed to increase first, which saturated and finally reduced strongly with illumination intensity. For illumination intensity of >10 sun, V_{oc} was observed to decrease, and this was attributed to heating due to concentrated light. These studies suggest that a better understanding of complex interplay between morphological changes and the chemistry of films as a function of temperature can lead to significant improvements in both the performance and operational stability. Copolymerization of P3HT with a little amount of 3,4-dihexyl-thiophene has been observed to enhance thermal stability of the solar cells [49]. The concept was to slightly reduce the regioregularity of P3HT with introduction of dialkyl monomers. This copolymer had stable nanophase BHJ morphology, which was evident by TEM images [49].

We studied morphological changes induced by thermal annealing in solar cells based on P3HT:PCBM and P3HT:ICBA active layers [50]. The two systems exhibit entirely different morphologies and the one based on P3HT:ICBA was more stable. The solar cells were subjected to multiple thermal annealing treatments at 140°C in a glove box and studied for photovoltaic performance. The effect of repeated annealing cycle on the cell efficiency is shown in Figure 5.8a. Where P3HT:PCBM solar cells exhibited a rapid reduction in

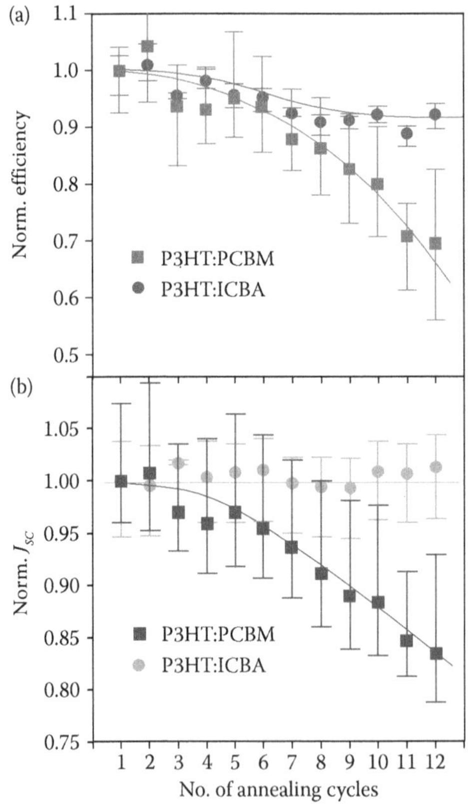

FIGURE 5.8
(a) Effect of subsequent thermal annealing on the efficiency of ITO/PEDOT:PSS/P3HT:ICBA/Ca/Ag and ITO/PEDOT:PSS/P3HT:PCBM/Ca/Ag solar cells. (b) The effect of thermal annealing on J_{sc} of the respective devices. (Reproduced with permission from P. Kumar et al., 2014. *Appl. Phys. Lett.* 104: 193905.)

efficiency (predominantly because of a correlated reduction in J_{sc}), the P3HT:ICBA solar cells exhibited only a modest reduction (<10%) in efficiency, which was consistent with no deterioration in J_{sc} as shown in Figure 5.8b. The spectral and morphological properties of P3HT blends with PCBM and ICBA are completely different [51]. Upon extended annealing the P3HT:PCBM system underwent phase segregation to form highly separated P3HT and PCBM domains, whereas ICBA remained miscible for a wide range of polymer compositions. This different behavior of ICBA could be simply attributed to its structure. ICBA is a bisadduct fullerene possessing two indene side groups, which hinders its crystallization or aggregation. The formation of large phase segregated regions is known to cause device degradation, consistent with the systematic decrease in η and J_{sc} as observed in Figure 5.8. For P3HT:ICBA system this phase segregation does not occur and thus there is effectively no loss in J_{sc} after repeated annealing cycles. These conclusions are further supported by the optical, atomic force microscopic (AFM), and photocurrent mapping images of the PSCs after 12 annealing cycles shown in Figure 5.9. The optical

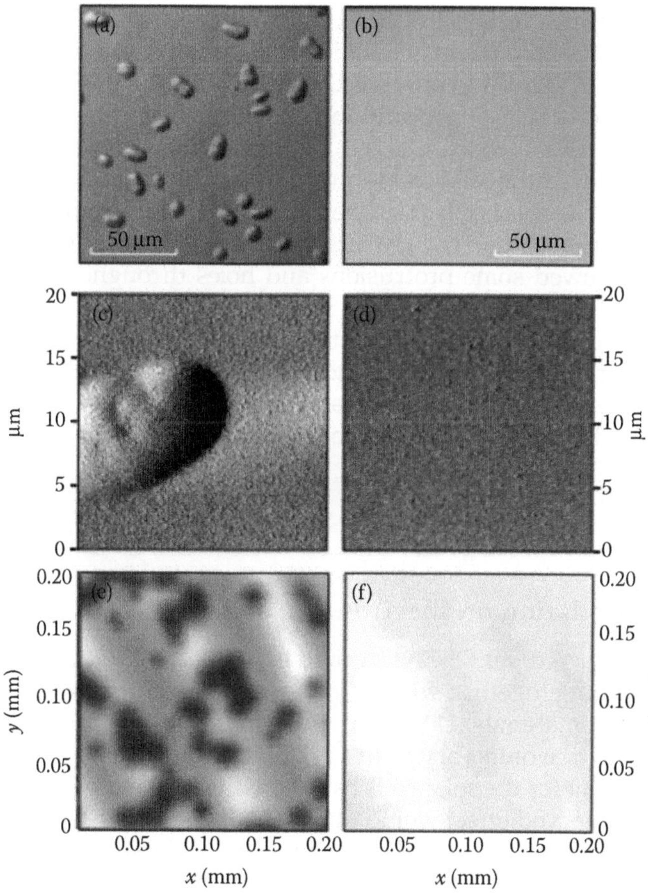

FIGURE 5.9
Optical (top panel), AFM (middle panel), and photocurrent mapping (bottom panel) images of ITO/PEDOT:PSS/P3HT:PCBM/Ca/Ag (left column) and ITO/PEDOT:PSS/P3HT:ICBA/Ca/Ag (right column) solar cells after 12 subsequent thermal annealing in the glove box. (Reproduced with permission from P. Kumar et al., 2014. *Appl. Phys. Lett.* 104: 193905.)

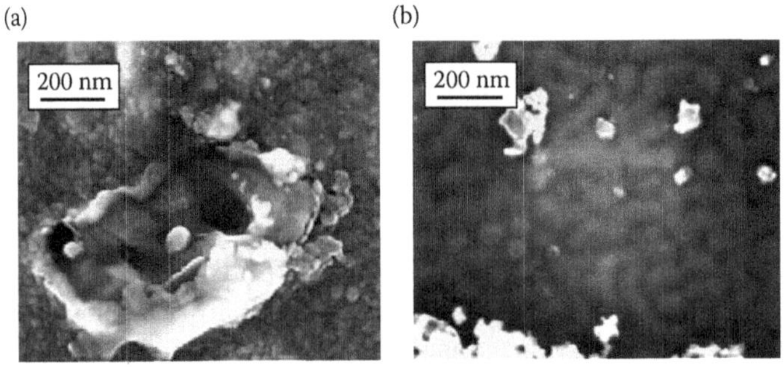

FIGURE 5.10
(a) SEM image of one of the holes observed in Al electrode and (b) SEM image of the area from where Al had disappeared in the solar cell illuminated in $^{18}O_2/N_2$ atmosphere. (Adapted with permission from K. Norrman, N. B. Larsen, and F. C. Krebs, 2006. *Sol. Energy Mater. Sol. Cells.* 90: 2793–2814.)

and AFM images (Figure 5.9a and c, respectively) clearly show the formation of large (~10 μm) PCBM aggregates in P3HT:PCBM blend film and these aggregates correspond to the regions of lower photocurrent generation (dark regions in Figure 5.9e). On the contrary, no aggregate formation was observed in P3HT:ICBA film even after 12 annealing cycles (Figure 5.9b and d) and a uniform photocurrent generation was observed across the cell (Figure 5.9f).

Norrman et al. employed the SEM technique to study the surface morphology of the top Al electrode in one of the solar cells [11]. The evaporated Al got deposited in grains with size ranging from ~10 to 100 nm. The SEM image of the solar cell illuminated in an $^{18}O_2/N_2$ atmosphere clearly showed some protrusions and holes throughout the Al electrode of degraded solar cells. Figure 5.10a shows the SME image of one of the holes in the Al surface, whereas Figure 5.10b shows the SEM image of the region where Al had disappeared due to degradation. In the area where Al had disappeared, the underlying organic layer was clearly visible, and a small protrusion was also observed there that suggested that the protrusion in Al could be originating from the underlying organic layers.

5.4 Effect of Degradation on Electronic Properties

Degradation in efficiency of an OSC with time is a collective effect of various degradation processes happening together, which deteriorate electronic as well as optical properties of the constituent materials. Degradation mechanisms and degradation profiles of photovoltaic parameters would vary with variation in test conditions. The degradation profiles may be different for the solar cells based on different materials, but in general the physics behind the observed effect would almost be the same. So I just take the example of the studies performed by me and my colleagues to understand the effect of degradation on electronic properties and the physics behind degradation [52]. Experiments were performed on the solar cells prepared in ITO/PEDOT:PSS/P3HT:PCBM/Al configuration, and to get detailed insight of degradation processes, one of the samples was exposed to ambient atmosphere (H_2O ~44%, O_2 ~19%, and T = ~300 K) for 5 min before deposition of the top Al electrodes and the other was kept unexposed. The rest of the procedures and preparation conditions were kept identical for all the samples. The sample with

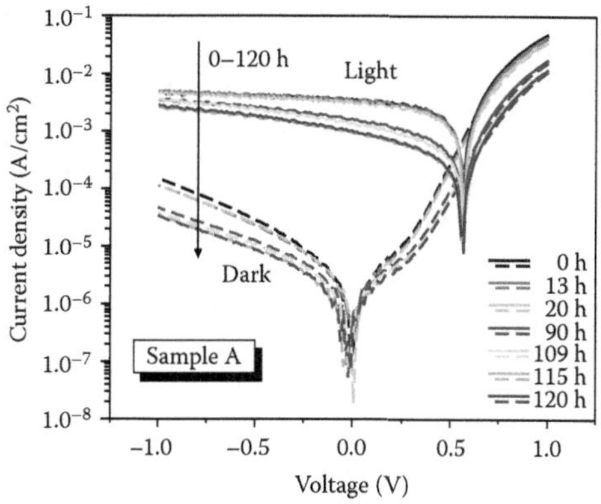

FIGURE 5.11
Dark (dashed) and illuminated (solid) *J–V* characteristics of the solar cell with active layer not exposed to ambient conditions. (Reproduced with permission from A. Gaur and P. Kumar, 2013. *Polym. Adv. Technol.* 24: 630–637.)

the unexposed active layer was designated as sample A, whereas that with the exposed active layer was designated as sample B. Exposure to ambience resulted in adsorption of some O_2 and H_2O molecules on the active layer that caused relatively faster degradation in cell performance. The degradation studies were performed according to the test procedure ISOS-D-1, which corresponds to shelf-life studies by keeping the cells in dark, ambient conditions (43 ± 3% rh, 27 ± 3°C) without any protection [53]. For photovoltaic performance evaluation, *J–V* characteristics of the samples were measured in dark and under illumination with time, wherefrom J_{sc}, V_{oc}, fill factor (*FF*), and the efficiency were extracted. The sample with the exposed active layer exhibited faster (exponential with time) degradation, whereas that with the unexposed active layer exhibited slow (linear with time) degradation. It was proposed that the linear degradation was because of degradation in the active layer only, whereas the exponential degradation was because of through degradation of the solar cell in the measurement period. Figure 5.11 shows the dark and illuminated *J–V* characteristics of sample A, measured at different time intervals, whereas those of sample B are shown in Figure 5.12. Current density was observed

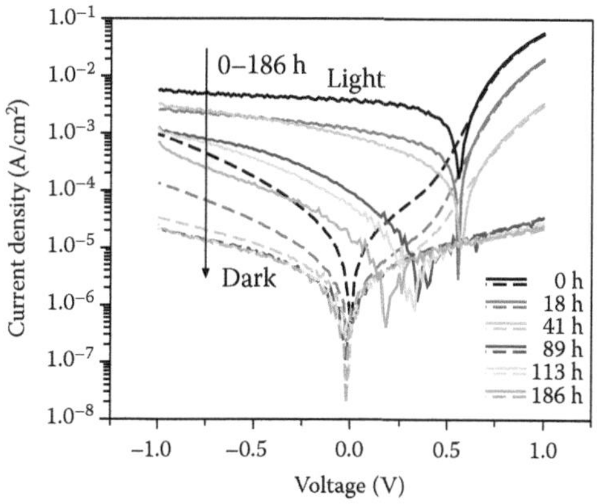

FIGURE 5.12
Dark (dashed) and illuminated (solid) *J–V* characteristics of the solar cell with active layer exposed to ambient conditions (Reproduced with permission from A. Gaur and P. Kumar, 2013. *Polym. Adv. Technol.* 24: 630–637.)

to decrease with time in both the reverse and forward bias; however, the degradation in sample B was relatively faster.

Initially, the dark *J–V* characteristics of sample A exhibited a rectification ratio of ~3×10^2 at ±1 V, which had no deterioration even after 120 h. It was because of the uniform degradation in current density in both the reverse and forward biases. Figure 5.13a shows degradation in J_{sc} and V_{oc}, whereas Figure 5.13b shows the degradation in *FF* and efficiency of sample A. J_{sc} was found to exhibit a linear decay, whereas V_{oc} had no remarkable degradation even after 120 h. In OSCs, V_{oc} is controlled by the difference in the highest occupied molecular orbital (HOMO) of the donor and the lowest unoccupied molecular orbital (LUMO) of the acceptor molecules and also by the difference in work functions of the two electrodes. No deterioration in V_{oc} of sample A, suggested no degradation in electrodes even up to ~120 h. Therefore, the degradation in J_{sc} is because of degradation of the active layer only, which was caused by the reaction of the active layer with O_2 and H_2O molecules, diffused through the top Al electrodes and uncovered area of the active layer [11]. *FF* decreased linearly with the slow rate of degradation because of the changes in electronic properties of the active layer. The charge carrier mobilities decreased, and new trap sites were created, which resulted in difficulty in extraction of the charge carriers, and as a result *FF* decreased. The efficiency first decayed almost linearly for the initial hours, but

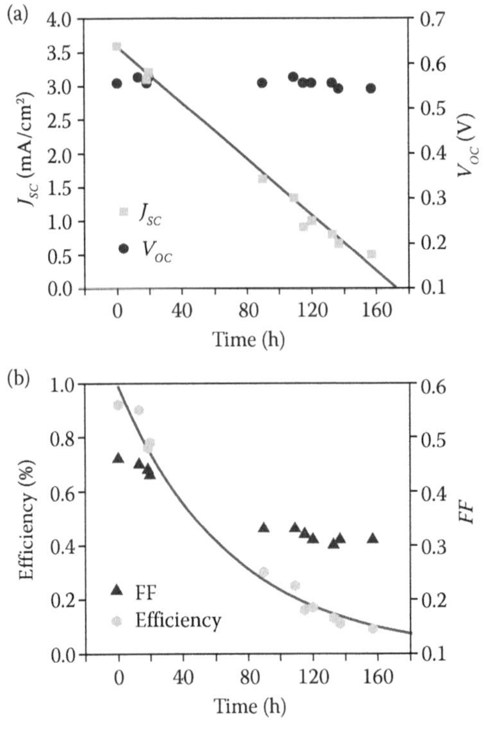

FIGURE 5.13
Degradation profiles for (a) J_{sc} and V_{oc} and (b) *FF* and efficiency of the solar cell with active layer not exposed to ambient conditions. Solid curves represent fitting of the data with linear variation for J_{sc} and exponential variation for efficiency. (Reproduced with permission from A. Gaur and P. Kumar, 2013. *Polym. Adv. Technol.* 24: 630–637.)

the best fit in the entire time range could be achieved only with exponential decay. The degradation was slow due to fewer amounts of O_2 and H_2O molecules.

Compared to sample A, sample B exhibited relatively faster degradation in both the dark and under illumination. In forward bias, the degradation in current density was relatively faster than that in reverse bias. Note that the solar cells were kept and tested under identical conditions, and the big differences in their stability was because of relatively larger contents of O_2 and H_2O molecules in sample B that came from the exposure of its active layer to the ambient air. Initially (0 h), the dark characteristic of sample B exhibited a rectification ratio of ~62 at ±1 V that reduced to 1 after ~113 h that corresponded to no rectification. When the rectification ratio reduced to 1, the current density followed ohmic conduction, which corresponded to leakage current through shunt R_p. No rectification in the dark shows elimination of the diode behavior and is because of degradation of electrode materials along with the active layer. It is important to note that even though the rectification had been eliminated, a photocurrent under illumination was still obtained. This could be because of diffusion or the difference of work functions of the two electrodes was still not zero, and an internal electric field existed, though it had reduced due to degradation in electrodes. It was important to note that the intersection point of dark and illuminated characteristics corresponds to V_{bi} in BHJ PSCs and reduction in the intersection voltage with time was because of reduction in built-in voltage (V_{bi}) of the cell [54,55]. Initially, the illuminated current density for $V \gg V_{bi}$ was more than that for $V < 0$. Interestingly, the situation changed after ~89 h, where current density for $V \gg V_{bi}$ became much less than that for $V < 0$, but still the illuminated current density was more than the dark current density, which suggested that photocurrent contribution was in the same direction as that in the dark for $V > V_{bi}$. This difference could be attributed to different charge transport mechanisms in forward and reverse bias of a degraded solar cell.

Figure 5.14a shows the variation in J_{sc} and V_{oc}, whereas variations in *FF* and efficiency of sample B are shown in Figure 5.14b. J_{sc} first decreased rapidly and then slowly with time and V_{oc} exhibited no degradation for initial 41 h, but it decreased linearly afterward. Reduction in V_{oc} was because of reduction in V_{bi} and reduction in V_{bi} was because of reduction in work function difference of the two electrodes. However, the possibility of reduction in the difference of HOMO of the donor and LUMO of the acceptor due to degradation of the active layer could also not be ignored. A reduction in the work function of Al has been observed due to interaction with oxygen [56], therefore, reduction in the work function difference was attributed to a reduction in the work function of the ITO/PEDOT:PSS electrode. Due to the hygroscopic nature of PEDOT:PSS (it absorbs water and loses its electronic properties), the work function decreases. Because of the acidic nature of PEDOT:PSS, it etches ITO and changes its properties. Therefore, changes in V_{oc} in the present case were attributed to the change in work function of the ITO/PEDOT:PSS electrode. *FF* decreased rapidly and then slowly with time and stabilized to ~0.27 after ~140 h. Similar to J_{sc}, the efficiency decayed first rapidly and then slowly with time. J_{sc} is controlled by both the active layer and electrode materials; therefore, the exponential decay in J_{sc} of sample B has been attributed to degradation of the complete cell, which includes both the physical and chemical degradations. As discussed earlier, O_2 exposure to P3HT increases charge carrier concentration and decreases the carrier mobility with enhanced trap density. The traps reduce the charge carrier mobility, disturb the internal electric field distribution, work as recombination centers, and affect the charge carrier transport.

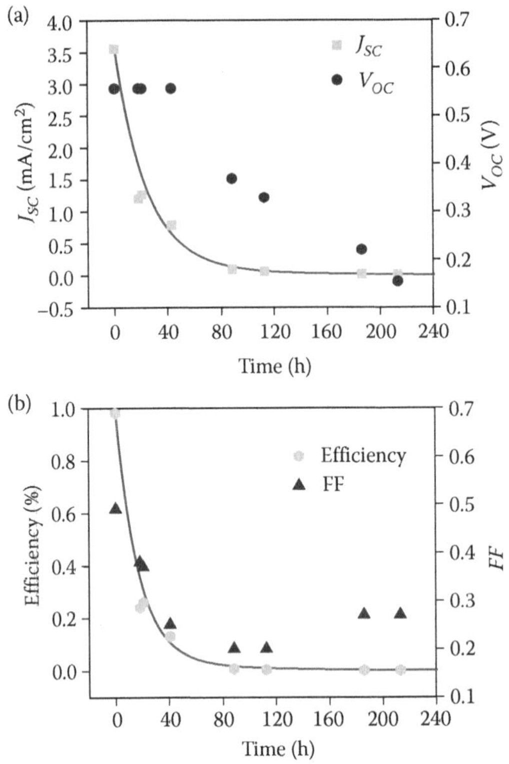

FIGURE 5.14
Degradation profiles for (a) J_{sc} and V_{oc} and (b) *FF* and efficiency of the solar cell with active layer exposed to ambient conditions. Solid curves represent fitting of the data with exponential variation for both the J_{sc} and efficiency. (Reproduced with permission from A. Gaur and P. Kumar, 2013. *Polym. Adv. Technol.* 24: 630–637.)

My colleagues and I also calculated the photocurrents extracted out of the devices (J'_{ph}) by subtracting the dark current density from current density under illumination at different time intervals [52]. Figure 5.15 shows the variation in J'_{ph} of sample A with applied voltage and its degradation with time, whereas variation in J'_{ph} for sample B is shown in the inset of Figure 5.15. Initially, J'_{ph} of sample A showed a linear variation in reverse bias (third quadrant) and nonlinear variation in forward bias (fourth quadrant). As the device started degrading, the magnitude of J'_{ph} decreased with time, but it still varied linearly in the third quadrant and became linear in the fourth quadrant as well. In every case J'_{ph} became zero at the same +ve voltage, beyond which it got reversed in direction. It is worth mentioning that this point corresponded to the V_{bi}. J_{ph} (V) varies with the internal electric field, and at $V = V_{bi}$, the internal electric field becomes zero, resulting in J_{ph} (V_{bi}) = 0 [54,55]. The change from nonlinear to linear variation in J'_{ph} in the fourth quadrant could be attributed to the increment in recombination losses that caused reduction in *FF*. Reversal in the direction of J'_{ph} was because of the change in direction of the effective electric field, which increased with voltage for $V > V_{bi}$. Since every time J'_{ph} became zero at $V = 0.63$ V, it suggested no deterioration in V_{bi}, which was controlled by the electrodes work functions. Therefore, it could be inferenced from the preceding results that even after 120 h, the electrodes (ITO and Al) exhibit no deterioration. This stability of electrodes was because of preparation of the device in a controlled atmosphere.

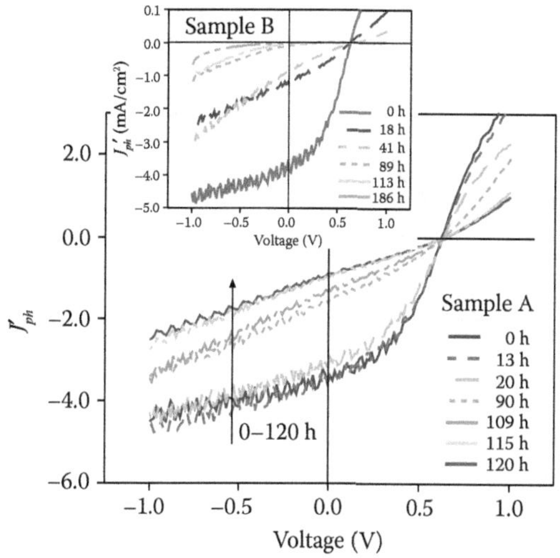

FIGURE 5.15
Variation of J'_{ph} with applied voltage at different time intervals for sample A. Inset shows the variation J'_{ph} for sample B. (Reproduced with permission from A. Gaur and P. Kumar, 2013. *Polym. Adv. Technol.* 24: 630–637.)

In the case of sample B, initially J'_{ph} showed linear variation in the third quadrant that changed to nonlinear after ~89 h. In the fourth quadrant, J'_{ph} changed from convex to linear and then to concave with time, leading to a great reduction in *FF*. *FF* is controlled by R_s and R_p, but from the models available in the literature, calculations showed that whatever the values of R_s and R_p, the characteristics might change from convex to linear but not to concave. Concave shape of the characteristics under illumination could be understood from different charge transport phenomenon, and the calculations showed that in this region, the current was governed by space charge limited phenomenon where J'_{ph} varied as V^m. The space charge limited phenomenon came from the traps and large difference in charge carrier mobilities. Therefore, the concave shape of the characteristics in the fourth quadrant could be attributed to SCLC caused by degradation in active layer. For initial hours, the voltage at which J'_{ph} became zero was the same, but after ~41 h, it started decreasing. The reduction in voltage corresponding to zero J'_{ph} was because of the reduction in V_{bi} with time. After ~41 h, for $V > V_{bi}$, the J'_{ph} got reversed in direction, but remained almost zero.

Analysis of R_s and R_p in the dark and under illumination for both samples showed that both R_s and R_p increase with time. The increase in R_s and R_p of sample B was relatively faster compared to that in sample A. Figure 5.16 shows the variation in R_p with time in the dark and under illumination for both samples. For sample A, R_p in the dark increased first slowly and then rapidly after ~89 h; however, for sample B, it increased first rapidly and then slowly afterward and tended to saturate. Compared to the dark, R_p decreased for both samples under illumination and showed almost no variation with time for sample A; however, in case of sample B, it increased rapidly after ~89 h. Different degradation rates in R_p for the two samples were because of ambience exposure to the active layer of sample B. Under illumination, the reduction in R_p could be understood in terms of photoinduced doping of the active layer, which is represented by the addition of photoinduced shunt resistance across the existing R_p in the dark. The increase in R_s was because of the increase

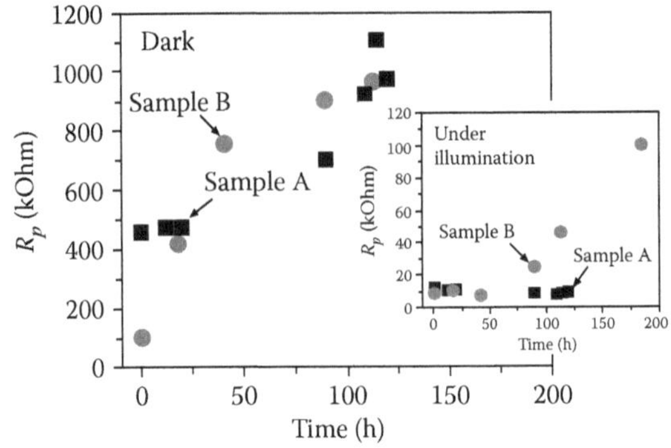

FIGURE 5.16
Variation in R_p with time for the two samples in the dark. Inset shows the variation in R_p with time under illumination for the same samples. (Reproduced with permission from A. Gaur and P. Kumar, 2013. *Polym. Adv. Technol.* 24: 630–637.)

in bulk resistivity due to reduction in charge carrier mobilities, formation of traps, and increase in contact resistance. However, the increase in R_p with time could be attributed to change in the active layer morphology and formation of dead zones, which reduced the leakage current. The increase in R_s and R_p reduced the charge carrier mobilities and *FF* first rapidly and then slowly afterward. It is important to note that R_s is more crucial compared to R_p in determination of *FF* and efficiency of a solar cell. Therefore, even after an increase in R_p, *FF* decreased due to increment in R_s. The imbalance in charge carrier mobilities would create space charge in the active layer that would reduce further the *FF*. More the imbalance in carrier mobilities less will be the *FF*. In a degraded solar cell, it was very difficult to find out to what extent the electron and hole mobilities changed. Some specifically designed experimental setup would be required for such studies. However, for simplification in calculations, consideration of similar degradation in both the charge carriers would be acceptable to a good approximation.

5.5 Tools for Testing Degradation Mechanisms in OSCs

For understanding the degradation mechanisms in OSCs a number of techniques have been employed that give the information about physical and chemical degradations. Jorgensen et al. carried out a detailed review on degradation and frequently used techniques for analysis of degradation mechanisms in OSCs [9]. Some of the techniques and their use in getting relevant information about degradation in OSCs have been tabulated in Table 5.1. Most of the techniques give limited information of degradation processes and to get detailed information, a series of experiments need to be performed. As some of the techniques are destructive, it is difficult to perform all experiments on the same sample, therefore a number of samples are required to get the next level of information. A brief introduction to some of these techniques and the information achieved from about degradation is given next.

TABLE 5.1

Some of the Techniques Used to Study and Analyze the Degradation Mechanisms in OSCs

Technique	Destructive (D)/ Nondestructive (ND) Technique	Bulk Analysis	Surface Analysis	Morphological Analysis	Depth Profiling	Chemical Analysis	2D Imaging
AFM	D		√	√			√
SEM	D		√	√			√
TEM	D			√			√
J–V measurement	ND	√					
Impedance spectroscopy	ND	√					
XPS	D		√		√	√	√
TOF-SIMS	D		√		√	√	√
UV–vis absorption spectroscopy	ND	√				√	
IR spectroscopy	ND	√				√	
X-ray reflectrometry	ND	√		√			
Photocurrent mapping	ND	√		√			√
Fluorescence microscopy	D	√				√	√

5.5.1 Atomic Force Microscopy (AFM)

AFM is an important tool for imaging the surfaces of different samples. It gives information of surface morphology and surface roughness on a nanoscale level. It has been used to study grain boundaries and microscopic pinholes in the top electrodes of OSCs. It has found special applications in investigating the surface degradation and morphological changes in OSCs. Degradation factors, such as electrode delamination, protrusion, and bursting of the top electrode, can be easily seen using AFM technique. AFM has a limited area resolution, but it has excellent lateral resolution. The surface imaging can be done in contact, noncontact, or tapping modes. In contact mode the cantilever tip is in direct and continuous contact of the sample during measurement. Direct contact scanning may damage the samples so it is preferably used for rigid samples. For soft samples the images are either taken in noncontact or tapping modes.

5.5.2 Scanning Electron Microscopy (SEM)

SEM is a very powerful tool of imaging and like AFM, it is used most often for surface analysis and morphological studies in OSCs. It is superior to AFM in terms of both the lateral resolution and broad area range analysis. For SEM analysis, the samples are scanned by a focused beam of high-energy (0.5–30 kV) electrons. The primary electrons interact with sample atoms and generate many low-energy secondary electrons. By measuring the intensity of the secondary electrons as a function of position, the corresponding image of the surface is achieved. A high-energy electron beam may damage the soft samples, therefore it is considered a destructive technique. However, for rigid samples it may be nondestructive. It is widely used for optimization of device efficiency by controlling the nanoscale bulk as well as surface morphologies. It has proved to be an excellent technique for cross-sectional analysis of OSCs. It is used for monitoring the nanoscale morphological changes happening in the bulk or surface of the solar cells during degradation. Like AFM it is used to study the grain boundaries, pinholes, protrusions, delamination, and bursting of top electrodes. Secondary electrons are also accompanied by backscattered high-energy electrons and x-rays generated by the primary electrons. Backscattered electrons give elemental information about the sample, and the energy dispersive x-ray (EDX) analysis gives more detailed chemical and elemental information of the sample. EDX analysis by reflection of x-rays generated from a tungsten anode x-ray tube has also been used to study degradation mechanisms in OSCs [57]. From these studies the authors proposed that an aluminum oxide layer is formed at the organic–Al interface and indium atoms diffuse from ITO to the active layer.

5.5.3 Transmission Electron Microscopy (TEM)

TEM is another very powerful technique to get information of nanoscale bulk morphology and distribution of nanostructures in OSCs. It is frequently used for optimization of OSC performance by analyzing and controlling the bulk morphology of different layers. It has also been used widely to study degradation mechanisms in OSCs. TEM is used to study the morphological changes or changes in nanostructures with time in bulk of solar cells. Like in SEM, a high-energy focused beam of electrons is used to analyze the samples. The signals generated by interaction of incident electrons with sample atoms are detected in the transmission mode and used to generate the corresponding image. Here also the high-energy electron beam may damage the samples, therefore it is also considered a destructive technique.

5.5.4 *J–V* Measurements

Measurement of *J–V* characteristics in the dark and under illumination with time is an important method to get information about degradation mechanisms in OSCs. The *J–V* characteristics are measured by applying a bias to the cell and measuring the corresponding current in dark and under illumination. The PCE and other photovoltaic parameters like J_{sc}, V_{oc}, and *FF* are measured from illuminated *J–V* characteristics. It is a direct and simple method to determine the degradation profiles of photovoltaic parameters. For degradation profiles and lifetime of the solar cells, the photovoltaic parameters are extracted from the *J–V* characteristics, measured time to time, and plotted as a function of time. Measuring the *J–V* characteristics we can also get lots of information about cells' electronic properties and their variation with time. Some examples are shown in Figures 5.11 through 5.16. Degradation in different photovoltaic parameters is related to degradation in different constituents of the solar cells. For example, degradation in J_{sc} is related to degradation in the active layer and contact resistance, degradation in V_{oc} is related to degradation in electrode materials and the active layer, whereas degradation in *FF* is related to degradation in contact resistance and the active layer. *J–V* measurements are nondestructive and most of the information is achieved by applying voltages in the range –1 V to +2 V, which is safe for OSCs. Application of high voltage would cause a high current to flow through the sample that may damage it.

5.5.5 Impedance Spectroscopy

Impedance spectroscopy is again a nondestructive technique that measures dielectric properties of the materials as a function of high frequencies. An OSC is actually a resistive-capacitive (RC) device, and impedance spectra are the plot of its real part of impedance versus imaginary impedance. The semicircle in the impedance spectra corresponds to the RC component wherefrom the capacitive values are calculated. Impedance spectroscopic analysis of degraded solar cells gives the information of variation in its RC and dielectric values. If a dielectric layer is developed at any interface like at the Al cathode, then it can be easily detected using this technique. Not only that, doping or depletion of any organic layer on degradation can also be detected [58,59].

5.5.6 X-Ray Photoelectron Spectroscopy (XPS)

XPS is a surface sensitive technique that gives information about chemical composition and electronic states of the elements in a sample by measuring binding energy of their core electrons. It is an elemental analysis technique where samples are bombarded by x-rays, which release core electrons from the atoms, and by measuring the energy of the ejected electrons the binding energy is calculated. It has a probe depth of up to ~10 nm into the material being analyzed. It is also used for depth profiling of various elements in OSCs. In-depth profiling the analyzed atoms are etched out by sputtering, so that the next atom is available for analysis. It is a simple and straightforward technique for quantitative elemental analysis of OSCs. It is widely used for understanding of degradation mechanisms in OSCs [4]. We performed degradation and regeneration studies on ITO/PEDOT:PSS/P3HT:PCBM/Ca/Ag solar cells and in order to probe the changes occurring at the Ca/Ag electrode interfaces, a detailed XPS depth profiling analysis of fresh, degraded, and regenerated solar cells was conducted [33]. The silver and calcium layers were directly identified by measuring the Ag and Ca 2p peaks, the active layer was identified from the C 1s peak, and the anode was tracked by looking at the indium $3d_{5/2}$ peak. In addition, by monitoring the S 2p signal and the carbon:sulfur (C:S) ratio, it was possible to monitor both the relative P3HT:PCBM concentration and differentiation between the active organic and PEDOT:PSS layers. Figure 5.17 shows the XPS elemental depth profile analysis of (a) fresh, (b) degraded, and (c) regenerated solar cells. After regeneration the In $3d_{5/2}$, O 1s, and Si 2p peaks at the anode interface got broaden and the anode interface became less distinct, indicating the diffusion of In and O_2 into active layer. An O 1s peak was also located at the cathode interface coincident with the position of Ca 1s peak. The O 1s peak increased upon degradation and upon regeneration that was consistent with oxidative degradation of the cathode. Figure 5.18 shows the S 2p spectrum at the peak of Ca interface signal in the depth profile (Figure 5.17) for the fresh, degraded, and regenerated devices. All three plots clearly showed the presence of two sulfur species, with the lower and higher energy fitted peaks corresponding to sulfide and P3HT components, respectively, indicating the presence of CaS at the interface. Upon degradation and subsequent regeneration a reduction in sulfur signal was observed and, more important, a change in the ratio of two peaks with sulfide contribution decreased systematically. This observation was consistent with the systematic loss of CaS, which could be explained by the initial formation of mixed hydroxythiol species and subsequent loss of H_2S to form $Ca(OH)_2$. XPS studies were also performed to probe the specification of oxidation products at the Ca/Ag interface. Upon degradation the O:Ca ratio systematically increased across the interface region suggesting the formation

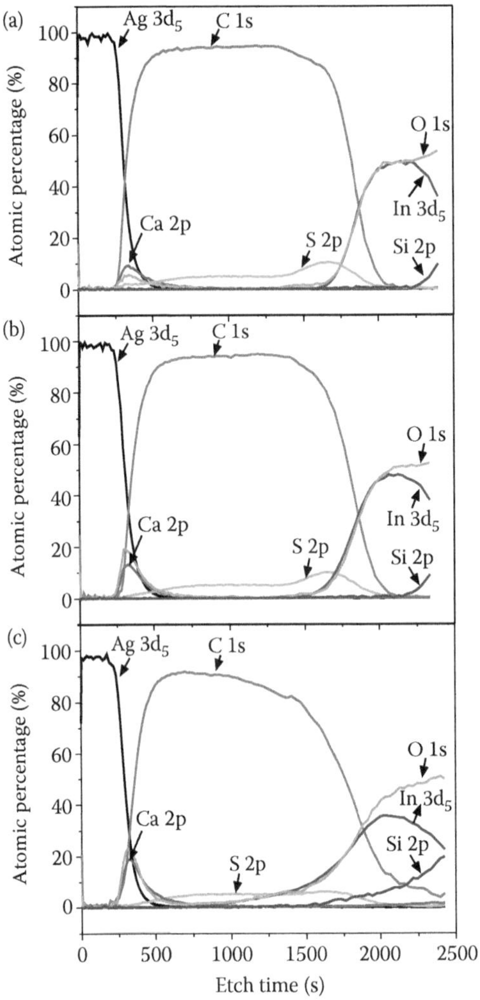

FIGURE 5.17
Elemental XPS depth profiles of (a) fresh, (b) degraded, and (c) regenerated OSCs with a Ca/Ag cathode. (Reproduced with permission from P. Kumar et al., 2014. *ACS Appl. Mater. Inter.* 6: 5281.)

of CaO and $Ca(OH)_2$. And upon regeneration, the O:Ca ratio reduced systemically across the interface region, suggesting thermal dehydration of $Ca(OH)_2$ to CaO.

5.5.7 Time-of-Flight Secondary Ion Mass Spectroscopy (TOF-SIMS)

TOF-SIMS is the most sensitive surface analysis technique and is widely used for chemical composition, imaging, and depth profiling of the solid samples. For analysis, the samples are bombarded with a focused high-energy primary ion beam, which ejects secondary ions from the sample. The primary ion source contains the ions of gaseous elements like $^{40}Ar^+$, Xe^+, $^{16}O^-$, $^{16}O_2^+$, $^{16}O_2^-$, or elemental ions like $^{133}Cs^+$, or ionized molecules like SF_5^+ and C_{60}^+. By collecting and analyzing the mass ratio of ejected secondary ions, the elemental and chemical compositions of the samples are determined. TOF-SIMS is a qualitative technique and has a probe depth of 1–2 nm and the area coverage up to some square

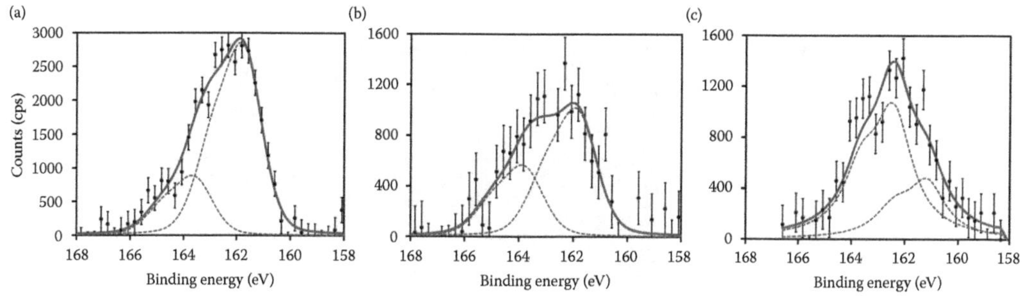

FIGURE 5.18
Elemental S 2p XPS spectra at the peak of the Ca interface signal in the depth profiles of (a) fresh, (b) degraded, and (c) regenerated OSCs with a Ca/Ag cathode. The S 2p peaks are fitted with a spin orbit splitting of 1.2 eV, and the error bars are calculated from the standard deviation of the differences between the experimental and fitted data across all of the energies of the S 2p spectra. (Reproduced with permission from P. Kumar et al., 2014. *ACS Appl. Mater. Inter.* 6: 5281.)

centimeters for surface imaging. For depth profiling the TOF-SIMS is accompanied with sputtering to remove the top analyzed elements. Unfortunately, the sputter bombardment generates fragment ions giving ionic information instead of molecular information. For example, the TOF-SIMS analysis of C_{60} or PCBM by primary ions only has the information of $C_mH_n^{\pm}$ ions along with $C_{60}^{\pm}$ molecular ions, but when sputtering is involved along with primary ions the secondary ions released for analysis consist of $C_mH_n^{\pm}$ ions only. Additionally, the sputtering causes mixing of interlayers, which makes the interfaces difficult to analyze. This effect becomes more prominent when the sputtering is done for a longer time. However, it is still a very important technique and has been extensively used to analyze and understand degradation mechanisms in OSCs. It has enabled researchers to monitor the effect of moisture, oxygen, and other spices on organic molecules and other constituents of the cell. It gives the information of molecular decomposition and formation of new entities. It is an important tool for surface imaging of devices. For surface imaging, the mass spectrum is obtained from different points on the surface and different species are designated by different colors. This gives the idea of in-plane distribution of species like oxidized molecules and oxygen uptakes. Norrman et al. used TOF-SIMS imaging to analyze the PEDOT:PSS layer in partial inverted devices (ITO/ZnO/P3HT:PCBM/PEDOT:PSS) without the top Ag electrode [47]. The devices were illuminated by 330 W/m^2 of AM1.5 irradiance at 65 ± 2°C for 3 weeks without encapsulation in dry $^{18}O_2$:N_2 and oxygen-free $H_2{}^{18}O$:N_2 (65 ± 2% rh), and then TOF-SIMS images of PEDOT:PSS layer were taken that are shown in Figure 5.19. Images a–d correspond to the device stored in dry $^{18}O_2$:N_2 atmosphere, whereas images e–h correspond to the device stored in an oxygen-free $H_2{}^{18}O$:N_2 atmosphere. Different images are for different species in the PEDOT:PSS layer.

Figure 5.19d clearly shows that the PEDOT:PSS layer exhibits a phase separation between PEDOT and PSS constituents. The dark regions in the image correspond to the PEDOT-rich phase, whereas gray and white regions correspond to the PSS-rich phase. Figure 5.19a shows the distribution of ^{16}O in PEDOT:PSS film. Since there was only ^{18}O in the atmosphere, the impressions of ^{16}O arise from the $-SO_3H$ group of PSS. Figure 5.19b is the image showing $^{18}O_2$ distribution in the PEDOT:PSS layer. The image clearly shows that $^{18}O_2$ is adsorbed and react with the PEDOT phase. Almost no adsorption of $^{18}O_2$ happens in the PSS phase. Figure 5.19c corresponds to $HS^{18}O_4$ imaging and shows that sulfur reacts with $^{18}O_2$ significantly. Figure 5.19h corresponds to the image for PSS species in PEDOT:PSS film

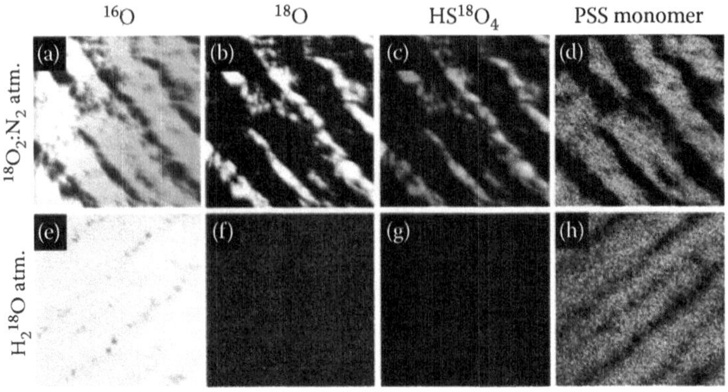

FIGURE 5.19
TOF-SIMS images (500 × 500 μm²) of PEDOT:PSS layer of partial inverted devices stored in dry $^{18}O_2$:N_2 (a–d) and oxygen free $H_2{}^{18}O$:N_2 (e–h) atmospheres for 3 weeks. The black region corresponds to 0% signal, the gray region represents some signal, and the white region represents 100% signal. (Reprinted with permission from K. Norrman et al., Degradation patterns in water and oxygen of an inverted polymer solar cell, *J. Am. Chem. Soc.* 132, 16883–16892. Copyright 2010 American Chemical Society.)

for the device kept under illumination in an oxygen-free $H_2{}^{18}O$:N_2 atmosphere. The phase separation between PEDOT and PSS was not as distinct as that in dry atmosphere (Figure 5.19d), however, it happened and $H_2{}^{18}O$ was observed to get adsorbed and reacted in a lesser extent but uniformly in both phases. The O_2 or H_2O uptake in PEDOT:PSS could be determined by comparing $^{18}O/^{16}O$ ratios. In an oxygen-free atmosphere the $^{18}O/^{16}O$ ratio in PEDOT:PSS film was 2%, whereas this ratio in dry atmosphere was 15%. These results suggest that oxygen diffuses mainly through the PEDOT phase and moisture has less impact on the inverted structure. It was also inferred that phase separation hampers the barrier properties of PEDOT:PSS causing O_2 transportation and oxidation at the interface.

Where TOF-SIMS imaging is used for in-plane analysis, the depth profiling is equally important for in-depth distribution of different elements or species in OSCs. Depth profiling by TOF-SIMS has been very useful for detection of diffusion of O_2, H_2O molecules, and other electrode materials across the OSCs. Norrman et al. performed water-induced degradation in MEH-PPV-based and P3HT-based polymer solar cells with the help of TOF-SIMS [47]. Analysis of degraded solar cells was performed both in-plane and in-depth to get three-dimensional information of degradation. The solar cells were prepared in ITO/PEDOT:PSS/active layer/Al configuration on glass substrates and the active layer was MEH-PPV or P3HT blend with PCBM. The solar cells were exposed to ~90% of isotopically labeled relative humidity in the dark and under illumination and studied for their different interfaces. For interface studies, different layers were peeled off from top to bottom, one by one to get different interfaces. For removal of Al electrode, an adhesive tape was applied on it and peeled off, thus giving the Al/active layer interface for analysis. The active layer was removed using a cotton bud soaked in chloroform, gently scrubbed around the surface thus exposing the active layer/PEDOT:PSS interface. Similarly the PEDOT:PSS layer was removed using a cotton bud soaked in water. Each interface was subjected to TOF-SIMS studies. Different solar cells were also subjected to depth profiling from the top Al electrode to the bottom glass substrate. Figure 5.20 shows the depth profile of MEH-PPV-based solar cells kept in the dark and under illumination for AlO_2^-, C_4^-, and $^{18}O/^{16}O$. The measurements were performed from the top Al electrode to underlying organic layers. AlO_2^- is a marker for Al electrode or aluminum oxide, C_4^- is the marker of

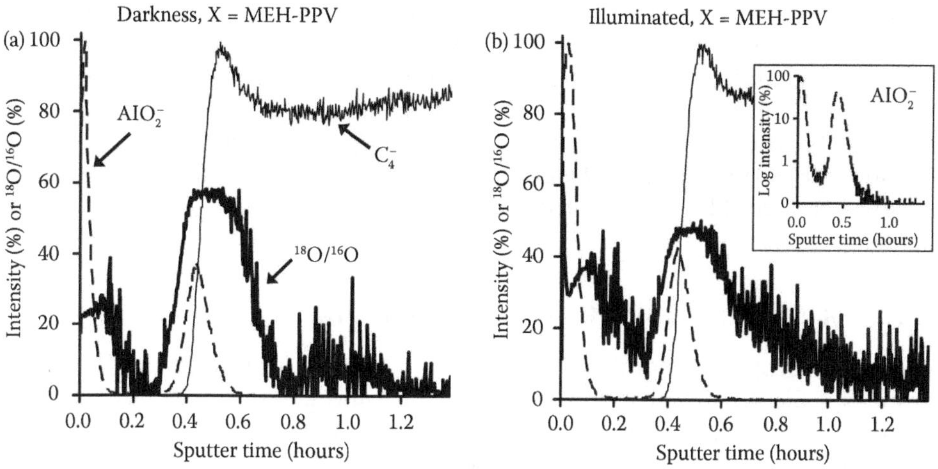

FIGURE 5.20
TOF-SIMS depth profiles of the devices exposed to ~90% relative humidity in the dark and under illumination. (Reprinted with permission from K. Norrman, S. A. Gevorgyan, and F. C. Krebs, Water induced degradation of polymer solar cells studied by $H_2{}^{18}O$ labeling, *ACS Appl. Mater. Inter.* 1, 102–112. Copyright 2009 American Chemical Society.)

active layer incorporating MEH-PPV and PCBM, whereas $^{18}O/^{16}O$ is the marker of moisture diffusion in the solar cells. AlO_2^- profiles show that aluminum gets oxidized from both sides and in the middle to some extent (see inset of Figure 5.20).

Norrman et al. observed that some particles are formed in different layers of solar cells and participate in performance degradation [47]. Formation of particles was confirmed from Figure 5.21, which shows the TOF-SIMS ion images of the sputtered hole in

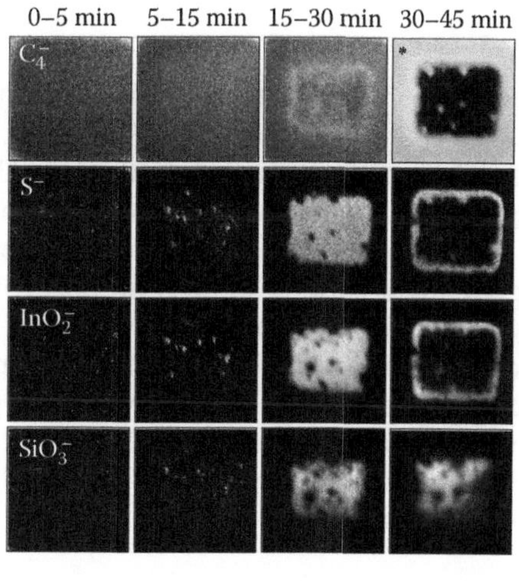

FIGURE 5.21
TOF-SIMS ion images of sputtered hole in illuminated Al/MEH-PPV:PCBM/PEDOT:PSS/ITO solar cell, measured from Al/MEH-PPV:PCBM interface. The black region represents no intensity, the gray region represents some intensity, and the white region represents maximum intensity. The asterisk (*) is for the logarithmic intensity scale. (Adapted with permission from K. Norrman, S. A. Gevorgyan, and F. C. Krebs, Water induced degradation of polymer solar cells studied by $H_2{}^{18}O$ labeling, *ACS Appl. Mater. Inter.* 1, 102–112. Copyright 2009 American Chemical Society.)

a MEH-PPV-based solar cell where the top Al electrode had been peeled off. C_4^- is the marker of active layer, S^- is the marker of PEDOT:PSS whereas InO_2^- and SiO_3^- are the markers for ITO and glass, respectively. The images taken at 5–15 min show the particles containing sulfur, indium, and silicon. In the images taken at 15–30 and 30–45 min other types of particles were observed (seen as dark spots) that contained carbon, sulfur, and indium. PEDOT:PSS and ITO were found to be involved in formation of these particle. However, it could not be detected if the active layer was also involved in particle formation.

5.5.8 Ultraviolet–Visible (UV–Vis) Absorption Spectroscopy

UV–vis spectroscopy is a nondestructive technique used to study degradation in optical properties like light absorption spectrum and band gap of semiconductors. Known intensity of monochromatic lights is made incident on the samples, and by detecting the transmitted intensity the amount of absorbed light at that light wavelength is calculated thus getting the complete absorption spectrum. The maximum wavelength absorbed by the semiconductor corresponds to its band gap. Photooxidation causes decomposition in chromophores of organic semiconductors. Chromophores are the hydrocarbon groups that give characteristic colors to the semiconductors. Decomposition of chromophores leads to loss in color of the semiconductor and it is known as photobleaching. The effect of photobleaching on optical absorption and band gap is studied by UV–vis spectroscopy. Dam et al. calculated the singlet oxygen degradation rate constant in a number of phenylenevinylene derivatives by measuring their relative absorbance with time [15]. The degradation rate was determined by plotting $\ln(A_t/A_0)$ as a function of elapsed photolysis time t. A_t and A_0 were absorbances of the materials at time t and $t = 0$, respectively. The plots were linear with time and their slopes gave the first-order degradation rate constants.

5.5.9 Infrared (IR) Spectroscopy

Like UV–vis absorption spectroscopy, IR spectroscopy is a nondestructive technique used to extract out the chemical information of materials. IR radiation is made incident on the materials and the corresponding absorbance is measured. Absorbed IR energy corresponds to vibrations or stretching of different bonds between different elements. Bond breaking or formation of new entities due to degradation would cause some of the IR absorption peaks to disappear and creation of some new absorption peaks. This technique has been used extensively to study degradation mechanisms in organic solar cells and Chambon et al. used it in combination with UV–vis spectroscopy to study degradation in MDMO-PPV:PCBM solar cells [2]. Similarly Cumpston et al. [17] and Neugebauer et al. [8] studied photochemical degradation in PPV-based OSCs using Fourier transform infrared (FTIR) spectroscopy. The studies showed reduction in vinylene bond signal and increment in carbonyl group signal upon oxidation.

5.5.10 X-Ray Reflectrometry

X-ray reflectrometry is a nondestructive technique to analyze interfaces in OSCs [60]. The samples are exposed to x-rays and the reflected x-rays give the information of film thickness, roughness, and electron density in the device. Paci et al. used this technique to study the degradation in MDMO-PPV:PCBM-based solar cells [57,61]. The reflected intensity had oscillations that corresponded to interference fringes from different layers. These studies helped to detect oxidation of Al at the Al–polymer interface and diffusion of indium atoms from ITO to the active layer.

5.5.11 Photocurrent Mapping

Photocurrent mapping is also known as light-beam-induced current (LBIC) mapping. It is an important tool to map the photocurrent response of a solar cell. The solar cells are illuminated by a light beam of 1 mm to 1 μm diameter, at specific positions and the corresponding photocurrents are measured. The solar cells are kept in short-circuit condition and the point-by-point scanning of the complete active area by a light beam gives a complete image of photocurrent generation. The setup includes a lock-in amplifier for registration of photocurrent and an x–y scanner that guides the light beam in the scanning area. The defects, dead zones, or the regions giving low photo-response are easily detected. For example, the dead zones will not give any photocurrent, and using different colors for different magnitudes of photocurrent, a photocurrent mapping image is achieved. This technique is widely used to study degradation in OSCs. Jeranko et al. fabricated MDMO-PPV:PCBM solar cells and studied their degradation using photocurrent mapping [62]. Figure 5.22 shows the photocurrent mapping image of a solar cell stored in the dark in

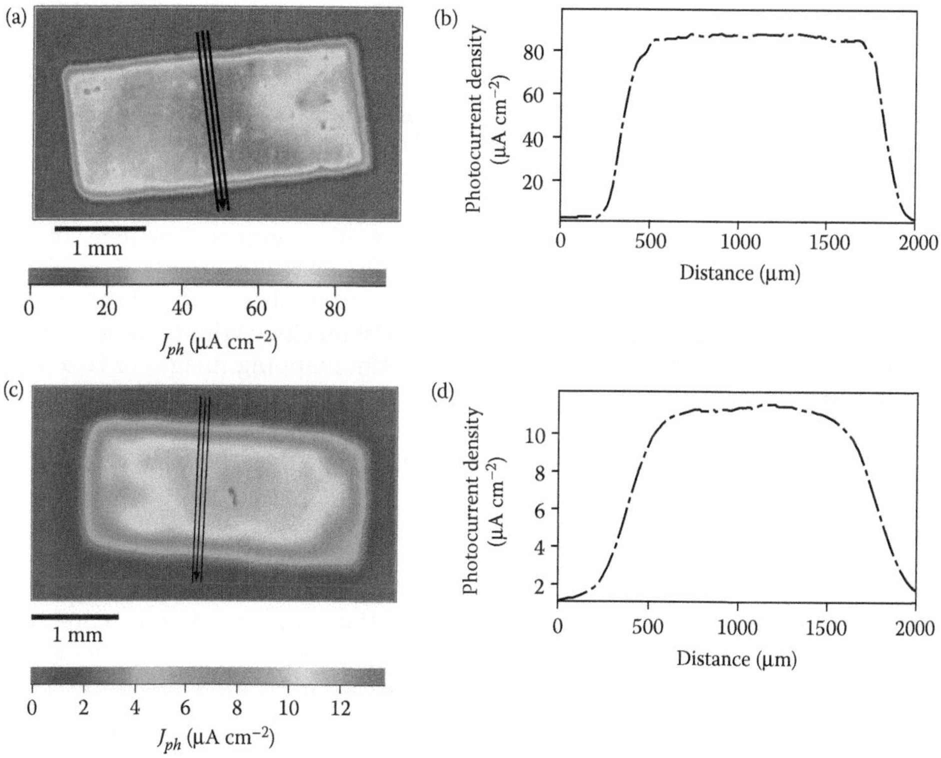

FIGURE 5.22
Photocurrent mapping images of a MDMO-PPV:PCBM-based OSC (a) taken after 1 day storage in dark in inert atmosphere and (c) taken after 2 weeks of air exposure. (b) and (c) show the cross-sectional photocurrent density of images (a) and (c), respectively, along the black arrow shown on the images. (Adapted with permission from T. Jeranko et al., 2004. *Sol. Energy Mater. Sol. Cells* 83: 247–262.)

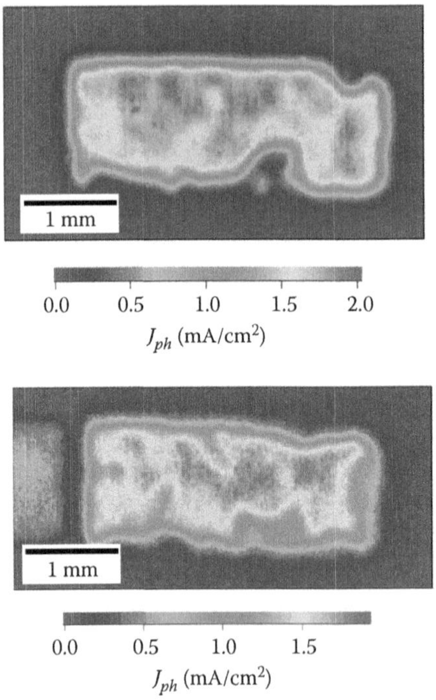

FIGURE 5.23
Photocurrent mapping images of two MDMO-PPV:PCBM-based OSCs taken after 2 weeks of air exposure. (Adapted with permission from T. Jeranko et al., 2004. *Sol. Energy Mater. Sol. Cells* 83: 247–262.)

inert atmosphere for 1 day and then after 2 weeks of air exposure. The deterioration in photocurrent at the edges could be clearly seen in the image taken after 2 weeks of air exposure. The corresponding cross-sectional photocurrent density has also been shown in the figure, which shows a lateral process of degradation along with transverse degradation in the device. Figure 5.23 shows the photocurrent mapping images of two MDMO-PPV:PCBM-based solar cells taken after 2 weeks of air exposure. Inhomogenity in the photocurrent could be seen clearly, which suggests the selective degradation in the solar cells. The high degree of degradation was observed at the edges, because of lateral diffusion of oxygen and moisture into the solar cells.

5.5.12 Fluorescence Microscopy

Fluorescence microscopy is an optical imaging technique that uses fluorescence or phosphorescence emitted from organic films to study their properties. Films under study are illuminated with light of a specific wavelength, which is absorbed by the films that emit a light of longer wavelength. When there are more than one species in the film, they could be easily detected through visualization of the contrast of the emitted fluorescence. This technique is widely used for degradation studies in OSCs. Norrman et al. studied degradation in ITO/PEDOT:PSS/C_{12}-PPV/C_{60}/Al solar cell using fluorescence microscopy [11]. The active material C_{12}-PPV was fluorescent, whereas C_{60} and degraded C_{60} were not. The samples were excited at 488 nm that gave the emission at >505 nm. Figure 5.24 shows the fluorescence microscopic images of ITO/PEDOT:PSS/C_{12}-PPV/C_{60}/Al OSCs illuminated in

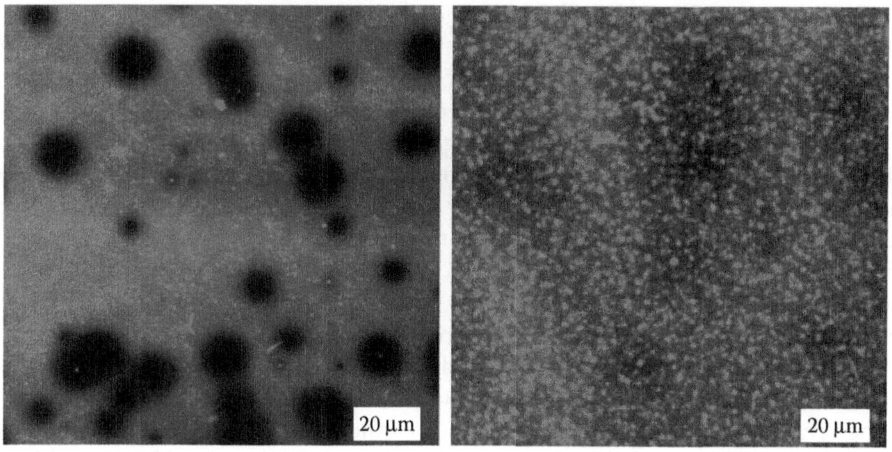

FIGURE 5.24
Fluorescence microscopic images of ITO/PEDOT:PSS/C_{12}-PPV/C_{60}/Al OSCs illuminated in $^{18}O_2/N_2$ atmosphere (left image) and stored in the dark in air (right image) taken from the transparent ITO side. (Adapted with permission from K. Norrman, N. B. Larsen, and F. C., Krebs, 2006. *Sol. Energy Mater. Sol. Cells* 90: 2793–2814.)

$^{18}O_2/N_2$ atmosphere and stored in the dark in air. It was observed that the film stored in the dark exhibited degradation in C_{60} only, whereas the film stored under illumination exhibited degradation in both the C_{12}-PPV and C_{60}. It was therefore concluded that illumination accelerates the oxidation or in other terms the degradation.

5.6 International Standards for Lifetime Testing of OSCs

Thorough understanding and solution of degradation in OSCs require proper measurement in performance variation with time. It is important to mention that different groups report the solar cell performance measured in different test conditions and it makes the interlaboratory comparison of cell performance very difficult. For rapid and actual development of OSC technology, the measurements should be accurate and reproducible at different laboratories. Therefore, for high reliability and accurate quantification of device performance, there has been a requirement of test standards for measurements and lifetime testing of OSCs. Other photovoltaic technologies encounter different kinds of degradation mechanisms (discussed in Chapter 2) that may not be applicable for OSCs, therefore new guidelines are required to set the testing standards for OSCs. The standards ought to be internationally accepted for aging and test procedures. Establishment of lifetime test standards would help in comparing the studies from different research laboratories and industries. In that direction, interlaboratory study would be helpful to explore the variance between measurements in various laboratories and would be helpful in establishing the test standards. Compared to other solar cell technologies, OSC is a highly diverse technology and its performance is affected by various parameters, therefore developing testing procedures for OSCs was very complex. However, some standards were developed in 2003 [63–65] for OSCs testing that were accepted by the solar cell community but were not followed that often because performance of OSCs was not very good in those days and the importance of standards for OSC testing was not really known. But now the need of such standards has become highly important. For easy adoption, the standards were required to be simple and acceptable to all. For this purpose, International Summit on Organic Photovoltaic Stability (ISOS) meetings were held for three consecutive years in 2008 in Denver, Colorado (ISOS-1), in 2009 in Amsterdam, The Netherlands (ISOS-2), and in 2010 in Roskilde, Denmark (ISOS-3), and after discussion some standard protocols for lifetime testing and international comparison of device performance were established [53,66–69].

Keeping in mind the issues of diversity in testing procedures by various research groups and their measurement capability, instead of one rigorous standard, three levels of testing procedures—Level 1 (Basic), Level 2 (Intermediate), and Level 3 (Advanced)—were recommended by the committee. Level 1 includes the handheld measurements using the simplest equipment with minimum guidelines. Level 2 requires equipment such as a source meter and calibrated solar simulator, which should be available at most laboratories. It includes the conditions and procedures suiting to most laboratories. Level 3 has a higher level of accuracy and sophistication, and requires additional equipment and protocols that may be out of the scope of a laboratory. Level 3 procedures are followed in certified labs and include a number of parameters for monitoring. In case a laboratory does not have sufficient facilities to meet any of these protocols, it was recommended that the laboratory measure the data with carefully noting the deviations from recommended protocols and report them along with the data. The climate conditions, latitude, and altitude should also be mentioned for ready comparison of geographical conditions. The ISOS protocols outline the parameters for various testing from shelf-life testing in the dark, outdoor testing, and laboratory weathering, to thermal cycling testing. It also suggests the guidelines for reporting the data. These protocols were not meant for test qualification but for intercomparison of laboratory results toward improving the reliability of reported data. A brief introduction to these standards along with the main test parameters like temperature, humidity, environment, light, and load, is given in Table 5.2. The testing condition ISOS-D-1 corresponds to the shelf-life testing, where devices are stored in the dark in ambient conditions and tested under natural sun light or solar simulator. ISOS-D-2 and 3 are controlled chamber testing and correspond to high temperature storage and damp heat testing, respectively, of the devices stored in dark. ISOS-O-1, 2, and 3 are the outdoor testing protocols; ISOS-L-1, 2, and 3 are laboratory-simulated testing protocols; and ISOS-T-1, 2, and 3 are thermal cycling testing protocols. As per the ISOS-O-1 protocols, the devices are kept outside the laboratory under natural sun and tested periodically indoors under a solar simulator. According to ISOS-O-2 protocols the devices are stored outside and testing is done outside under natural sunlight. As per the ISOS-O-3 norms, the devices are stored and tested outside, but for more accuracy in measurement, the devices are also tested indoors under a solar simulator. Outdoor testing is more straightforward, whereby devices are exposed to actual operating conditions and performance is tested periodically either outdoors in situ under natural sunlight or indoors under a solar simulator. Outdoor testing is more accurate giving the actual lifetime of solar cells in operating conditions, and in situ measurement is more important as it takes into account the effect of temperature, illumination intensity, wind, cloud, and seasonal variation of testing conditions on the device performance. Table 5.3 lists other testing protocols and equipment requirements for dark and outdoor testing. Details on other testing protocols can be found in Reese et al. [69]. These protocols are now being followed internationally [52,66,70].

TABLE 5.2

Protocols Recommended by ISOS for Lifetime Testing of OSCs

Test Type	**Dark**			**Outdoor**		
Test ID	**ISOS-D-1**	**ISOS-D-2**	**ISOS-D-3**	**ISOS-O-1**	**ISOS-O-2**	**ISOS-O-3**
Light	None	None	None	Sunlight	Sunlight	Sunlight
Temperature	Ambient[a]	65/85°C	65/85°C	Ambient	Ambient	Ambient
Humidity	Ambient	Ambient	85% RH[b]	Ambient	Ambient	Ambient
Environment	Ambient	Oven	Env. chamber[c]	Outdoor	Outdoor	Outdoor
Light source	Solar simulator or sunlight	Solar simulator	Solar simulator	Solar simulator	Sunlight	Solar simulator and sunlight
Load	None	None	None	MPP[d] or none	MPP or none	MPP

	Laboratory Weathering Testing			**Thermal Cycling**		
Test ID	**ISOS-L-1**	**ISOS-L-2**	**ISOS-L-3**	**ISOS-T-1**	**ISOS-T-2**	**ISOS-T-3**
Light	Simulator	Simulator	Simulator	None	None	None
Temperature	Ambient	65/85°C	65/85°C	RT[e] to 85°C	RT to 85°C	–40 to 85°C
Humidity	Ambient	Ambient	~50% RH	Ambient	Ambient	~50% RH
Environment	Light only	Light and temp.	Light, temp., RH	Hot plate/oven	Oven/Env. chamber	Env. chamber
Light source	Solar simulator	Solar simulator	Solar simulator	Solar simulator or sunlight	Solar simulator	Solar simulator
Load	MPP or none	MPP or none	MPP	None	None	None

	Solar–Thermal–Humidity Cycling		
Test ID	**ISOS-LT-1**	**ISOS-LT-2**	**ISOS-LT-3**
Light	Simulator	Simulator	Simulator
Temperature	Linear or step ramping between RT and 65°C	Linear ramping between 5 and 65°C	Linear ramping between –25 and 65°C
Humidity	Monitored, uncontrolled	Monitored, controlled at 50% RH beyond 40°C	Monitored, controlled at 50% RH beyond 40°C
Environment	Weathering chamber	Env. chamber with solar simulator	Env. chamber with solar simulator and freezing
Light source	Solar simulator	Solar simulator	Solar simulator
Load	MPP or none	MPP or none	MPP or none

Source: Adapted with permission from M. O. Reese et al., 2011. *Sol. Energy Mater. Sol. Cells* 95: 1253–1267.
[a] Ambient, 25°C/50% RH in general or 27°C/65% RH in tropical countries according to ISO 291 (2008).
[b] RH, relative humidity.
[c] Env. chamber, environmental chamber.
[d] MPP, maximum power point.
[e] RT, room temperature.

5.7 Lifetime Calculation

Degradation curves are the reflection of degradation mechanisms happening inside the devices. It may be linear, exponential, or a combination of different functions of time [26,71]. The lifetime of OSCs is defined in terms of T_{80} or T_{50}, which correspond to

TABLE 5.3

Other Testing Protocols and Equipment Required for Dark and Outdoor Testing of OSCs

		Dark		
		ISOS-D-1	**ISOS-D-2**	**ISOS-D-3**
Testing protocol	Temperature and humidity	Monitor ambient values	Control and monitor sample temp. and monitor RH	Control and monitor sample temp. and RH
	Measurement intervals	Daily to weekly	Daily to weekly	Daily to weekly
	Characterization temp.	Monitor	Monitor	According to IEC 60904-3/ASTM E948 (i.e., 25 ± 2/25 ± 1°C)
	Characterization irradiance	Monitor	Control (800–1100 W/m^2)	Control (800–1100 W/m^2)
	IPCE (at T_0, T_{80}, T_{50}, T_s, and T_{s80})[a]	Optional	Optional	Measure
Output	Time/date	Report	Report	Report
	Light source irradiance level	Report	Report	Report
	Performance parameters	Report J_{sc} and V_{oc}	Report J_{sc}, V_{oc}, FF, PCE/MPP, and J–V optional	Report J_{sc}, V_{oc}, FF, PCE/MPP, and J–V optional
Required equipment	Light source	Solar simulator (close to AM1.5G)	Solar simulator with AM1.5G	Solar simulator with AM1.5G
	Temperature monitoring	Ambient temp. measuring unit	Sensor with ±2°C accuracy	Sensor with ±2°C accuracy
	Humidity monitoring	Ambient RH measuring unit	Sensor with ±5% accuracy	Sensor with ±5% accuracy
	J–V measurement	Computer interfaced source measure unit	Computer interfaced source measure unit	Computer interfaced source measure unit
	Storage	Drawer, etc.	Oven, Env. chamber	Env. chamber
		Outside		
		ISOS-O-1	**ISOS-O-2**	**ISOS-O-3**
Testing protocol	Solar irradiance	Monitor	Monitor	Monitor
	Temperature and humidity	Monitor ambient values	Monitor NOCT[b] and ambient RH	Monitor NOCT and ambient RH
	Measurement intervals	Daily to weekly	15 min to 1 h	Outside 15 min to 1 h, inside weekly or monthly
	Characterization temp.	Monitor sample temp. on backside	Monitor sample temp. on backside	Monitor sample temp. on backside
	Characterization irradiance	Monitor	Monitor	Monitor
	Wind speed monitoring	Optional	Optional	Monitor and the system should be installed ca. 1.2 m to the east or west of the devices under test and 0.7 m higher

(Continued)

TABLE 5.3 (*Continued*)

Other Testing Protocols and Equipment Required for Dark and Outdoor Testing of OSCs

		Outside		
		ISOS-O-1	ISOS-O-2	ISOS-O-3
	IPCE (at T_0, T_{80}, T_{50}, T_s, T_{s80})	Optional	Optional	Report
	Note data taken in the range	Optional	Ambient temperature outside of range 20 ± 15°C, Irradiance below 400 W/m^2	Ambient temperature outside of range 20 ± 15°C, irradiance below 400 W/m^2, wind speeds outside of range 1 ± 0.75 m/s
Output	Location/time	Report latitude, longitude and date	Report latitude, longitude, and date	Report latitude, longitude, and date
	Performance parameters	Report J_{sc} and V_{oc} (*FF* and PCE/MPP and/or full JVs if possible)	Report J_{sc}, V_{oc}, *FF*, PCE/MPP, and *J–Vs* optional	Report J_{sc}, V_{oc}, *FF*, PCE/MPP, and *J–Vs* optional
	Description of measurement protocol and setup	Report	Report	Report
	Load type	Report	Report	Report
Required equipment	Light source	Solar simulator (close to AM1.5G)	Sunlight under clear sky conditions	Sunlight under clear sky conditions and solar simulator with AM1.5G
	Temperature monitoring	Ambient temp. measuring unit	Sensor with ± 2°C accuracy	Sensor with ± 2°C accuracy
	Humidity monitoring	Ambient RH measuring unit	Sensor with ± 5% accuracy	Sensor with ± 5% accuracy
	Irradiance monitoring	Pyranometer and/or photodiode	Irradiance monitoring unit like pyranometer	Photodiode for solar simulator and sunlight irradiance monitoring
	J–V measurement	Computer interfaced source measure unit	Computer interfaced source measure unit	Computer interfaced source measure unit

Source: Adapted with permission from M. O. Reese et al., 2011. *Sol. Energy Mater. Sol. Cells* 95: 1253–1267.

[a] T_0, 0 time after device fabrication; T_s, time for second test after device fabrication; T_{80}, time at which device efficiency reaches to 80% of its initial value; T_{50}, time at which device efficiency reaches to 50% of its initial value; and T_{s80}, time of second test at which device efficiency reaches to 80% of its initial value.

[b] NOCT, nominal operating cell temperature.

the time when efficiency reaches to 80% or 50%, respectively, of its initial value. The half-life of a solar cell corresponds to T_{50}. Accelerated lifetime testing is a traditional process to predict the lifetime of organic electronic devices under a specific test condition. The most commonly used accelerated test conditions for organic devices are high temperature, temperature gradients, power extremes, voltage extremes, humidity, radiation, mechanical shock and stress, relaxation tests, vibrational tests, and their combinations. The accelerated test conditions increase the rate of degradation of the devices. In accelerated testing, the change in a property of the device due to degradation is measured with time. Once the accelerated lifetime is achieved, the actual lifetime is predicted mathematically using different models. Some of the important models used in the accelerated lifetime testing are the Arrhenius equation, the Eyring equation, and the Power law. For the first-order kinetic degradation process, the degradation constant of the Arrhenius equation is given by

$$k_{\text{deg}} = A\exp\left(\frac{-E_a}{kT}\right), \tag{5.9}$$

where A is a constant, which depends on the materials, degradation mechanism, and test conditions; and E_a is the activation energy for the degradation process to get through. The kinetics of this process is strongly temperature dependent. The Eyring model is an extension of Arrhenius model and also takes into account the influence of test parameters other than temperature, such as illumination intensity and external bias.

Preparation in highly inert environment and protection from ambient conditions result in longer life for OSCs. Schuller et al. [72] fabricated OSCs based on the MDMOPPV:PCBM active layer sandwiched between ITO/PEDOT:PSS and Ca/Ag electrodes, sealed in inert atmosphere with glass cover. The temperature accelerated lifetime analysis was performed. For comparison, Al was also used as a cathode electrode, but the electrode was not observed to have any effect on the degradation mechanism. The devices were illuminated with a halogen lamp at ~35 mW/cm^2. Four different temperatures—40°C, 60°C, 85°C, and 105°C—were chosen for acceleration lifetime testing. The J_{sc} was observed to decrease linearly $[J_{sc}(t) = J_{sc}(0)(1 - k_{\text{deg}}t)]$ with time t. k_{deg} was determined from the linear fit of the experimental data and the acceleration factor (K) for the individual temperature was determined from the Arrhenius model, according to which

$$K = \frac{k_{\text{deg}} - T_2}{k_{\text{deg}} - T_1} = \exp\left(\frac{E_a}{k}\left(\frac{1}{T1} - \frac{1}{T2}\right)\right), \tag{5.10}$$

where T_1 and T_2 are the experimental temperatures. The devices at 85°C exhibited less than 20% change in J_{sc} for 1000 h and the experimental data of degradation showed excellent agreement with the Arrhenius model.

De Bettignies et al. [73] carried out lifetime testing measurements on the optimized P3HT:PCBM solar cells and calculated the accelerated lifetime using the Arrhenius model. A significant difference was noted in the degradation with Ca/Ag and LiF/Al electrodes. For the initial 25 h, the degradation was independent of the nature of the cathode, but after that the LiF/Al cathode led to faster degradation in the cells. They studied in detail the degradation of J_{sc}, V_{oc}, and FF, and presented a degradation mechanism. The studies on J_{sc} were in agreement to the linear reduction of J_{sc} with time, where from the degradation constant was determined, and to obtain the acceleration factor (K), they used the Arrhenius model. De Bettignies et al. calculated the degradation factor at 60°C and extrapolated as if the cells were working at 25°C. J_{sc} exhibited less than 15% reduction after 200 h at 60°C, which corresponded to 1000 h at 25°C. Krebs et al. [74] observed a shelf life of almost 1 year with only 35% loss in the initial efficiency. Degradation of OSCs usually happens with different time constants for initial and afterward degradation. Degradation processes that take place in initial hours become less important afterward and the processes that take place in later stages remain less important in initial hours. To make OSCs capable enough to serve the society, more work in this field is urgently required.

References

1. F. C. Krebs and K. Norrman, 2007. Analysis of failure mechanism for stable organic photovoltaic during 10000 h of testing, *Prog. Photovolt.: Res. Appl.* 15: 697–712.
2. S. Chambon, A. Rivaton, J. L. Gardette, and M. Firon, 2007. Photo- and thermal degradation of MDMO-PPV:PCBM blends, *Sol. Energy Mater. Sol. Cells* 91: 394–398.
3. A. G. Erlat, B. M. Henry, C. R. M. Grovenor, A. G. D. Briggs, R. J. Chater, and Y. Tsukahara, 2004. Mechanism of water vapor transport through PET/AlO_xN_y gas barrier films, *J. Phys. Chem. B* 108: 883–890.
4. K. Norrman, M. V. Madsen, S. A. Gevorgyan, and F. C. Krebs, 2010. Degradation patterns in water and oxygen of an inverted polymer solar cell, *J. Am. Chem. Soc.* 132: 16883–16892.
5. X. Yang, J. K. J. van Duren, R. A. J. Janssen, M. A. J. Michels, and J. Loos, 2004. Morphology and thermal stability of the active layer in poly(*p*-phenylenevinylene)/methanofullerene plastic photovoltaic devices, *Macromolecules* 37: 2151–2158.
6. S. A. Gevorgyan and F. C. Krebs, 2008. Bulk heterojunctions based on native polythiophene, *Chem. Mater.* 20: 4386–4390.
7. D. E. Gallardo, C. Bertoni, S. Dunn, N. Gaponik, and A. Eychmüller, 2007. Cathodic and anodic material diffusion in polymer/semiconductor–nanocrystal composite devices, *Adv. Mater.* 19: 3364–3367.
8. H. Neugebauer, C. J. Brabec, J. C. Hummelen, and N. S. Sariciftci, 2000. Stability and photodegradation mechanisms of conjugated polymer/fullerene plastic solar cells, *Sol. Energy Mater. Sol. Cells* 61: 35–42.
9. M. Jørgensen, K. Norrman, and F. C. Krebs, 2008. Stability/degradation of polymer solar cell, *Sol. Energy Mater. Sol. Cells* 92: 686–714.
10. M. Jørgensen, K. Norrman, S. A. Gevorgyan, T. Tromholt, B. Andreasen, and F. C. Krebs, 2012. Stability of polymer solar cells, *Adv. Mater.* 24: 580–612.
11. K. Norrman, N. B. Larsen, and F. C. Krebs, 2006. Lifetimes of organic photovoltaics: Combining chemical and physical characterization techniques to study degradation mechanisms, *Sol. Energy Mater. Sol. Cells* 90: 2793–2814.
12. M. Lira-Cantu, K. Norrman, J. W. Andreasen, and F. C. Krebs, 2006. Oxygen release and exchange in niobium oxide MEHPPV hybrid solar cells, *Chem. Mater.* 18: 5684–5690.
13. K. Norrman and F. C. Krebs, 2005. Chemical degradation mechanisms of organic photovoltaics studied by TOF-SIMS and isotopic labeling, *Proc. SPIE* 5938: 59380D.
14. R. D. Scurlock, B. Wang, P. R. Ogilby, J. R. Sheats, and R. L. Clough, 1995. Singlet oxygen as a reactive intermediate in the photodegradation of an electroluminescent polymer, *J. Am. Chem. Soc.* 117: 10194–10202.
15. N. Dam, R. D. Scurlock, B. Wang, L. Ma, M. Sundahl, and P. R. Ogilby, 1999. Singlet oxygen as a reactive intermediate in the photodegradation of phenylenevinylene oligomers, *Chem. Mater.* 11: 1302–1305.
16. R. F. Bianchi, D. T. Balogh, M. Tinani, R. M. Faria, and E. A. Irene, 2004. Ellipsometry study of the photo-oxidation of poly[(2-methoxy-5-hexyloxy)-*p*-phenylenevinylene], *J. Polym. Sci. B: Polym. Phys.* 42: 1033–1041.
17. B. H. Cumpston and K. F. Jensen, 1998. Photooxidative stability of substituted poly(phenylene vinylene) (PPV) and poly(phenylene acetylene) (PPA), *J. Appl. Polym. Sci.* 69: 2451–2558.
18. F. Padinger, T. Fromherz, P. Denk, C. J. Brabec, J. Zettner, T. Hierl, and N. S. Sariciftci, 2001. Degradation of bulk heterojunction solar cells operated in an inert gas atmosphere: A systematic study, *Synth. Metals* 121: 1605–1606.
19. J. Alstrup, K. Norrman, M. Jørgensen, and F. C. Krebs, 2006. Lifetimes of organic photovoltaics: Design and synthesis of single oligomer molecules in order to study chemical degradation mechanisms, *Sol. Energy Mater. Sol. Cells* 90: 2777–2792.
20. K. Norrman, J. Alstrup, M. Jørgensen, and F. C. Krebs, 2006. Lifetimes of organic photovoltaics: Photooxidative degradation of a model compound, *Surf. Interface Anal.* 38: 1302–1310.

21. M. G. Matturro, R. P. Reynolds, R. V. Kastrup, and C. F. Pictroski, 1986. Thioozonide decomposition. Sulfur and oxygen atom transfer. Evidence for the formation of a carbonyl O-sulfide intermediate, *J. Am. Chem. Soc.* 108: 2775–2776.
22. J. M. Zhuo, L. H. Zhoa, R. Q. Png, L. Y. Wong, P. J. Chia, J. C. Tang, S. Sivaramakrishnan et al., 2009. Direct spectroscopic evidence for a photodoping mechanism in polythiophene and poly(bithiophene-alt-thienothiophene) organic semiconductor thin films involving oxygen and sorbed moisture, *Adv. Mater.* 21: 4747–4752.
23. J. Schafferhans, A. Baumann, A. Wagenpfahl, C. Deibel, and V. Dyakonov, 2010. Oxygen doping of P3HT:PCBM blends: Influence on trap states, charge carrier mobility and solar cell performance, *Org. Electron.* 11: 1693–1700.
24. L. Luer, H. J. Egelhaaf, D. Oelkrug, G. Cerullo, G. Lanzani, B. H. Huisman, and D. de Leeuw, 2004. Oxygen-induced quenching of photoexcited states in polythiophene films, *Org. Electron.* 5: 83–89.
25. M. S. A. Abdou, F. P. Orfino, Y. Son, and S. Holdcroft, 1997. Interaction of oxygen with conjugated polymers: Charge transfer complex formation with poly(3-alkylthiophenes), *J. Am. Chem. Soc.* 119: 4518–4524.
26. P. Kumar, A. Sharma, and D. P. Singh, 2013. Effect of voltage sweep direction on the performance evaluation of P3HT:PCBM solar cells, *Prog. Photovolt.: Res. Appl.* 21: 950.
27. A. M. Peiro, C. Colombo, G. Doyle, J. Nelson, A. Mills, and J. R. Durant, 2006. Photochemical reduction of oxygen adsorbed to nanocrystalline TiO_2 films: A transient absorption and oxygen scavenging study of different TiO_2 preparations, *J. Phys. Chem. B* 110: 23255–23263.
28. A. Mills, G. Doyle, A. M. Peiro, and J. Durrant, 2006. Demonstration of a novel, flexible, photocatalytic oxygen-scavenging polymer film, *J. Photochem. J. Photobiol. A: Chem.* 177: 328–331.
29. A. M. Peiro, G. Doyle, A. Mills, and J. R. Durrant, 2005. Freestanding polymer–metal oxide nanocomposite films for light-driven oxygen scavenging, *Adv. Mater.* 17: 2365–2368.
30. M. Lira-Cantu, K. Norrman, J. W. Andreasen, N. Casan-Pastor, and F. C. Krebs, 2007. Detrimental effect of inert atmospheres on hybrid solar cells based on semiconductor oxides, *J. Electrochem. Soc.* 154: B508–B513.
31. D. C. Olson, J. Piris, R. T. Collins, S. E. Shaheen, and D. S. Ginley, 2006. Hybrid photovoltaic devices of polymer and ZnO nanofiber composites, *Thin Solid Films* 496: 26–29.
32. W. J. E. Beek, M. M. Wienk, M. Kemerink, X. Yang, and R. A. J. Janssen, 2005. Hybrid zinc oxide conjugated polymer bulk heterojunction solar cells, *J. Phys. Chem. B* 109: 9505–9516.
33. P. Kumar, C. Bilen, K. Feron, N. C. Nicolaidis, Bill B. Gong, X. Zhou, W. J. Belcher, and P. C. Dastoor, 2014. Comparative degradation and regeneration of polymer solar cells with different electrodes, *ACS Appl. Mater. Inter.* 6: 5281.
34. T. G. Devi and M. P. Kannan, 2007. X-ray diffraction (XRD) studies on the chemical states of some metal species in cellulosic chars and the Ellingham diagrams, *Energy Fuels* 21: 596–601.
35. R. B. Getman, Y. Xu, and W. F. Schneider, 2008. Thermodynamics of environment-dependent oxygen chemisorption on Pt(111), *J. Phys. Chem. C* 112: 9559–9572.
36. J. Nishnaga, T. Aihara, H. Yamagata, and Y. Horikoshi, 2005. Mechanical and optical characteristics of Al-doped C_{60} films, *J. Cryst. Growth* 278: 633–637.
37. M. Logdlund and J. L. Bredas, 1994. Theoretical studies of the interaction between aluminum and poly(*p*-phenylenevinylene) and derivatives, *J. Chem. Phys.* 101: 4357.
38. F. Bebensee, M. Schmid, H. P. Steinruck, C. T. Campbell, and J. M. Gottfried, 2010. Toward well-defined metal–polymer interfaces: Temperature-controlled suppression of subsurface diffusion and reaction at the calcium/poly(3 hexylthiophene) interface, *J. Am. Chem. Soc.* 132: 12163–12165.
39. P. Peumans, V. Bulovic, and S. R. Forrest, 2000. Efficient photon harvesting at high optical intensities in ultrathin organic double-heterostructure photovoltaic diodes, *Appl. Phys. Lett.* 76: 2650–2652.
40. Q. L. Song, F. Y. Li, H. Yang, H. R. Wu, X. Z. Wang, W. Zhou, J. M. Zhao, X. M. Ding, C. H. Huang, and X. Y. Hou, 2005. Small molecule organic solar cells with improved stability, *Chem. Phys. Lett.* 416: 42–46.

41. H. R. Wu, Q. L. Song, M. L. Wang, F. Y. Li, H. Yang, Y. Wu, C. H. Huang, X. M. Ding, and X. Y. Hou, 2007. Stable small molecule organic solar cells with 1,3,5-tris(2-*N*-phenylbenzimidazolyl) benzene as an organic buffer, *Thin Solid Films* 515: 8050–8053.
42. C. J. Brabec, S. E. Shaheen, C. Winder, and N. S. Sariciftci, 2002. Effect of LiF/metal electrodes on the performance of plastic solar cells, *Appl. Phys. Lett.* 80: 1288.
43. F. C. Krebs and H. Spanggaard, 2005. Significant improvement of polymer solar cell stability, *Chem. Mater.* 17: 5235–5237.
44. M. P. de Jong, L. J. van Ijzendoorn, and M. J. A. de Voigt, 2000. Stability of the interface between indium-tin-oxide and poly(3,4-ethylenedioxythiophene)/poly(styrenesulfonate) in polymer light-emitting diodes, *Appl. Phys. Lett.* 77: 2255.
45. K. Kawanoa, R. Pacios, D. Poplavskyy, J. Nelson, D. D. C. Bradley, and J. R. Durrant, 2006. Degradation of organic solar cells due to air exposure, *Sol. Energy Mater. Sol. Cells* 90: 3520–3530.
46. X. Yang, J. Loos, S. C. Veenstra, W. J. H. Verhees, M. M. Wienk, J. M. Kroon, M. A. J. Michels, and R. A. J. Janssen, 2005. Nanoscale morphology of high-performance polymer solar cells, *Nano Lett.* 5: 579–583.
47. K. Norrman, S. A. Gevorgyan, and F. C. Krebs, 2009. Water induced degradation of polymer solar cells studied by $H_2{}^{18}O$ labeling, *ACS Appl. Mater. Inter.* 1: 102–112.
48. T. Tromholt, E. A. Katz, B. Hirsch, A. Vossier, and F. C. Krebs, 2010. Effects of concentrated sunlight on organic photovoltaics, *Appl. Phys. Lett.* 96: 73501.
49. K. Sivula, C. K. Luscombe, B. C. Thompson, and J. M. J. Frechet, 2006. Enhancing the thermal stability of polythiophene: Fullerene solar cells by decreasing effective polymer regioregularity, *J. Am. Chem. Soc.* 128: 13988–13989.
50. P. Kumar, C. Bilen, K. Feron, X. Zhou, W. J. Belcher, and P. C. Dastoor, 2014. Enhanced regeneration of degraded polymer solar cells by thermal annealing, *Appl. Phys. Lett.* 104: 193905.
51. Y. H. Lin, Y. T. Tsai, C. C. Wu, C. H. Tsai, C. H. Chiang, H. F. Hsu, J. J. Lee, and C. Y. Cheng, 2012. Comparative study of spectral and morphological properties of blends of P3HT with PCBM and ICBA, *Org. Electron.* 13: 2333–2341.
52. A. Gaur and P. Kumar, 2013. Effect of degradation on electronic properties of polymer solar cells, *Polym. Adv. Technol.* 24: 630–637.
53. S. A. Gevorgyan, A. J. Medford, E. Bundgaard, S. B. Sapkota, H. F. Schleiermacher, B. Zimmermann, U. Würfel et al., 2011. An inter-laboratory stability study of roll-to-roll coated flexible polymer solar modules, *Sol. Energy Mater. Sol Cells* 95: 1398–1416.
54. P. Kumar, S. C. Jain, H. Kumar, S. Chand, and V. Kumar, 2009. Effect of illumination intensity and temperature on open circuit voltage in organic solar cells, *Appl. Phys. Lett.* 94: 183505.
55. P. Kumar, H. Kumar, S. C. Jain, P. Venkatesu, S. Chand, and V. Kumar, 2009. Effect of active layer thickness on open circuit voltage in organic photovoltaic devices, *Jpn. J. Appl. Phys.* 48: 121501.
56. R. L. Wells and T. Fort, 1972. Interaction of oxygen with clean aluminum surfaces by measurement of work function changes, *Surf. Sci.* 33: 172–178.
57. B. Paci, A. Generosi, V. R. Albertini, P. Perfetti, B. de Bettignies, M. Firon, J. Leroy, and C. Sentein, 2006. Controlling photoinduced degradation in plastic photovoltaic cells: A time-resolved energy dispersive x-ray reflectometry study, *Appl. Phys. Lett.* 89: 043507.
58. C. Melzer, V. V. Krasnikov, and G. Hadziioannou, 2003. Organic donor/acceptor photovoltaics: The role of C60/metal interfaces, *Appl. Phys. Lett.* 82: 3101.
59. B. Pradhan and A. J. Pal, 2005. Role of dielectric properties in organic photovoltaic devices, *Chem. Phys. Lett.* 416: 327–330.
60. J. W. Andreasen, S. A. Gevorgyan, C. M. Schlepütz, and F. C. Krebs, 2008. Applicability of X-ray reflectometry to studies of polymer solar cell degradation, *Sol. Energy Mater. Sol. Cells* 92: 793–798.
61. B. Paci, A. Generosi, V. R. Albertini, P. Perfetti, B. de Bettignies, M. Firon, J. Leroy, and C. Sentein, 2005. In situ energy dispersive x-ray reflectometry measurements on organic solar cells upon working, *Appl. Phys. Lett.* 87: 194110.
62. T. Jeranko, H. Tributsch, N. S. Sariciftci, and J. C. Hummelen, 2004. Patterns of efficiency and degradation of composite polymer solar cells, *Sol. Energy Mater. Sol. Cells* 83: 247–262.

63. S. C. Tse, H. H. Fong, and S. K. So, 2003. Electron transit time and reliable mobility measurements from thick film hydroxyquinoline-based organic light-emitting diode, *J. Appl. Phys.* 94: 2033.
64. H. Fong, K. Lun, and S. So, 2002. Extrinsic electron traps in tris(8-hydroxyquinoline) aluminium, *Jpn. J. Appl. Phys.* 41: L1122.
65. A. J. Mozer and N. S. Sariciftci, 2004. Negative electric field dependence of charge carrier drift mobility in conjugated, semiconducting polymers, *Chem. Phys. Lett.* 389: 438–442.
66. D. M. Tanenbaum, M. Hermenau, E. Voroshazi, M. T. Lloyd, Y. Galagan, B. Zimmermann, M. Hosel et al., 2012. The ISOS-3 inter-laboratory collaboration focused on the stability of a variety of organic photovoltaic devices, *RSC Adv.* 2: 882–893.
67. G. Teran-Escobar, D. M. Tanenbaum, E. Voroshazi, M. Hermenau, K. Norrman, M. T. Lloyd, Y. Galagan et al., 2012. On the stability of a variety of organic photovoltaic devices by IPCE and in situ IPCE analyses-the ISOS-3 inter-laboratory collaboration, *Phys. Chem. Chem. Phys.* 14: 11824.
68. R. Rosch, D. M. Tanenbaum, M. Jørgensen, M. Seeland, M. Barenklau, M. Hermenau, E. Voroshazi et al., 2012. Investigation of the degradation mechanisms of a variety of organic photovoltaic devices by combination of imaging techniques—the ISOS-3 inter-laboratory collaboration, *Energy Environ. Sci.* 5: 6521.
69. M. O. Reese, S. A. Gevorgyan, M. Jørgensen, E. Bundgaard, S. R. Kurtz, D. S. Ginley, D. C. Olson et al., 2011. Consensus stability testing protocols for organic photovoltaic materials and devices, *Sol. Energy Mater. Sol. Cells* 95: 1253–1267.
70. V. Turkovic, S. Engmann, N. Tsierkezos, H. Hoppe, U. Ritter, and G. Gobsch, 2014. Long-term stabilization of organic solar cells using hindered phenols as additives, *ACS Appl. Mater. Interfaces* 6: 18525–18537.
71. F. C. Krebs, J. E. Carle, N. Cruys-Bagger, M. Andersen, M. R. Lilliedal, M. A. Hammond, and S. Hvidt, 2005. Lifetimes of organic photovoltaics: Photochemistry, atmosphere effects and barrier layers in ITO-MEHPPV:PCBM-aluminium devices, *Sol. Energy Mater. Sol. Cells* 86: 499–516.
72. S. Schuller, P. Schilinsky, J. Hauch, and C. J. Brabec, 2004. Determination of the degradation constant of bulk heterojunction solar cells by accelerated lifetime measurements, *Appl. Phys. A*79: 37–40.
73. R. De Bettignies, J. Leroy, M. Firon, and C. Sentein, 2006. Accelerated lifetime measurements of P3HT:PCBM solar cells, *Synt. Met.* 156: 510–513.
74. F. C. Krebs, 2006. Encapsulation of polymer photovoltaic prototypes, *Sol. Energy Mater. Sol. Cells* 90: 3633–3643.

6

How to Prevent Degradation in Organic Solar Cells

6.1 Introduction

Although the efficiency of organic solar cells (OSCs) has improved to a great extent and has gone beyond the value required to give competition to conventional solar cells, they are not stable enough. Degradation and poor stability are the biggest problems keeping OSCs away from commercialization. If OSCs are to replace silicon (Si) solar cells, prevention from decline in their performance with time is must. For prevention from degradation the understanding of degradation is of prime importance. The mechanisms controlling degradation in OSCs have been discussed in the previous chapter. Modification in chemical structures, oxidation of constituents, cross-linking of polymers and their chain scissions are some of the important factors causing degradation in OSCs. This chapter will focus on the methods suitable for prevention of degradation in OSCs. Before covering the methods adopted to prevent degradation in OSCs, let us first discuss the prevention of degradation in dye-sensitized solar cells (DSSCs) that will help us in understanding the prevention of degradation in OSCs.

6.2 Prevention of Degradation in Dye-Sensitized Solar Cells (DSSCs)

To prevent degradation in DSSCs, encapsulation is primarily important. The encapsulation must ensure that there is neither leakage nor evaporation of the liquid electrolyte. The sealing materials should introduce a high diffusion barrier for electrolyte constituents (including I_2) out of the cell and a high barrier for diffusion of any external entity like H_2O and O_2 molecules into the cell. Presence of O_2 is more harmful than H_2O for stability of DSSCs. Under high illumination, the effect of O_2 becomes more severe as it attacks all organic molecules in the cell. Usually a hotmelt Surlyn tape is used to seal the DSSCs but it is not suitable when devices operate at very high temperatures, therefore sealing of the electrolyte should be done using materials of high melting points. For example, Hinsch et al. developed a technique where low-melting-temperature glass frit was used and the sealing was done at 650°C [1]. The composition of glass frit was adjusted such a way that it matched with the thermal expansion coefficient of glass slides. This sealing led to more thermally stable solar cells. The stability of DSSCs also depends on the chemical composition of electrolytes and additives.

Usually the electrolyte mediums are very sensitive to ultraviolet (UV) light and for better stability of solar cells, UV filters should be used. Presence of UV light also initiates TiO_2-mediated oxidation processes that cause degradation in organic compounds including dyes and solvents. A dramatic improvement in the cell stability was achieved in UV light when MgI_2 was added to the electrolyte medium [1]. Improved stability by addition of MgI_2 has been attributed to faster reduction of holes by iodine, formation of an MgO surface layer, and prevention of side reactions. Pure nitrile-based electrolyte solvents like acetonitrile and proprionitrile have shown high UV stability. Solvents having chemically bonded oxygen are less stable. Purity of solvents is also important to ensure high stability in DSSCs. Preparation of DSSCs with a suitable method and use of inexpensive UV filters with adequate encapsulation can result in products stable enough for various low-power applications [2]. TiO_2/dye/electrolyte interfaces are very important from a stability point of view. The presence of H_2O molecules in the electrolyte solution increases the numbers of protons in the cell that may get adsorbed on the TiO_2 surface and increase the surface charge [3]. The surface charge on TiO_2 would affect its Fermi level and as a result the open-circuit voltage (V_{oc}). The addition of pyridine derivatives buffers the protons and imparts better stability to DSSCs [4]. Pyridine may also get adsorbed on the TiO_2 surface and prevent adsorption of other entities, which may affect the TiO_2 Fermi level. Carboxylic acids have also been quite effective that way [5]. Some groups have also used metal-oxide insulating layers like Al_2O_3 on TiO_2 to stabilize it [6–8]. Use of an insulating layer also prevents the back-electron transfer from TiO_2 to the electrolyte. An improved method for production of Pt-coated counter electrode would be very helpful in achieving high stability. Deposition of Pt on counter electrode should be strong enough that it is not released from the counter electrode. As a replacement to liquid electrolyte medium, there have been many reports of solid-state mediums using organic/inorganic hole transport materials, nanocomposite materials, gel, and polymer electrolytes [9–16]. Polymer electrolytes have gained more interest due to ease in fabrication, low cost, and better stability. The quasi-solid-state mediums have also shown great stability in DSSCs [17,18]. In quasi-solid-state mediums the framework materials offer liquid-like channels for transportation of I^-/I_3^- ions [19,20].

6.3 Prevention of Degradation in Organic Solar Cells (OSCs)

Degradation in OSCs can be differentiated into active layer degradation and electrode degradation. To improve the stability of OSCs, one should control both kinds of degradation. The possibility of degradation in OSCs can be circumvented by protection from oxygen and moisture and insertion of suitable buffer layers, which could inhibit diffusion of different materials and external entities like O_2 and H_2O. However, photobleaching/photodegradation and thermally induced variations would cause even the protected devices to degrade during operation in direct sunlight. Therefore, the devices and materials should be more robust and insensitive to the entities affecting their performance. This can be achieved via molecular engineering, device engineering, and proper encapsulation of devices all together.

6.3.1 Molecular Engineering

The organic materials themselves should be air stable and hard enough to retard the diffusion of external entities like in poly(3-carboxydithiophene) (P3CT), which is prepared by thermocleavage of its soluble precursor and is a polyester derivative [21]. Gevorgyan et al. prepared bulk-heterojunction (BHJ) solar cells using poly-(3-(2-methylhexan-2-yl)-oxy-carbonyldithiophene) (P3MHOCT) and fullerene derivatives [22]. The curing at 200°C converted P3MHOCT to more rigid P3CT, which was converted to polythiophene when subjected to 310°C. The chemical structure of P3CT is shown in Figure 6.1. The device with polythiophene exhibited slow decay over 500 h of continuous illumination in nitrogen atmosphere. In the context of longer stability of polymer solar cells, Manceau et al. did synthesize thermocleavable polymers with enhanced stability [23]. Petersen et al. synthesized and studied the thermocleavable esters of low band gap monomers and polymers based on diphenyl-dithienylthieno-pyrazine [24]. The best performing material was subjected to lifetime studies in four different environments (dry nitrogen, dry oxygen, humid nitrogen, and ambient atmosphere). The best stability was observed in nitrogen, whereas nearly the same degree of stability was observed in dry oxygen. The solar cells based on P3CT exhibited extremely high stability compared to those based on P3HT, but the efficiency was low compared to P3HT and even native thiophene [25]. It is not simple to make BHJ solar cells out of P3CT, as it requires cleavage at about 200°C, which is too high for BHJ with PCBM. For simplicity in fabrication and high performance, the molecules should be easy to process, and for that purpose the use of thermocleavable materials is not a very good idea. The molecules should possess organic groups that are less sensitive to oxygen and moisture. For example, the thiophene derivatives are less prone to reaction with oxygen and moisture than easily oxidizable vinylene derivatives. The fullerene bisadducts like ICBA and NCBA impart better stability to the active layer compared to PCBM [26–28,29]. The photostability of PBDTTPD has been observed to be more than P3HT and PCDTBT [30];

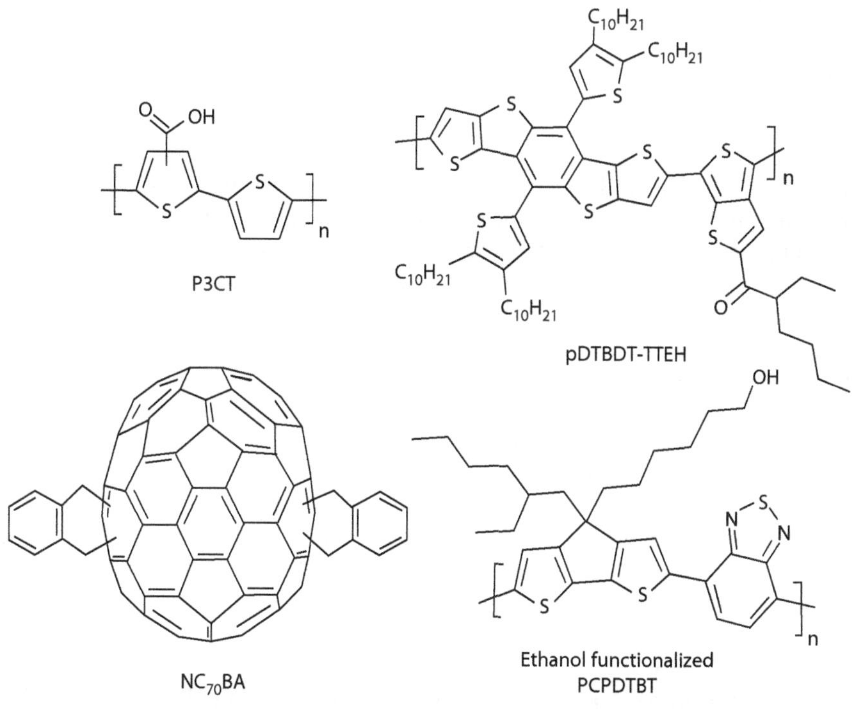

FIGURE 6.1
Chemical structures of P3CT, pDTBDT-TTEH, $NC_{70}BA$, and ethanol functionalized PCPDTBT.

this has been attributed to the presence of oxy-alkyl side chains [31], which are supposed to be more photostable than alkyl side chains [32].

The importance of low band gap polymers toward achieving high efficiency was discussed in Chapter 3. Achievement of the latest high efficiencies in OSCs has been possible because of low band gap polymers [33–36]. Keeping the advantage of low band gap polymers in achieving high power conversion efficiency (PCE), new low band gap polymers are being designed and synthesized toward achieving high stability in OSCs [37]. Highly conjugated polymers with planer backbone are expected to be very crucial not only for high efficiency but for high stability as well. Some of the promising low band gap polymers have been synthesized by copolymerization of electron-rich donor units like benzo[1,2-*b*;4,5-*b*′] dithiophene (BDT) and acceptor units like diketopyrrolopyrrole (DPP), *N*-alkylthieno[3,4-*c*] pyrrole-4,6-dione (TPD), benzothiadiazole (BZ), and thieno[3,4-b]dithiophene (TT) [38–41]. Shin et al. designed and synthesized copolymers having dithienobenzodithiophene (DTBDT) donor derivatives and 2-ethyl-1-(thieno[3,4-*b*]thiophen-2-yl)hexan-1-one (TTEH) and *N*-octyl-2,7-dithia-5-azacyclopenta[*a*]pentalene-4,6-dione (DTPD) acceptors to give, respectively, the pDTBDT-TTEH and pDTBDT-DTPD copolymers, which had a band gap of ~1.6 eV [37]. The chemical structure of the pDTBDT-TTEH polymer is shown in Figure 6.1. To test the stability of solar cells, the synthesized pDTBDT-TTEH and pDTBDT-DTPD polymers were blended with $PC_{71}BM$ acceptor in a 1:1.5 wt% ratio and the solar cells were prepared in ITO/TiO_2/polymer:$PC_{71}BM$/MoO_3/Ag inverted geometry. The solar cells based on pDTBDT-TTEH:$PC_{71}BM$ outperformed those based on pDTBDT-DTPD:$PC_{71}BM$ active layer and exhibited maximum PCEs of 6.16% and 4.44%, respectively. The solar cells were stored in air in the dark without encapsulation. pDTBDT-TTEH devices were observed to be highly air stable and pDTBDT-DTPD devices exhibited relatively rapid degradation. pDTBDT-TTEH-based solar cells retained more than 95% of their initial efficiency for over 1000 h without any encapsulation, whereas those based on pDTBDT-DTPD came down to ~30% of their initial efficiency in the same conditions. This difference in stability of solar cells was attributed to different morphological stabilities of the active layers. pDTBDT-DTPD solar cells exhibited strong aggregation and phase separation, which resulted in significant loss in short-circuit current density (J_{sc}) with time. It is important to mention that morphological stability is strongly related to film crystallinity. Blending of $PC_{71}BM$ in pDTBDT-TTEH retains the initial morphology and optimized phase separation of the active layer, however, blending of $PC_{71}BM$ in pDTBDT-DTPD deters its crystallinity and aggregates are formed. Therefore, the highly crystalline and packed nanodomains in pDTBDT-TTEH:$PC_{71}BM$ active layer inhibit air infiltration and prevent the active layer from degradation, which does not happen with amorphous and aggregated film of pDTBDT-DTPD:$PC_{71}BM$. Though the two polymers had the same planer fused ring system of DTBDT, the difference in their crystallinity comes from the TTEH and DTPD side groups. The TTEH side group leads to a planer polymer backbone, whereas DTPD forms a highly twisted backbone. These results suggest that the long planer backbones in polymers are beneficial for not only high efficiency but for high stability as well. pDTBDT-TTEH:$PC_{71}BM$-based solar cells also exhibited better stability compared to well-studied PTB7:$PC_{71}BM$ solar cells, and that was again because of less crystallinity of PTB7:$PC_{71}BM$ compared to pDTBDT-TTEH:$PC_{71}BM$ active layer [37].

Meng et al. synthesized a new C_{70} derivative dihydronaphthyl-based C_{70} bisadduct ($NC_{70}BA$) and explored it as a better acceptor compared to PCBM for OSCs applications [29]. Blending of $NC_{70}BA$ with P3HT showed not only higher PCE but also higher thermal stability compared to P3HT:PCBM composite. The chemical structure of $NC_{70}BA$ is shown in Figure 6.1. P3HT:$NC_{70}BA$-based solar cells exhibited 5.95% PCE. $NC_{70}BA$ had

0.2 eV higher LUMO (lowest unoccupied molecular orbital) than PCBM, which resulted in higher V_{oc} and improved light absorption in visible range. The decomposition temperature of $NC_{70}BA$ was measured to be as high as 415°C. PCBM is known to form crystallites in the active layer after thermal annealing and the efficiency decreases gradually upon heating [42]. On the contrary, $NC_{70}BA$ has an amorphous nature and reduces thermally induced crystallization. The solar cells were heated at 150°C for 20 h, where the efficiency of those based on P3HT:PCBM reduced from 3.82% to 1.75% the efficiency of solar cells based on P3HT:$NC_{70}BA$ reduced from 5.95% to 4.89% only [29]. Reduction in thermally induced crystallization of active layers has also been observed on blending with ICBA [42]. Reduction in thermally induced crystallization leads to higher thermal stability in OSCs. Thermally induced degradation has also been minimized via cross-linking of polymer and fullerenes [43–46]. Introduction of certain functional groups on the polymers may cross-link and freeze the optimum morphology after thermal treatment or exposure to UV light. Freezing of the morphology provides enhanced thermal stability to the active layer. Kim et al. functionalized P3HT copolymers with bromine and cross-linked them via UV exposure [46]. The devices after photo-cross-linking were highly stable but this method is rather harsh and morphology of the active layer is unpredictable therefore not very good.

Obviating the cross-linking approach, Bertho et al. improved the thermal stability of OSCs via changing the intrinsic structure of the side chains of poly(3-alkylthiphene) (P3AT) and P3HT [47,48]. Usually for optimum morphology and donor–acceptor phase separation, the active layers are annealed at high temperatures, but annealing beyond the glass transition temperature (T_g) accelerates phase separation and forms large crystalline regions that lead to unoptimum morphology and deterioration in cell performance. Annealing beyond T_g has also been observed to develop a polymer layer at the active layer and the top electrode interface that causes deterioration in device performance [49]. The temperature of devices may also increase during direct operation in sunlight, therefore high thermal stability of the active layer morphology is highly important for long-term stability of OSCs, and high T_g materials may fulfill this goal. High T_g polymers inhibit demixing of polymer and fullerene by reducing diffusion of PCBM and results in a rigid polymer:fullerene matrix [50]. Kester et al. employed the high T_g strategy along with side chain functionalization of PCPDTBT by ester and alcohol moieties toward achieving high thermal stability [51]. The ester and alcohol moieties were inserted on the 4,4-dialkyl-4H-cyclopenta[2,1-b;3,4-b′] dithiophene (CPDT) side chains. Modification of side chains did not affect the solid-state optoelectronic properties of PCPDTBT but increased T_g and reduced the tendency of aggregation, which led to enhanced stability. The alcohol functionalized PCPDTBT showed the best stability results. The chemical structure of alcohol functionalized PCPDTBT is shown in Figure 6.1. Side chain modification of PCPDTBT reduced the crystallization and demixing of polymer:fullerene blends upon storage at high temperatures that imparted longer lifetime to devices. Sometimes thermal treatment of high T_g polymers may also convert them into a highly rigid form, giving long life to the devices like P3CT, discussed earlier. A similar study on a thiophene derivative possessing carboxylic acid function on every second thiophene ring led to improved stability of solar cells on thermal treatment, and by using a thermal acceleration factor of 4 between 25°C and 72°C and indoor illumination intensities of 50–100 Wm^{-2}, the operational lifetimes of >20,000 h were anticipated under the inert condition [52]. Purity of materials is also very important for achieving high stability in OSCs [30]. Matekar et al. improved long-term stability of PBDTTPD-based solar cells by purifying it and removing the organic impurities [30]. The devices with high molecular weight degraded faster compared to those based on low molecular weight PBDTTPD, however purification of high molecular weight polymer improved the lifetime of devices dramatically. Similarly Kong et al. reported significant improvement in the stability of PCPDTBT solar cells by selectively extracting out the high molecular weight PCPDTBT and removing its trap embedded components [53]. Extraction of trap-free high molecular weight PCPDTBT resulted in enhanced PCE and reduction in initial burn-in losses that led to longer stability in solar cells.

Photobleaching is an important degradation phenomenon in organic semiconductors that corresponds to loss of optical absorption and their color due to photochemical reactions in the semiconductor upon light illumination. The photochemical reactions cause decomposition of chromophores and formation of inferior compounds. For enhanced longevity of OSCs, the organic molecules and the active layers should be chemically robust with reduced photobleaching rate. Mateker et al. investigated the photobleaching in a number of organic molecules including polymers, small molecules, and fullerenes, and found that film morphology of a material affects the photobleaching rate greatly, where amorphous films bleach faster than crystalline films [54]. A study on the effect of film density on its photobleaching rate showed that the stability increases with an increase in film density. Figure 6.2 shows the photobleaching rate of different materials as a function of density of their corresponding films. T1 and X2 on the graph are two solution-processed small molecules. The films having density more than 1.5 g/cm^3 photobleach slowly compared to less dense films. This relationship is not absolute, but some changes in the side groups might cause discrepancy, like in case of P3HT and MDMO-PPV. The densities of regiorandom P3HT and MDMO-PPV are almost the same, but their photostability differ by almost two orders of magnitude. Also the density of amorphous film of rubrene is

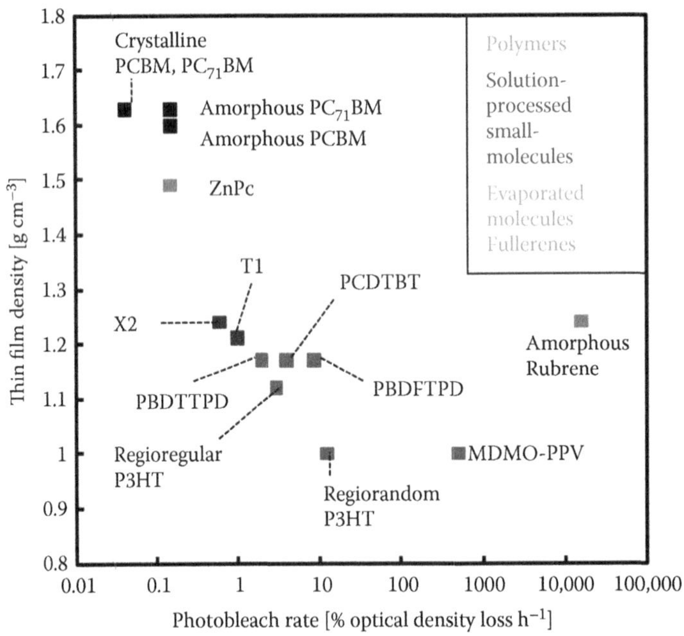

FIGURE 6.2
Photobleaching rate of different materials as a function of density of their films. (Reprinted with permission from W. R. Mateker et al., Molecular packing and arrangement govern the photooxidative stability of organic photovoltaic materials, *Chem. Mater.* 27, 6345–6353. Copyright 2015 American Chemical Society.)

more than the polymers P3HT and MDMO-PPV, but its rate of photobleaching is comparatively very high. Also PCDTBT, PBDTTPD, and PBDFTPD have similar film densities, but they differ in their photostability. Mateker et al. also investigated the effect of polymer backbone planarity on their chemical stability [54]. It was found that the planer backbone polymers were more photostable compared to those with twisted and less ordered backbones. Twisting in the backbone causes disruption in delocalization of π-electrons, which weakens the bond and increases chemical reactivity. PBDTTPD has almost planer backbone, which is expected from the interaction between oxygen on TPD and hydrogen on BDT units [54]. PBDTTPD was found to be more photostable compared to its rather twisted furan derivative PBDFTPD. Figure 6.3 shows decay in the normalized optical density of PBDTTPD and PBDFTPD with time. PBDTTPD exhibits much slower photodegradation compared to PBDFTPD. The chemical structures of PBDTTPD and PBDFTPD are also shown. For PBDTTPD, the BDT and TPD units are in the same plane, whereas BDF and TPD units in PBDFTPD are in different planes. It is, therefore, concluded that molecular engineering is very important, and for highly stable solar cells the active materials should be designed in such a way that they possess high T_g and planer backbone, and the films made of them should be dense and crystalline with controlled grain size.

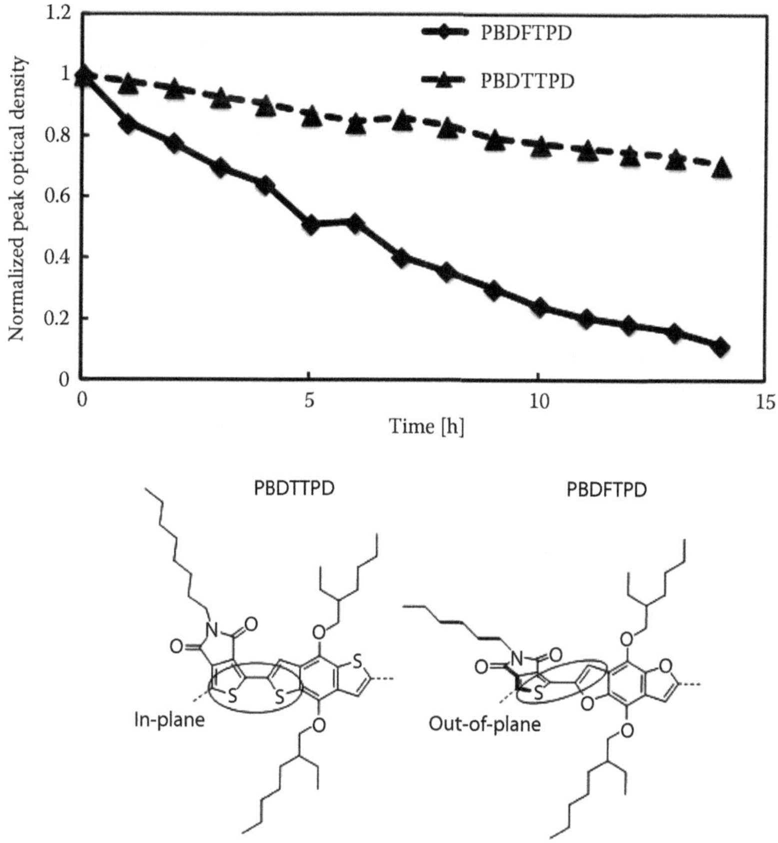

FIGURE 6.3
Photodegradation in normalized peak optical density of PBDTTPD and PBDFTPD with time (top) and their chemical structures (bottom). (Reprinted with permission from W. R. Mateker et al., Molecular packing and arrangement govern the photooxidative stability of organic photovoltaic materials, *Chem. Mater.* 27, 6345–6353. Copyright 2015 American Chemical Society.)

6.3.2 Device Engineering

For longer lifetime, the solar cells should be prepared in a way that the active layer and other constituents are not affected by the factors causing their degradation. For example, the electrode degradation could be minimized by using rather stable metal electrodes and suitable interface layers. ITO coated with a thin layer of PEDOT:PSS is usually used as anode in normal OSCs. PEDOT:PSS is an important material that helps in achieving high PCE, but because of its hygroscopic and acidic nature it has serious detrimental effects on the stability of solar cells therefore it needed a suitable replacement. Initially the replacement of PEDOT:PSS with a suitable hole transport layer was unsuccessful, but later the transition metal oxides like vanadium oxide (V_2O_5), tungsten oxide (WO_3), nickel oxide (NiO), and molybdenum oxide (MoO_3) were reported to be the successful replacements for PEDOT:PSS [55–58]. For normal devices, cathode is usually Al with a thin buffer layer of a low work function metal like Ca and Ba [42,59,60]. Ca and Ba are very sensitive materials to H_2O and O_2, and even Al exhibits oxidation upon ambient exposure. For enhanced stability of OSCs, application of thin buffer layers of TiO_2 or ZnO before Al has been found a good option for the replacement of Ca [35,59,61,62]. Usually the crystalline TiO_2 films are formed at temperatures above 450°C, which is not suitable for deposition on polymer films, therefore the sol–gel method is used to prepare thin films of titanium suboxide (TiO_x), which works as a good electron transport and injection layer. Lee et al. [63] observed that the introduction of a TiO_x layer between the organic active layer and the cathode results in excellent air stability. Optimum thickness of the TiO_x layer before the cathode also works as the optical spacer and improves the photovoltaic performance of OSCs [64], but here the TiO_x layer acted as a shield to prevent the intrusion of oxygen and moisture into the active layer and improved the stability of the unsealed device by approximately 2 orders of magnitude. TiO_x has a photocatalysis nature and inherent oxygen deficiencies, which make it a good material for protection from O_2 and H_2O. Due to oxygen deficiencies, it takes oxygen from the polymer and reduces its oxidation. Figure 6.4 shows the retardation and absorption of oxygen in TiO_x.

If we have air-stable OSCs where no encapsulation is required, the problem of short lifetime of OSCs will be solved. This requires highly stable materials and the device geometry, where adjacent materials are compatible to each other in terms of deposition and chemical

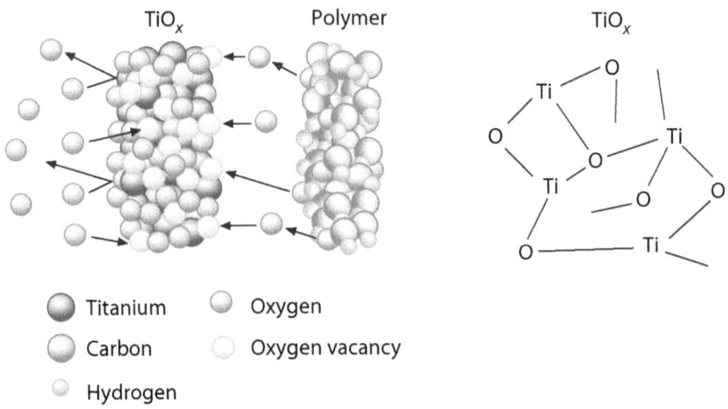

FIGURE 6.4
TiO_x, protecting the polymer layer from oxygen and moisture. The chemical structure of TiO_x is also shown. (Data from K. Lee et al., 2007. *Adv. Mater.* 19: 2445–2449.)

reactions. Inverted device geometry has shown great potential in this direction, and highly stable and efficient solar cells have been fabricated [35,36,65–68]. Inverted geometry and the progress made in the field have already been discussed in Chapter 3. Inverted geometry employs air-stable buffer and transport layers giving high stability to devices. Inverted geometry is also highly compatible to large-scale manufacturing of OSCs, as it can be processed in air without any vacuum processes. It does not possess the sensitive materials, such as Ca, Ba, Mg, or Al, like in normal geometry. In inverted geometry metal oxides such as ZnO and TiO_x are used as electron transport layers (ETL), whereas PEDOT:PSS or transition metal oxides such as V_2O_5, WO_3, NiO, and MoO_3 are used as hole transport layers. The top electrode is generally a high work function Ag or Au metal. Where the solar cells based on PEDOT:PSS degrade to 20% of initial efficiency in 200 h, MoO_3-based solar cells retain up to 70% of their initial efficiency up to 1000 h [56]. You et al. used MoO_3/Ag as the top anode contact, whereas TiO_2, ZnO, and TiO_2:Cs nanoparticles were used as ETLs with different active layers [67]. TiO_2, TiO_2:Cs, and ZnO were solution processed, whereas MoO_3/Ag were thermally evaporated in vacuum chamber. All ETLs had similar work function ~4.2 eV, close to LUMO of PCBM, and the performance of inverted solar cells was as good as normal solar cells with corresponding active layers. The performance of TiO_2:Cs-based solar cells showed inferior performance compared to those based on ZnO and was attributed to inferior electrical transport and interfacial morphology compared to those with ZnO. ZnO showed very stable performance, however, all inverted solar cells exhibited very high stability compared to normal geometry. Vacuum deposition comes out to be a very costly and time-consuming process also it is not compatible to roll-to-roll (R2R) processing, therefore efforts are being made toward solution processing of transition metal oxides [68–72].

Chen et al. used solution processable vanadium suboxides (VO_x) with the sol–gel method as the hole transport layer (HTL) in inverted OSCs [69]. P3HT and alternating poly(thiophene-phenylene-thiophene)-(2,1,3-benzothiadiazole) (a-PTPTBT) were used as donor polymers with PCBM acceptor in the active layers. The solar cells were prepared on ITO-coated glass substrates, where ZnO was used over ITO as ETL, and V_2O_5 and VO_x were used as HTLs over the active layers with an Ag top anode. Replacement of V_2O_5 by VO_x did not affect the PCE much in P3HT:PCBM-based solar cells but improved the stability tremendously. VO_x film was prepared from spin coating of vanadium triisopropoxide (VTIPO), whereas V_2O_5 film was prepared by spin coating of V_2O_5 powder suspended in isopropanol. Figure 6.5 shows PCE degradation graphs of solar cells based on V_2O_5 and VO_x HTLs. For comparison, the degradation profiles of inverted solar cells based on PEDOT:PSS HTL and normal P3HT:PCBM solar cells are also shown. Before degradation studies, the solar cells were encapsulated in a glove box using UV-curing glue and glass substrates. The enhanced stability with VO_x has been attributed to the smooth continuous film of VO_x compared to the rather rough film of V_2O_5. The solution-processed VO_x film formed a dense organic–inorganic hybrid layer, imparting high protection from ambient.

Kanwat and Jang reported extremely stable OSCs on doping of PEDOT:PSS with tungsten oxide (WO_x) and using it as an anode buffer layer [68]. The solar cells were made of P3HT:ICBA active layer with PEDOT:PSS and PEDOT:PSS:WO_x anode buffer layers. The cathode buffer layer was lithium-doped zinc oxide (LZO). All layers were prepared by spin coating from their respective solutions except the top Al electrode, which was deposited by thermal evaporation in vacuum chamber. The solar cells without WO_x doping in PEDOT:PSS exhibited ~4.1% PCE, whereas those with WO_x-doped PEDOT:PSS showed ~4.6% PCE. Improvement in the photovoltaic performance was because of improved J_{sc}. Doping of WO_x in PEDOT:PSS, increased its hole transport properties and reduced series

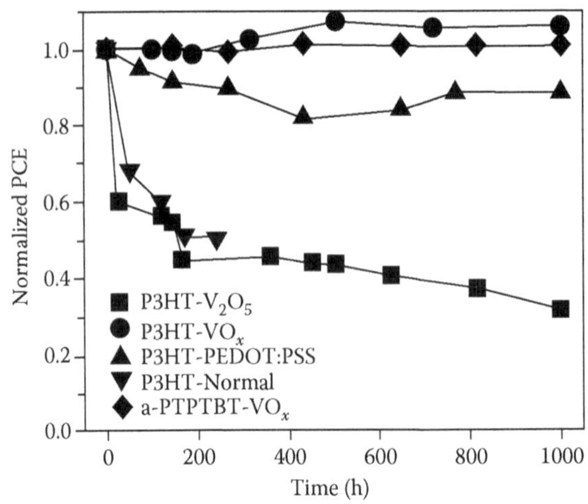

FIGURE 6.5
Degradation profiles of PCE of inverted solar cells based on different polymers and HTLs. For comparison the profile of normal P3HT solar cell has also been shown. (Redrawn with permission from C. P. Chen, Y. D. Chen, and S. C. Chuang, 2011. *Adv. Mater.* 23: 3859–3863.)

resistance and leakage current. Doping of PEDOT:PSS with WO_x not only improved the photovoltaic performance of solar cells but also tremendously enhanced their stability. More important, where the hole transport property of PEDOT:PSS decreased drastically with time, the hole transport property of WO_x-doped PEDOT:PSS improved with aging and has been attributed to variation in WO_x chemical states. Figure 6.6 shows the degradation profiles of normalized PCE of inverted OSCs. The solar cells with WO_x-doped PEDOT:PSS HTL exhibited extremely high stability with only about 5% degradation in initial efficiency after 5000 h in air, whereas in the same time those based on a PEDOT:PSS interlayer degraded to less than 25% of their initial efficiency in the same testing conditions.

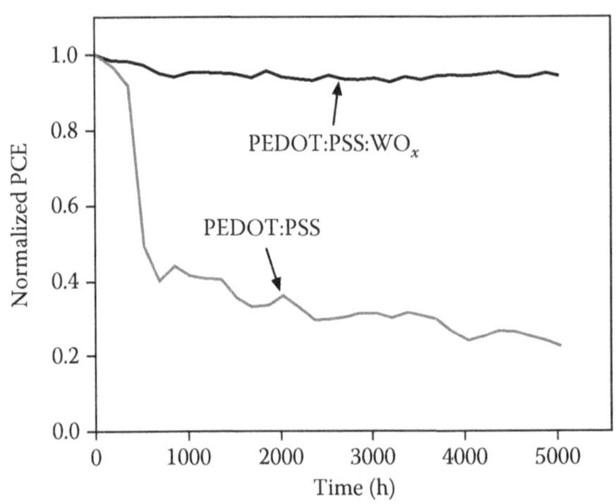

FIGURE 6.6
Degradation profiles of normalized PCE of inverted OSCs based on P3HT:ICBA active layer with PEDOT:PSS and PEDOT:PSS:WO_x hole buffer layers. (Modified from A. Kanwat and J. Jang, 2014. *J. Mater. Chem.* C 2: 901–907.)

It is worth mentioning that these devices were encapsulated in a glove box prior to degradation studies. From x-ray photoelectron spectroscopy (XPS) and time-of-flight secondary ion mass spectroscopy (TOF-SIMS) studies, it was found that doping of WO_x in PEDOT:PSS reduces the oxidation of P3HT at the metal–organic interface.

Since the acidic nature of PEDOT:PSS is detrimental to the stability of OSCs, Meng et al. used pH-neutral PEDOT:PSS as HTL and achieved great improvement in the cell stability [73]. PEDOT:PSS could be neutralized by adding 0.2 M NaOH solution to it [74]. However, the solar cell with pH-neutral PEDOT:PSS exhibited PCE of 3.44%, which was much lower compared to 6.28% from acidic PEDOT:PSS. The active layers were made of PCDTBT:PC_{71}BM blend [73]. When pH-neutral PEDOT:PSS film was subjected to both the UV-ozone and oxygen plasma treatments, the efficiency increased to 6.6%. This improvement in the cell efficiency has been attributed to reduction in energy offset between HTL and the active layer that increased V_{oc} and reduced charge carrier recombination. Neutral PEDOT:PSS stops etching of ITO and slows diffusion into the active layer that leads to enhanced stability of devices. Before degradation studies, the solar cells were encapsulated in inert atmosphere. After 7 days storage in ambient conditions, where the performance of acidic PEDOT:PSS-based solar cell decreased by 20% of their initial value, the performance of that based on neutral PEDOT:PSS decreased by only 9%. After 50 days of storage the performance of solar cell based on acidic PEDOT:PSS decreased by 45%, but that of based on neutral PEDOT:PSS decreased by only 20%.

Insertion of interface layers at different places can not only improve the performance of solar cells but also enhance their stability. Interface layers block the diffusion of O_2 and H_2O molecules and other entities like metal electrode atoms that inhibit various chemical reactions in the devices and lead to higher stability [75]. Interface layers can also absorb UV component from solar radiation and reduce the photooxidation of active layers. Selection of a suitable interface layer with optimized thickness is very important for enhanced performance. Therefore, interface engineering has become an important field of research in OSCs for enhanced stability [76–85]. The interface layer could be of metal oxides, polymers, small molecules, graphene oxide, or fullerene derivatives. A detailed report on interface engineering could be found in Zeng et al. [86]. For device engineering towards enhanced stability, other methods could also be employed to improve stability of the active layer. For example, we doped the active P3HT:PCBM layer with 5,6,11,12-tetraphenylnaphthacene (rubrene) and observed enhancement in its thermal stability [87]. Doping of rubrene makes the P3HT:PCBM system more robust and stable in thermal, electrochemical and morphological aspects. Turkovic et al. achieved enhanced long-term stability using hindered phenol-based antioxidants as stabilizing additives in OSCs [88]. They used a number of hindered phenols additives in P3HT:PCBM solar cells and the one octadecyl 3-(3,5-di-tert-butyl-4-hydroxyphenyl)-propionate improved stability and increased the accumulated power generated (APG) by the cell by a factor of 3 compared to that without any additive. The stabilized performance was attributed to reduced exciton recombination via the hydrogen donation mechanism in combination with radical scavenging.

6.3.3 Encapsulation

Encapsulation of devices in a nonreactive environment could be one of the ways to prevent these devices from ambient O_2 and H_2O molecules. However, it is not a permanent solution of degradation, as the degradation happens in vacuum or inert atmosphere as well as due to photooxidation and interdiffusion of device constituents. Therefore, the best option would be to solve the problem of internal degradation first via application of materials that are resistant to electrochemical and photochemical reactions. Then encapsulation would further give them protection from external entities like O_2 and H_2O molecules. Encapsulation is important and even in the present scenario 3–5 years of lifetime is easily achievable with proper encapsulation. The encapsulation is done by a barrier film that is employed over the active area of the cell with an adhesive. The barrier films could also work as UV filters and protect them from scratches. UV light filtration would reduce photooxidation of the active layer [86]. Since soda-lime glass (the base substrate) itself filters the UV light of wavelengths <300 nm, to cut higher wavelengths additional UV filter can also be used. Barrier materials and adhesive materials play a crucial role in achieving high stability in organic devices.

6.3.3.1 Barrier Material

Various organic and inorganic materials could be used as barrier materials for encapsulation of OSCs, however, the barrier materials should possess a very low water vapor transmission rate (WVTR) and oxygen transmission rate (OTR). To achieve high protection from ambient, the WVTR of the barrier material should be around 10^{-6} g/m^2/day and OTR should be around 10^{-3} cm^3/m^2/day. A good barrier material should have high stability under UV exposure and no photocorrosion should take place. Since glass is a very high barrier material for O_2 and H_2O transmissions (with zero WVTR and OTR) it is the most suitable barrier material for encapsulation but it is not suitable for encapsulation of flexible R2R modules. Thick metal plates can also be used as a good barrier material. Encapsulation should be done in a way that it does not peel off the electrode or active layer materials and does not deteriorate the cell performance. For testing and operation, the contacts should be taken out of the encapsulation carefully or the contact on the cell may be taken away from the active area and left unencapsulated. F. C. Krebs used a very simple method to encapsulate P3HT:PCBM solar cells on glass substrates, where he used Al back plate as the barrier material [89]. The encapsulation was done in glove box, and a glass fiber reinforced thermosetting epoxy (known as prepreg) was used as an adhesive to stick the sealing Al plate with device. The prepreg was applied at the edges of the Al back plate and placed on a hotplate at 80°C. When the sealant became soft upon heating, the solar cell was carefully placed on it. For proper bonding of the solar cell on glass and the back Al plate, a weight was placed on top of the assembly. The assembly was left on the hotplate under weight for 12 h for thermosetting of the epoxy. The solar cell exhibited more than 1 year of shelf-life. Zimmermann et al. encapsulated P3HT:PCBM-based inverted solar cells with a glass plate, which was bonded to the solar cell using a glue [90]. The solar cells were subjected to continuous illumination under a sulfur plasma lamp at 1000 W/m^2 and after 1500 h the solar cells maintained 90% of their initial efficiency. This stability corresponded to 1.5 years under sun light exposure. For the solar cells on glass, the encapsulation from the top side only is sufficient, however, the solar cells on plastic sheets require complete encapsulation from both sides. For encapsulation of solar cells on glass, the barrier material could be transparent or opaque to visible light, as it does not face the incident light; but for flexible solar cells, the barrier material facing illumination is required to be transparent to allow the incident light to enter and be absorbed in the active layer. To retain other attractions of OSCs, encapsulation should be lightweight and cost effective.

For encapsulation of R2R modules and to maintain their flexibility, the barrier materials should also be flexible, therefore glass or metal plates are not suitable for this purpose. Encapsulation of printed flexible solar cells require high barrier materials and usually polyethylene terephthalate (PET) sheets or plastic foils with ultrahigh barrier to O_2 and H_2O

transmissions are used for this purpose. Since very high barrier PET sheets have been difficult to achieve, the devices are being encapsulated with reasonably good barrier sheets. The barrier films used in food packing industries could also be used for encapsulation of OSCs. Moro et al. used a flexible barrier film with WVTR of 1.2×10^{-3} g/m^2/day to encapsulate P3HT:PCBM solar cells prepared on glass substrates and got a shelf-life of around 3500 h [91]. Lungenschmied et al. prepared P3HT:PCBM solar cell modules on both the glass and ultrahigh barrier plastic foil from NOVA-PLASMA and sealed them with the same ultrahigh barrier foils in a highly inert atmosphere and achieved the lifetime of more than 6000 h (T_{50}, @ 50% of initial efficiency) [92]. A barrier foil was applied on both kinds of modules with a two-component epoxy adhesive, which maintained appreciable flexibility after curing. Hauch et al. prepared P3HT:PCBM solar cells on PET sheets and encapsulated with transparent barrier films with WVTR of 0.03 g/m^2/day at 38°C/100% RH [93]. Figure 6.7 shows the schematic cross section view of the encapsulated flexible solar cell by Hauch et al. The modules were exposed to outdoor conditions for about 14 months and a performance evaluation was done time to time. The lifetime testing was carried out in direct outdoor exposure as well as under accelerated testing at 60°C at 1 sun in the laboratory. The outdoor testing was done under load at maximum power point, and even after 1 year the modules exhibited only around 20% loss in their output power. Interestingly, when the modules were placed on the rooftop, the performance increased by 40% immediately after placement. This increase in efficiency was because of the increase in *FF*, whereas J_{sc} exhibited almost no change even after 14 months of testing. A similar effect on *FF* was also observed for the cells tested in laboratory under accelerating condition. Figure 6.8 shows the evaluation studies by Hauch et al. [93] on the normalized output power of the rooftop modules as a function of time. The environmental temperature at the time of measurement has also been shown. From the accelerated lifetime testing in the laboratory, the life of the modules was estimated to be roughly around 1 year of outdoor exposure at 42°C lateral altitude.

Krebs et al. employed an R2R laminator to encapsulate R2R-processed flexible solar modules [94]. The encapsulation was done in air using a barrier foil and 3M 467 MPF adhesive. First, the adhesive was applied on the barrier foil and then that foil was laminated onto the solar cells. The encapsulation was done from both sides of the solar module. To avoid damage to the active area the lamination was done first on the back side and then on the front side. Instead of sticking barrier film to the solar cells with glue, thermal lamination

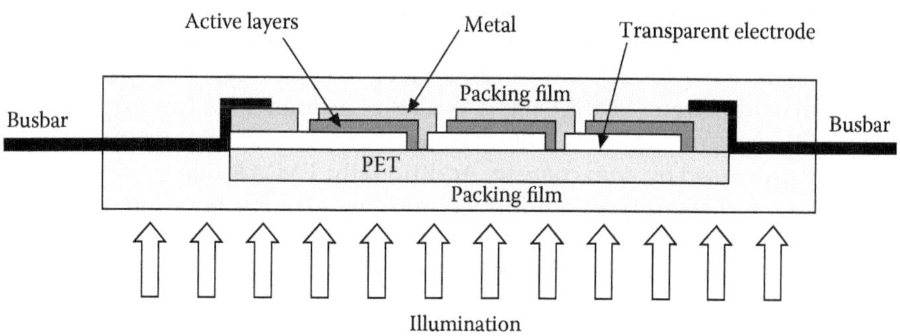

FIGURE 6.7
Schematic cross-sectional view of the encapsulated flexible solar module. (Redrawn with permission from J. A. Hauch et al., 2008. *Sol. Energy Mater. Sol. Cells*, 92: 727–731.)

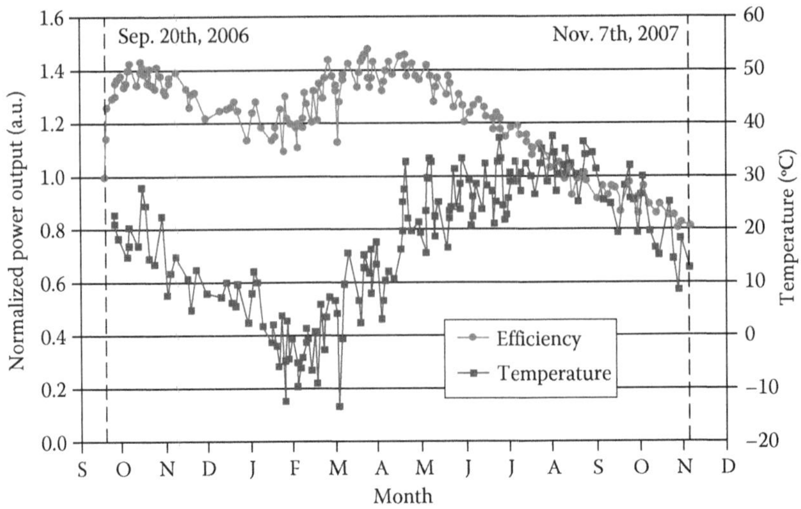

FIGURE 6.8
Normalized output power of a module kept at the rooftop as a function of time. The temperatures of the environment at the time of measurements are also shown. (Reproduced with permission from J. A. Hauch et al., 2008. *Sol. Energy Mater. Sol. Cells* 92: 727–731.)

and coatings of different barrier materials can also be done for encapsulation. The coating of encapsulating materials could be done by spin coating, thermal evaporation, or vapor phase deposition. There are several reports where oxide materials like SiO_2, SiO_x, Al_2O_3, etc. have been used as protecting layers against degradation for organic devices [95–97]. But such coatings are time consuming, costly, and cause some losses in flexibility of solar cells.

6.3.3.2 Adhesive Material

An adhesive for encapsulation of organic devices should be solvent free and electrically insulating. The adhesive material should seal the solar cells and should not react with active area if it comes in contact with. The adhesive itself should also have very low WVTR and OTR. Some of the polymers such as poly(methyl methacrylate) (PMMA) and copolymer of tetrafluoroethylene and 2,2-bis-trifluoromethyl-4,5-difluoro-1,3-dioxole, {P(TFE-PDD)}, have also been used for encapsulation of organic devices. They could also be used along with oxide materials in bilayer protective structures or as adhesives to stick the encapsulating barrier materials [98–101]. In bilayer protective structures, PMMA or P(TFE-PDD) are coated with oxide layers. PMMA has a WVTR of around 1.7×10^{-3} g/m^2/day and OTR of around 6×10^{-3} cm^3/m^2/day. It is a flexible, transparent, and cost-effective material that could be easily processed by spin coating. Additionally, PMMA filters UV radiation below 300 nm and has good stability in outdoor conditions [101]. For bonding of glass sheet on solar cells using PMMA, it could be spin coated on a glass sheet and then put on the top of the solar cell. Annealing of PMMA at ~100°C for some time would solidify it, and both the solar cell and encapsulating glass sheet would bond together. However, PMMA does not provide a very strong bonding between encapsulating glass and solar cells, therefore usually epoxy glues are more frequently applied as bonding adhesives. The bilayer polymer/oxide encapsulation is complicated and causes loss in flexibility. Epoxy encapsulation is relatively much easier and cost effective compared to bilayer PMMA/SiO_x encapsulation.

The epoxy glues may be two-component or single-component glues, which could be hardened by UV light exposure, thermal treatment, or room temperature storage for a longer time. Epoxy Technology Inc., Epoxies Etc, and 3M, United States; DELO, Germany; Ossila Limited, UK; and Kommerling, Germany are some of the companies that offer a variety of epoxies, adhesives, and glues for encapsulation of OSCs. Along with the adhesives, the companies provide their user manuals and data sheets for proper usage. Some adhesives are pressure sensitive and can be easily applied onto barrier materials and solar cells by lamination. The adhesives could be printed or applied by different means and cured after encapsulation. Thermosetting adhesives could be applied to solar cells and barrier materials and hardened by thermal annealing.

6.4 Regeneration of Solar Cells

Regeneration of degraded solar cells is an important field of research not only from the perspective of better understanding of degradation mechanisms but it will also make solar cell technologies more cost effective. Regeneration of solar cells has a great significance toward solution of degradation problems. Only a few research groups have thought in this direction, therefore only a few reports on the regeneration of solar cells are available in the literature. Regeneration in solar cells is basically reversal of one or more degradation phenomena (reversible degradation phenomena) with some external treatment. However, one kind of treatment may not be suitable for other types of solar cells, and it entirely depends upon the degradation phenomena of that particular solar cell. Not all the degradation phenomena are reversible, therefore 100% regeneration is almost impossible. In some cases none of the phenomena is reversible and such solar cells could not be regenerated. For regeneration in DSSCs, let us take an example by Zhang et al. where they fabricated DSSCs using N719 dye and a quasi-solid-state electrolyte medium having 0.1 M LiI, 0.05 M I_2, 0.6 M 1,2-dimethyl-3-propylimidazolium iodide, and 2 wt% poly(vinylidene fluoride) (PVDF), which also exhibited degradation with time. It is worth mentioning that in DSSCs, the electrolyte mediums play a very critical role in their stability, therefore electrolyte mediums are being replaced by relatively stable quasi-solid-state and solid-state mediums [17,18]. During degradation studies Zhang et al. realized that the degradation in such quasi-solid-state DSSCs was because of blockage of I^-/I_3^- ion transport channels at the TiO_2/dye/electrolyte interface. Therefore, they applied a positive electrical pulsed bias of about 2.5 V on the cell to open the blocked channels and got success in regenerating the degraded DSSCs [102]. Up to 30.7% of the initial cell efficiency was recovered by application of controlled double pulse, and the recovery could last for more than 28 days. Due to opening of the blocked channels, the recovery in efficiency was mainly from the recovery in J_{sc}, which could be improved by increasing the pulse height to a certain limit. Too high pulse would cause very high current to flow through and it would damage the cell therefore the pulse height was kept under limit. This study suggested that blocking of I^-/I_3^- ion conducting channels was the key factor causing degradation in such solar cells.

For regeneration of OSCs, let us take the example of work done by my group [42,103]. Attempts were made to regenerate OSCs by thermal treatment. The initial attempts were carried out on the solar cells based on a P3HT:PCBM active layer, and Ca/Ag or Ca/Al cathodes. PEDOT:PSS-coated ITO electrodes were used as the anode. Completely degraded solar cells were subjected to thermal annealing in inert atmosphere. The solar cells with

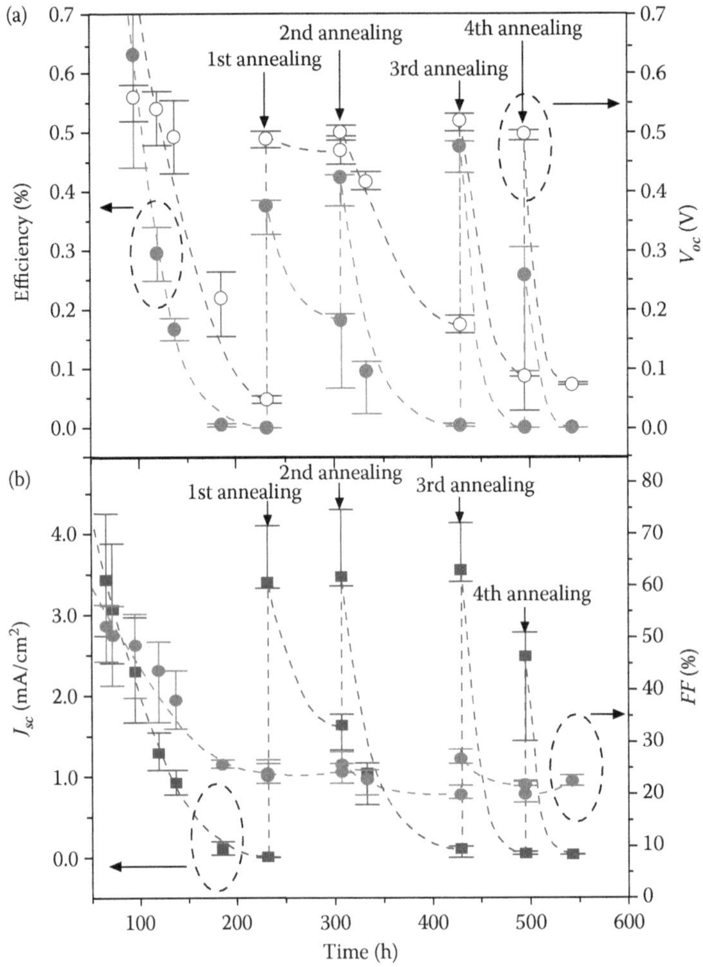

FIGURE 6.9
Effect of repeated heating cycles in an inert atmosphere on fully degraded devices with Ca/Ag cathodes: (a) variation of η and V_{oc} and (b) variation of J_{sc} and *FF*. Dashed lines are drawn to guide the eye to the trends in data. (Reprinted with permission from P. Kumar et al., Comparative degradation and regeneration of polymer solar cells with different electrodes, *ACS Appl. Mater. Inter.* 6, 5281–5289. Copyright 2014 American Chemical Society.)

Ca/Al cathode exhibited no improvement after annealing, whereas the solar cells with Ca/Ag cathode exhibited a considerable recovery in performance. Indeed, these solar cells could be subjected to many degradation and regeneration cycles. Figure 6.9a shows the degradation profiles of efficiency (η) and V_{oc} following repeated heating cycles in an inert atmosphere for fully degraded solar cells with a Ca/Ag cathode, which were stored in the dark between measurements. Figure 6.9b shows the corresponding profiles of J_{sc} and *FF*. It is clear from Figure 6.9 that each heating cycle leads to a significant increase in V_{oc} and J_{sc}, resulting in a reproducible 12%–15% recovery in η in every regeneration cycle. For comparison, Figure 6.10 shows the effect of heating cycles on J_{sc} and V_{oc} of degraded solar cells with Ca/Al cathodes. Interestingly, despite the fact that the solar cells with Ca/Al cathodes were functionally fully degraded after 230 h of exposure, V_{oc} retained the same value as

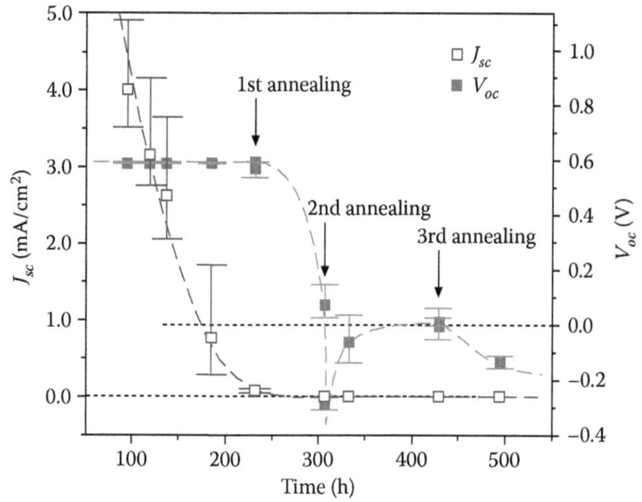

FIGURE 6.10
Effect of thermal annealing at different time intervals and further degradation in J_{sc} and V_{oc} of the solar cells with Ca/Al cathodes. Dashed lines are drawn to guide the eye to the trends in data. (Reprinted with permission from P. Kumar et al., Comparative degradation and regeneration of polymer solar cells with different electrodes, *ACS Appl. Mater. Inter.* 6, 5281–5289. Copyright 2014 American Chemical Society.)

the as-prepared devices. However, after the first heating cycle, V_{oc} degraded very quickly and J_{sc} was further reduced. After the second heating cycle the illuminated *J–V* characteristics started passing through the second quadrant, which resulted into negative V_{oc} and J_{sc}, and could be attributed to further degradation.

Figure 6.11 shows the external quantum efficiency (EQE) spectra of degraded devices with Ca/Ag and Ca/Al electrodes before and after the first heating cycle. As expected,

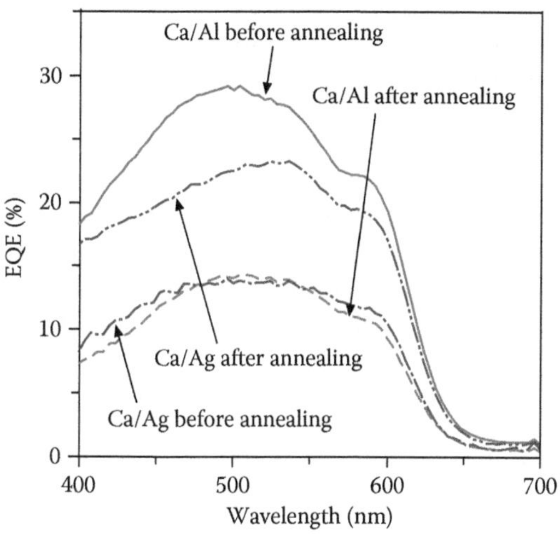

FIGURE 6.11
Effect of thermal annealing on the EQE of the degraded solar cells with Ca/Al and Ca/Ag electrodes. (Reprinted with permission from P. Kumar et al., Comparative degradation and regeneration of polymer solar cells with different electrodes, *ACS Appl. Mater. Inter.* 6, 5281–5289. Copyright 2014 American Chemical Society.)

devices with Ca/Ag electrodes exhibited an overall increase in EQE after the first heating cycle, whereas those with Ca/Al electrodes exhibited reduction in EQE, consistent with the observed variation in J_{sc}. Moreover, there was no significant change to the shape of the EQE plots, both before and after the heating cycle and for the different electrode materials. This observation suggested that improved response upon heating a degraded device does not arise from a change in the crystallinity of the P3HT material or from a change in the relative contribution from the two components in the active layer, both of which would otherwise change the shape of the EQE plots [104]. To probe further the regeneration of solar cells with Ca/Ag electrodes, the dark and illuminated *J–V* characteristics before and after each heating cycle were analyzed and the characteristics are shown in Figure 6.12. It can be seen that for the devices measured after each heating cycle, the illuminated *J–V* characteristics intersect their corresponding dark *J–V* characteristics at a common voltage (0.56 V), whereas the characteristics measured before each heating cycle did not exhibit a common intersection point. The intersection point of the dark and illuminated *J–V* characteristics is related to the built-in voltage (V_{bi}) of the devices, and thus the common intersection point indicates that the device recovers to the same V_{bi} after every heating cycle.

Since V_{bi} is determined mainly by the interfacial properties at the electrodes [105,106] and that the recovery is only observed for the Ca/Ag electrode devices, the results suggest that the device recovery is primarily driven by the changes at the Ca/Ag electrode interface. For detailed understanding, XPS studies were performed and it was observed that vacuum deposition of Ca onto the P3HT:PCBM active layer results in the formation of CaS islands at the Ca/active layer interface. Vacuum deposition of the Ag layer on top of the Ca acts as a barrier to oxygen and water, while the Ca layer prevents the formation of AgS. Even for the "fresh" sample, oxygen and water diffusion occurs (either upon exposure to ambient air or from entrapped water), and partial oxidation of the remaining calcium electrode takes place. Upon degradation in a moist air atmosphere (moderate RH), oxygen and water diffusion continues, and the calcium layer gets fully converted to oxide and then to a mixed oxide/hydroxide layer. As the Ca becomes fully oxidized (with more complete conversion to $Ca(OH)_2$), then oxidation of the Ag (or Al) electrode occurs. The native Ag oxide is thin (1–2 nm) due to self-passivation of the oxide growth, and a complete layer was not obtained since electronic inversion of the device function was not observed. Al forms a thicker oxide that also acts to passivate the cathode surface. In addition, the gradual conversion of CaS into CaO and subsequently to $Ca(OH)_2$ via formation of mixed hydroxythiol species (Ca(SH)-(OH)) takes place. Finally, upon regeneration at 140°C in the glove box, there were two thermodynamic processes that occurred. First, the silver oxide started to convert back to Ag, as evidenced by the recovery of V_{oc}. Second, under a nitrogen atmosphere, the CaO/$Ca(OH)_2$ equilibrium temperature is lowered to ~150°C, and thus the equilibrium is shifted to favor the dehydration of $Ca(OH)_2$. In the case of the Ca/Al cathode, however, the aluminum oxidation is irreversible, and hence no recovery of device performance was possible. Hence, the regeneration in Ca/Ag based solar cells arose primarily due to reduction of Ag_2O back to Ag. However, for the Ca/Al cathode, degradation and regeneration both result in irreversible oxidation of the Al electrode to produce a layer of insulating Al_2O_3, which inhibits the extraction of charge carriers and the performance deteriorates.

In order to further improve the regeneration, we replaced P3HT:PCBM with P3HT:ICBA. The solar cells with a P3HT:ICBA active layer and Ca/Ag cathode demonstrated ~50% restoration over several degradation/regeneration cycles upon thermal annealing in inert atmosphere [42]. Enhanced regeneration with P3HT:ICBA compared to P3HT:PCBM

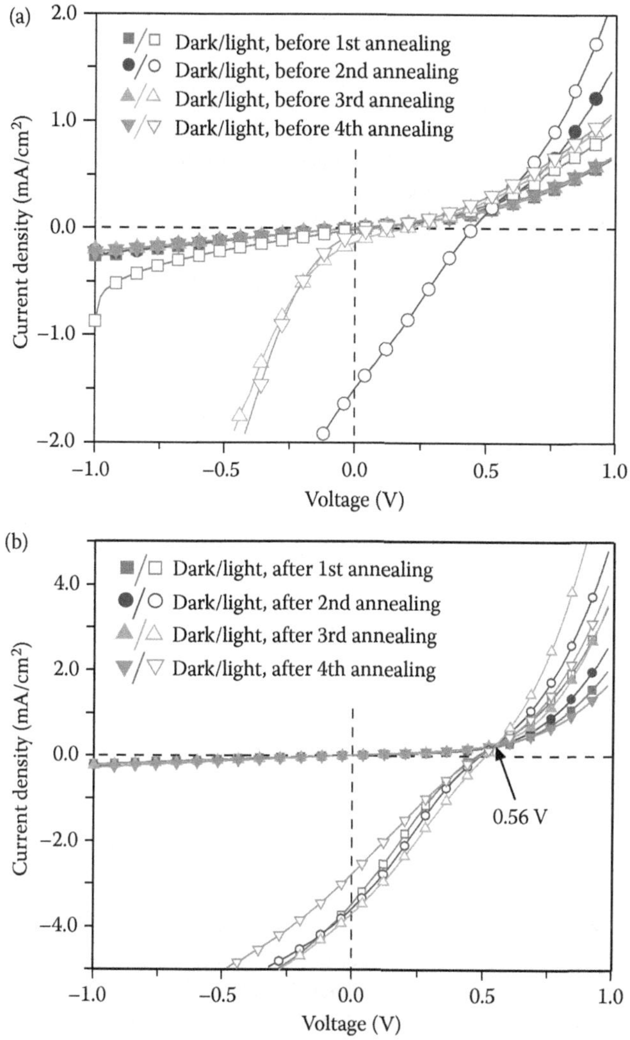

FIGURE 6.12
(a) Dark (solid) and illuminated (open) *J–V* characteristics of a typical solar cell with Ca/Ag electrodes before each heating cycle. (b) Dark (solid) and illuminated (open) *J–V* characteristics of solar cell with Ca/Ag electrodes after each heating cycle (140°C for 5 min). (Reprinted with permission from P. Kumar et al., Comparative degradation and regeneration of polymer solar cells with different electrodes, *ACS Appl. Mater. Inter.* 6, 5281–5289. Copyright 2014 American Chemical Society.)

active layer was because of its better thermal stability. The degradation occurs via a combination of three primary pathways: (1) cathodic oxidation, (2) active layer phase segregation, and (3) anodic diffusion. The primary degradation pathway in these devices is limited to anodic diffusions. We achieved recovery in device performance due to reversal of cathodic oxidation and robust active layer. Work is currently underway to explore alternative anode systems that will mitigate the third degradation pathway and hence fulfill the potential of the demonstrated regeneration process. As such, this work offered the tantalizing prospect of OSC design that could be fully regenerated through thermal processing.

References

1. A. Hinsch, J. M. Kroon, R. Kern, I. Uhlendorf, J. Holzbock, A. Meyer, and J. Ferber, 2001. Long-term stability of dye-sensitized solar cells, *Prog. Photovolt.: Res. Appl.* 9: 425.
2. H. Pettersson and T. Gruszecki, 2001. Long-term stability of low-power dye-sensitised solar cells prepared by industrial methods, *Sol. Energy Mater. Sol. Cells* 70: 203–212.
3. R. Memming, 2001. *Semiconductor Electrochemistry*, Weinheim, Germany: Wiley-VCH.
4. R. Kern, N. Van Der Burg, G. Chmiel, J. Ferber, G. Hasenhindl, A. Hinsch, R. Kinderman et al., 2000. Long term stability of dye-sensitized solar cells for large area power applications, *Opto-Electron. Rev.* 8: 284.
5. T. S. Kang, K. H. Chun, J. S. Hong, S. H. Moon, and K. J. Kim, 2000. Enhanced stability of photocurrent-voltage curves in Ru(II)-dye-sensitized nanocrystalline TiO_2 electrodes with carboxylic acids, *J. Electrochem. Soc.* 147: 3049–3053.
6. A. Kay and M. Gratzel, 2002. Dye-sensitized core–shell nanocrystals: Improved efficiency of mesoporous tin oxide electrodes coated with a thin layer of an insulating oxide, *Chem. Mater.* 14: 2930.
7. E. Palomares, J. N. Clifford, S. A. Haque, T. Lutz, and J. R. Durrant, 2003. Control of charge recombination dynamics in dye sensitized solar cells by the use of conformally deposited metal oxide blocking layers, *J. Am. Chem. Soc.* 125: 475.
8. B. A. Gregg, F. Pichot, S. Ferrere, and C. L. Fields, 2001. Interfacial recombination processes in dye-sensitized solar cells and methods to passivate the interfaces, *J. Phys. Chem. B* 105: 1422.
9. I. Chung, B. Lee, J. He, R. P. H. Chang, and M. G. Kanatzidis, 2012. All-solid-state dye-sensitized solar cells with high efficiency, *Nature* 485: 486–489.
10. H. Wang, X. Zhang, F. Gong, G. Zhou, and Z.-S. Wang, 2012. Novel ester-functionalized solid-state electrolyte for highly efficient all-solid-state dye-sensitized solar cells, *Adv. Mater.* 24: 121–124.
11. T. Stergiopoulos, I. M. Arabatzis, G. Katsaros, and P. Falaras, 2002. Binary polyethylene oxide/titania solid-state redox electrolyte for highly efficient nanocrystalline TiO_2 photoelectrochemical cells, *Nano Lett.* 2: 1259–1261.
12. P. Wang, S. M. Zakeeruddin, J. E. Moser, M. K. Nazeeruddin, T. Sekiguchi, and M. Gratzel, 2003. A stable quasi-solid-state dye-sensitized solar cell with an amphiphilic ruthenium sensitizer and polymer gel electrolyte, *Nat. Mater.* 2: 402–407.
13. N. Hirata, J. E. Kroeze, T. Park, D. Jones, S. A. Haque, A. B. Holmes, and J. R. Durrant, 2006. Interface engineering for solid-state dye-sensitised nanocrystalline solar cells: The use of an organic redox cascade, *Chem. Commun.* 5: 535–537.
14. M. Kaneko and T. Hoshi, 2003. Dye-sensitized solar cell with polysaccharide solid electrolyte, *Chem. Lett.* 32: 872–873.
15. Q. B. Meng, K. Takahashi, X. T. Zhang, I. Sutanto, T. N. Rao, O. Sato, and A. Fujishima, 2003. Fabrication of an efficient solid-state dye-sensitized solar cell, *Langmuir* 19: 3572–3574.
16. L. Wang, S. Fang, Y. Lin, X. Zhou, and M. Li, 2005. A 7.72% efficient dye sensitized solar cell based on novel necklace-like polymer gel electrolyte containing latent chemically cross-linked gel electrolyte precursors, *Chem. Commun.* 45: 5687–5689.
17. Z. Yu, D. Qin, Y. Zhang, H. Sun, Y. Luo, Q. Meng, and D. Li, 2011. Quasi-solid-state dye-sensitized solar cell fabricated with poly (β-hydroxyethyl methacrylate) based organogel electrolyte, *Energy Environ. Sci.* 4: 1298–1305.
18. Q. Yu, C. Yu, F. Guo, J. Wang, S. Jiao, S. Gao, H. Li, and L. Zhao, 2012. A stable and efficient quasi-solid-state dye-sensitized solar cell with a low molecular weight organic gelator, *Energy Environ. Sci.* 5: 6151–6155.
19. A. Hagfeldt, G. Boschloo, L. Sun, L. Kloo, and H. Pettersson, 2010. Dye-sensitized solar cells, *Chem. Rev.* 110: 6595–6663.

20. Y. Yang, C. H. Zhou, S. Xu, H. Hu, B. L. Chen, J. Zhang, S. J. Wu, W. Liu, and X. Z. Zhao, 2008. Improved stability of quasi-solid-state dye-sensitized solar cell based on poly (ethylene oxide)–poly (vinylidene fluoride) polymer-blend electrolytes, *J. Power Sources* 185: 1492–1498.
21. J. S. Liu, E. N. Kadnikova, Y. X. Liu, M. D. McGehee, and J. M. J. Frechet, 2004. Polythiophene containing thermally removable solubilizing groups enhances the interface and the performance of polymer-titania hybrid solar cells, *J. Am. Chem. Soc.* 126: 9486–9487.
22. S. A. Gevorgyan and F. C. Krebs, 2008. Bulk heterojunctions based on native polythiophene, *Chem. Mater.* 20: 4386–4390.
23. M. Manceau, M. Helgesen, and F. C. Krebs, 2010. Thermocleavable polymers: Materials with enhanced photochemical stability, *Poly. Degrad. Stab.* 95: 2666–2669.
24. M. H. Petersen, S. A. Gevorgyan, and F. C. Krebs, 2008. Thermocleavable low band gap polymers and solar cells therefrom with remarkable stability toward oxygen, *Macromolecules* 41: 8986–8994.
25. F. C. Krebs, 2005. Design and applications of polymer solar cells with lifetimes longer than 1000 hours, *Proc. SPIE* 5938: 59380Y.
26. P. Kumar and S. Chand, 2012. Recent progress and future aspects of organic solar cells, *Prog. Photovolt: Res. Appl.* 20: 377.
27. Y. H. Lin, Y. T. Tsai, C. C. Wu, C. H. Tsai, C. H. Chiang, H. F. Hsu, J. J. Lee, and C. Y. Cheng, 2012. Comparative study of spectral and morphological properties of blends of P3HT with PCBM and ICBA, *Org. Electron.* 13: 2333–2341.
28. V. Turkovic, S. Engmann, D. A. M. Egbe, M. Himmerlich, S. Krischok, G. Gobsch, and H. Hoppe, 2014. Multiple stress degradation analysis of the active layer in organic photovoltaics, *Sol. Energy Mater. Sol. Cells* 120: 654–668.
29. X. Meng, W. Zhang, Z. Tan, Y. Li, Y. Ma, T. Wang, L. Jiang, C. Shu, and C. Wang, 2012. Highly efficient and thermally stable polymer solar cells with dihydronaphthyl-based [70] fullerene bisadduct derivative as the acceptor, *Adv. Funct. Mater.* 22: 2187–2193.
30. W. R. Mateker, J. D. Douglas, C. Cabanetos, I. T. Sachs-Quintana, J. A. Bartelt, E. T. Hoke, A. E. Labban, P. M. Beaujuge, J. M. J. Fréchet, and M. D. McGehee, 2013. Improving the long-term stability of PBDTTPD polymer solar cells through material purification aimed at removing organic impurities, *Energy Environ. Sci.* 6: 2529–2537.
31. A. Tournebize, J. L. Gardette, C. Taviot-Guého, D. Bégué, M. A. Arnaud, C. Dagron-Lartigau, H. Medlej, R. C. Hiorns, S. Beaupré, and M. Leclerc, 2015. Is there a photostable conjugated polymer for efficient solar cells? *Polym. Degrad. Stab.* 112: 175.
32. H. S. Silva, A. Tournebize, D. Bégué, H. Peisert, T. Chassé, J. L. Gardette, S. Therias, A. Rivaton, and R. C. Hiorns, 2014. A universal route to improving conjugated macromolecule photostability, *RSC Adv.* 4: 54919–54923.
33. Y. Liu, J. Zhao, Z. Li, C. Mu, W. Ma, H. Hu, K. Jiang, H. Lin, H. Ade, and H. Yan, 2014. Aggregation and morphology control enables multiple cases of high-efficiency polymer solar cells, *Nat. Commun.* 5: 5293.
34. L. Dou, J. You, J. Yang, C. C. Chen, Y. He, S. Murase, T. Moriarty, K. Emery, G. Li, and Y. Yang, 2012. Tandem polymer solar cells featuring a spectrally matched low-bandgap polymer, *Nat. Photon.* 6: 180–185.
35. J. You, L. Dou, K. Yoshimura, T. Kato, K. Ohya, T. Moriarty, K. Emery et al., 2013. A polymer tandem solar cell with 10.6% PCE, *Nat. Commun.* 4: 1446.
36. C. C. Chen, W. H. Chang, K. Yoshimura, K. Ohya, J. You, J. Gao, Z. Hong, and Y. Yang, 2014. An efficient triple-junction polymer solar cell having a power conversion efficiency exceeding 11%, *Adv. Mater.* 26: 5670–5677.
37. N. Shin, H. J. Yun, Y. Yoon, H. J. Son, S. Y. Ju, S. K. Kwon, B. S. Kim, and Y. H. Kim, 2015. Highly stable polymer solar cells based on poly(dithienobenzodithiophene-co-thienothiophene), *Macromolecules* 48: 3890–3899.
38. W. Li, W. S. C. Roelofs, M. M. Wienk, and R. A. J. Janssen, 2012. Enhancing the photocurrent in diketopyrrolopyrrole-based polymer solar cells via energy level control, *J. Am. Chem. Soc.* 134: 13787–13795.

39. A. Najari, S. Beaupré, P. Berrouard, Y. Zou, J.-R. Pouliot, C. Lepage-Pérusse, and M. Leclerc, 2011. Synthesis and characterization of new thieno[3,4-c]pyrrole-4,6-dione derivatives for photovoltaic applications, *Adv. Funct. Mater.* 21: 718–728.
40. Y. R. Cheon, Y. J. Kim, J. J. Ha, M. J. Kim, C. E. Park, and Y. H. Kim, 2014. TPD-based copolymers with strong interchain aggregation and high hole mobility for efficient bulk heterojunction solar cells, *Macromolecules* 47: 8570–8577.
41. H. J. Yun, Y. J. Lee, S. J. Yoo, D. S. Chung, Y. H. Kim, and S. K. Kwon, 2013. Synthesis of poly(benzothiadiazole-*co*-dithienobenzodithiophenes) and effect of thiophene insertion for high-performance polymer solar cells, *Chem. Euro. J.* 19: 13242–13248.
42. P. Kumar, C. Bilen, K. Feron, X. Zhou, W. C. Belcher, and P. C. Dastoor, 2014. Enhanced regeneration of degraded polymer solar cells by thermal annealing, *Appl. Phys. Lett.* 104: 193905.
43. G. Griffini, J. D. Douglas, C. Piliego, T. W. Holcombe, S. Turri, J. M. J. Fréchet, and J. L. Mynar, 2011. Long-term thermal stability of high-efficiency polymer solar cells based on photocrosslinkable donor–acceptor conjugated polymers, *Adv. Mater.* 23: 1660–1664.
44. F. Ouhib, M. Tomassetti, J. Manca, F. Piersimoni, D. Spoltore, S. Bertho, H. Moons et al., 2013. Thermally stable bulk heterojunction solar cells based on cross-linkable acrylate-functionalized polythiophene diblock copolymers, *Macromolecules* 46: 785–795.
45. G. Brotas, J. Farinhas, Q. Ferreira, R. Rodrigues, I. L. Martins, J. Morgado, and A. J. Charas, 2014. Synthesis, characterization, and applications in photovoltaic cells of oxetane-functionalized P3HT derivatives, *J. Polym. Sci. A: Polym. Chem.* 52: 652–663.
46. B. J. Kim, Y. Miyamoto, B. Ma, and J. M. J. Frechet, 2009. Photocrosslinkable polythiophenes for efficient, thermally stable, organic photovoltaics, *Adv. Funct. Mater.* 19: 2273–2281.
47. S. Bertho, B. Campo, F. Piersimoni, D. Spoltore, J. D'Haen, L. Lutsen, W. Maes, D. Vanderzande, and J. Manca, 2013. Improved thermal stability of bulk heterojunctions based on side-chain functionalized poly(3-alkylthiophene) copolymers and PCBM, *Sol. Energy Mater. Sol. Cells* 110: 69–76.
48. J. Kesters, S. Kudret, S. Bertho, N. Van den Brande, M. Defour, B. Van Mele, H. Penxten et al. 2014. Enhanced intrinsic stability of the bulk heterojunction active layer blend of polymer solar cells by varying the polymer side chain pattern, *Org. Electron.* 15: 549–562.
49. I. T. Sachs-Quintana, T. Heumuller, W. R. Mateker, D. E. Orozco, R. Cheacharoen, S. Sweetnam, C. J. Brabec, and M. D. McGehee, 2014. Electron barrier formation at the organic-back contact interface is the first step in thermal degradation of polymer solar cells, *Adv. Funct. Mater.* 24: 3978–3985.
50. J. Vandenbergh, B. Conings, S. Bertho, J. Kesters, D. Spoltore, S. Esiner, J. Zhao et al., 2011. Thermal stability of poly[2-methoxy-5-(2′-phenylethoxy)-1,4-phenylenevinylene] (MPE-PPV):fullerene bulk heterojunction solar cells, *Macromolecules* 44: 8470–8478.
51. J. Kesters, P. Verstappen, J. Raymakers, W. Vanormelingen, J. Drijkoningen, J. D'Haen, J. Manca, L. Lutsen, D. Vanderzande, and W. Maes, 2015. Enhanced organic solar cell stability by polymer (PCPDTBT) side chain functionalization, *Chem. Mater.* 27: 1332–1341.
52. F. C. Krebs and H. Spanggaard, 2005. Significant improvement of polymer solar cell stability, *Chem. Mater.* 17: 5235–5237.
53. J. Kong, S. Song, M. Yoo, G. Y. Lee, O. Kwon, J. K. Park, H. Back et al., 2014. Long-term stable polymer solar cells with significantly reduced burn-in loss, *Nat. Commun.* 5: 5688.
54. W. R. Mateker, T. Heumueller, R. Cheacharoen, I. T. Sachs-Quintana, M. D. McGehee, J. Warnan, P. M. Beaujuge, X. Liu, and G. C. Bazan, 2015. Molecular packing and arrangement govern the photo-oxidative stability of organic photovoltaic materials, *Chem. Mater.* 27: 6345–6353.
55. G. Dennler, M. C. Scharber, and C. J. Brabec, 2009. Polymer–fullerene bulk-heterojunction solar cells, *Adv. Mater.* 21: 1323–1338.
56. Y. Sun, C. J. Takacs, S. R. Cowan, J. H. Seo, X. Gong, A. Roy, and A. J. Heeger, 2011. Efficient, air-stable bulk heterojunction polymer solar cells using MoO_x as the anode interfacial layer, *Adv. Mater.* 23: 2226–2230.
57. S. Chen, J. R. Manders, S. W. Tsang, and F. So, 2012. Metal oxides for interface engineering in polymer solar cells, *J. Mater. Chem.* 22: 24202–24212.

58. J. H. Huang, T. Y. Huang, H. Y. Wei, K. C. Ho, and C. W. Chu, 2012. Wet-milled transition metal oxide nanoparticles as buffer layers for bulk heterojunction solar cells, *RSC Adv.* 2: 7487–7491.
59. N. Nicolaidis, B. Vaughan, C. J. Mulligan, G. Bryant, T. Zillger, B. Trnovec, A. C. Hübler et al., 2014. Solution processable interface materials for nanoparticulate organic photovoltaic devices, *Appl. Phys. Lett.* 104: 43902.
60. C. J. Brabec, A. Cravino, D. Meissner, N. S. Sariciftci, T. Fromherz, M. T. Rispens, L. Sanchez, and J. C. Hummelen, 2001. Origin of the open circuit voltage of plastic solar cells, *Adv. Funct. Mater.* 11: 374–380.
61. C. J. Brabec, U. Scherf, and V. Dyakonov, 2011. *Organic Photovoltaics: Materials, Device Physics, and Manufacturing Technologies*, John Wiley & Sons, Germany.
62. J. You, C. C. Chen, Z. Hong, K. Yoshimura, K. Ohya, R. Xu, S. Ye, J. Gao, G. Li, and Y. Yang, 2013. 10.2% power conversion efficiency polymer tandem solar cells consisting of two identical subcells, *Adv. Mater.* 25: 3973–3978.
63. K. Lee, J. Y. Kim, S. H. Park, S. H. Kim, S. Cho, and A. J. Heeger, 2007. Air-stable polymer electronic devices, *Adv. Mater.* 19: 2445–2449.
64. J. Y. Kim, K. Lee, N. E. Coates, D. Moses, T. Q. Nguyen, M. Dante, and A. J. Heeger, 2007. Efficient tandem polymer solar cells fabricated by all-solution processing, *Science* 317: 222–225.
65. Z. He, C. Zhong, S. Su, M. Xu, H. Wu, and Y. Cao, 2012. Enhanced power-conversion efficiency in polymer solar cells using an inverted device structure, *Nat. Photon.* 6: 591–595.
66. T. Yang, M. Wang. C. Duan, X. Hu, L. Huang, J. Peng, F. Huang, and X. Gong, 2012. Inverted polymer solar cells with 8.4% efficiency by conjugated polyelectrolyte, *Energy Environ. Sci.* 5: 8208–8214.
67. J. You, C. C. Chen, L. Dou, S. Murase, H. S. Duan, S. A. Hawks, T. Xu et al., 2012. Metal oxide nanoparticles as an electron-transport layer in high-performance and stable inverted polymer solar cells, *Adv. Mater.* 24: 5267–5272.
68. A. Kanwat and J. Jang, 2014. Extremely stable organic photovoltaic incorporated with WO_x doped PEDOT:PSS anode buffer layer, *J. Mater. Chem. C* 2: 901–907.
69. C. P. Chen, Y. D. Chen, and S. C. Chuang, 2011. High-performance and highly durable inverted organic photovoltaics embedding solution-processable vanadium oxides as an interfacial hole-transporting layer, *Adv. Mater.* 23: 3859–3863.
70. R. Lampande, G. W. Kim, J. Boizot, Y. J. Kim, R. Pode, and J. H. Kwon, 2013. A highly efficient transition metal oxide layer for hole extraction and transport in inverted polymer bulk heterojunction solar cells, *J. Mater. Chem. A* 1: 6895–6900.
71. Z. Tan, L. Li, C. Cui, Y. Ding, Q. Xu, S. Li, D. Qian, and Y. Li, 2012. Solution-processed tungsten oxide as an effective anode buffer layer for high-performance polymer solar cells, *J. Phys. Chem. C* 116: 18626–18632.
72. H. Q. Wang, N. Li, N. S. Guldal, and C. J. Brabec, 2012. Nanocrystal V_2O_5 thin film as hole-extraction layer in normal architecture organic solar cells, *Org. Electron.* 13: 3014–3021.
73. Y. Meng, Z. Hu, N. Ai, Z. Jiang, J. Wang, J. Peng, and Y. Cao, 2014. Improving the stability of bulk heterojunction solar cells by incorporating pH-neutral PEDOT:PSS as the hole transport layer, *ACS Appl. Mater. Inter.* 6: 5122–5129.
74. H. Kim, S. Nam, H. Lee, S. Woo, C. S. Ha, M. Ree, and Y. Kim, 2011. Influence of controlled acidity of hole-collecting buffer layers on the performance and lifetime of polymer:fullerene solar cells, *J. Phys. Chem. C* 115: 13502–13510.
75. M. O. Reese, A. J. Morfa, M. S. White, N. Kopidakis, S. E. Shaheen, G. Rumbles, and D. S. Ginley, 2008. Pathways for the degradation of organic photovoltaic P3HT:PCBM based devices, *Sol. Energy Mater. Sol. Cells* 92: 746–752.
76. S. H. Park, A. Roy, S. Beaupre, S. Cho, N. Coates, J. S. Moon, D. Moses, M. Leclerc, K. Lee, and A. J. Heeger, 2009. Bulk heterojunction solar cells with internal quantum efficiency approaching 100%, *Nat. Photon.* 3: 297–302.
77. M. Jørgensen, K. Norrman, S. A. Gevorgyan, T. Tromholt, B. Andreasen, and F. C. Krebs, 2012. Stability of polymer solar cells, *Adv. Mater.* 24: 580–612.

78. H. L. Yip, S. K. Hau, N. S. Baek, and A. K. Y. Jen, 2008. Self-assembled monolayer modified ZnO/metal bilayer cathodes for polymer:fullerene bulk-heterojunction solar cells, *Appl. Phys. Lett.* 92: 193313.
79. H. L. Yip, S. K. Hau, and A. K. Y. Jen, 2009. Interface engineering of stable, efficient polymer solar cells, *SPIE Newsletter*, doi: 10.1117/2.1200905.1666.
80. J. Liu, M. Durstock, and L. Dai, 2014. Graphene oxide derivatives as hole- and electron-extraction layers for high-performance polymer solar cells, *Energy Environ. Sci.* 7: 1297–1306.
81. S. H. Kim, C. H. Lee, J. M. Yun, Y. J. Noh, S. S. Kim, S. Lee, S. M. Jo, H. I. Joh, and S. I. Na, 2014. Fluorine-functionalized and simultaneously reduced graphene oxide as a novel hole transporting layer for highly efficient and stable organic photovoltaic cells, *Nanoscale* 6: 7183–7187.
82. G. Kakavelakis, D. Konios, E. Stratakis, and E. Kymakis, 2014. Enhancement of the efficiency and stability of organic photovoltaic devices via the addition of a lithium-neutralized graphene oxide electron-transporting layer, *Chem. Mater.* 20: 5988–5993.
83. W. Y. Tan, R. Wang, M. Li, G. Liu, P. Chen, X. C. Li, S. M. Lu, H. L. Zhu, Q. M. Peng, and X. H. Zhu, 2014. Lending triarylphosphine oxide to phenanthroline: A facile approach to high-performance organic small-molecule cathode interfacial material for organic photovoltaics utilizing air-stable cathodes, *Adv. Funct. Mater.* 24: 6540–6547.
84. C. J. Brabec, S. E. Shaheen, C. Winder, and N. S. Sacriciftci, 2002. Effect of LiF/metal electrodes on the performance of plastic solar cells, *Appl. Phys. Lett.* 80: 1288–1289.
85. L. Yang, H. Xu, H. Tian, S. Yin, and F. Zhang, 2010. Effect of cathode buffer layer on the stability of polymer bulk heterojunction solar cells, *Sol. Energy Mater. Sol. Cells* 94: 1831–1834.
86. H. Zeng, X. Zhu, Y. Liang, and X. Guo, 2015. Interfacial layer engineering for performance enhancement in polymer solar cells, *Polymer* 7: 333–372.
87. A. Gaur and P. Kumar, 2013. Studies on stability of bi-functional P3HT:PCBM:rubrene optoelectronic devices, *Appl. Phys. A* 111: 877–887.
88. V. Turkovic, S. Engmann, N. Tsierkezos, H. Hoppe, U. Ritter, and G. Gobsch, 2014. Long-term stabilization of organic solar cells using hindered phenols as additives, *ACS Appl. Mater. Interfaces* 6: 18525–18537.
89. F. C. Krebs, 2006. Encapsulation of polymer photovoltaic prototypes, *Sol. Energy Mater. Sol. Cells* 90: 3633–3643.
90. B. Zimmermann, U. Wurfel, and M. Niggemann, 2009. Longterm stability of efficient inverted P3HT:PCBM solar cells, *Sol. Energy Mater. Sol. Cells* 93: 491–496.
91. L. Moro, N. M. Rutherford, R. J. Visser, J. A. Hauch, C. Klepek, P. Denk, P. Schilinsky, and C. J. Brabec, 2006. Barix multilayer barrier technology for organic solar cells, *Proc. SPIE* 6334: 63340M.
92. C. Lungenschmied, G. Dennler, H. Neugebauer, S. N. Sariciftci, M. Glatthaar, T. Meyer, and A. Meyer, 2007. Flexible, long-lived, large-area, organic solar cells, *Sol. Energy Mater. Sol. Cells* 91: 379–384.
93. J. A. Hauch, P. Schilinsky, S. A. Choulis, R. Childers, M. Biele, and C. J. Brabec. 2008. Flexible organic P3HT:PCBM bulk-heterojunction modules with more than 1 year outdoor lifetime, *Sol. Energy Mater. Sol. Cells* 92: 727–731.
94. F. C. Krebs, T. Tromholt, and M. Jørgensen, 2010. Upscaling of polymer solar cell fabrication using full roll-to-roll processing, *Nanoscale* 2: 873–886.
95. C. Maibohm, J. R. Brewer, H. Sturm, F. Balzer, and H. G. Rubahn, 2006. Bleaching and coating of organic nanofibers, *J. Appl. Phys.* 100: 054304.
96. P. Madakasira, K. Inoue, R. Ulbricht, S. B. Lee, M. Zhou, J. P. Ferraris, and A. A. Zakhidov, 2005. Multilayer encapsulation of plastic photovoltaic devices, *Synth. Met.* 155: 332.
97. F. Zhang, H. Zhu, W. Yang, Z. Wu, H. Qi, H. He, Z. Fan, and J. Shao, 2008. Al_2O_3/SiO_2 films prepared by electron-beam evaporation as UV antireflection coatings on 4H-SiC, *Appl. Surf. Sci.* 254: 3045–3048.
98. D. Vollath, I. Lamparth, and D. V. Szabó, 2002. Fluorescence from coated nanoparticles, *Mater. Res. Soc. Symp. Proc.* 703: 303.

99. D. Vollath, D. V. Szabó, and S. Schlabach, 2004. Oxide/polymer nanocomposites as new luminescent materials, *J. Nanopart. Res.* 6: 181–191.
100. Z. G. Wang, X. T. Zu, S. Zhu, X. Xiang, L. M. Fang, and L. M. Wang, 2006. Origin of luminescence from PMMA functionalized nanoparticles, *Phys. Lett. A* 350: 252–257.
101. L. Tavares, J. K. Hansen, H. G. Rubahn, and H. Sturm, 2010. Reduced bleaching in organic nanofibers by bilayer polymer/oxide coating, *J. Appl. Phys.* 107: 103521.
102. X. Zhang, X. Huang, and H. Jiang, 2013. Recovering degraded quasi-solid-state dye-sensitized solar cells by applying electrical pulses, *Phys. Chem. Chem. Phys.* 15: 6864–6869.
103. P. Kumar, C. Bilen, K. Feron, N. C. Nicolaidis, B. B. Gong, X. Zhou, W. J. Belcher, and P. C. Dastoor, 2014. Comparative degradation and regeneration of polymer solar cells with different electrodes, *ACS Appl. Mater. Inter.* 6: 5281–5289.
104. N. Nicolaidis, B. Routley, J. Holdsworth, W. Belcher, X. Zhou, and P. Dastoor, 2011. Fullerene contribution to photocurrent generation in organic photovoltaic cells, *J. Phys. Chem. C* 115: 7801–7805.
105. J. Luo, H. Wu, C. He, A. Li, W. Yang, and Y. Cao, 2009. Enhanced open-circuit voltage in polymer solar cells, *Appl. Phys. Lett.* 95: 43301.
106. T. Zhang, J. Huang, F. Hea, L. Chen, G. Niu, J. Pan, and Q. Song, 2013. The effect of built-in field on the interface exciton recombination and dissociation in N-N type organic solar cells, *Sol. Energy Mater. Sol. Cells* 112: 73–77.

7

Roll-to-Roll Organic Solar Cells

7.1 Introduction

Usually thin films of different materials in polymer solar cells are prepared by spin coating of their solutions. Spin coating is used almost exclusively but is not suitable for high-volume production, as it causes significant loss of the coating materials. Additionally several unintentional variations in processing may get introduced that are sometime difficult to identify and they affect the cell performance. For cost-effective large-scale production of polymer solar cells, slot-die coating, gravure coating, knife-over-edge coating, offset coating, spray coating, and printing techniques (such as inkjet printing, pad printing, and screen printing) are the most suitable techniques. F. C. Krebs presented a detailed review on various printing and coating techniques and their applications in fabrication of polymer solar cells [1]. For high-speed large-area roll-to-roll (R2R) processing of polymer solar cells, slot-die coating and screen printing are the most useful techniques. In R2R fabrication, the substrate is a roll of long wound flexible sheet, which is known as a web, and is unrolled for coating and rolled again after coating. The solutions being used in printing techniques are normally called ink. After completion of one process on the roll, it is transferred to another unwinder for the next step. The coatings steps could be discrete, which means each step has a separate unwinder and rewinder, or integrated, whereby the steps are followed one by another in a single unwinding and rewinding process. The flexible sheet could be of polyethylene terephthalate (PET), polyethylene naphthalate (PEN), polylactic acid (PLA), kapton, polyethylene (PE), polypropylene (PP), or nylon. Kapton has excellent thermal and mechanical properties, but it is too expensive for general purpose. PLA is biodegradable and may be good for disposal of dead organic solar cells (OSCs), but it has low resistivity against organic solvents and poor thermal properties. Compared to other flexible sheets, PET is found to have overall better properties to be a substrate material and is used most often in R2R production of OSCs. In R2R coating machines, the substrates enter from one side and after coating of different layers, the complete solar modules come out from the other side. Screen printing can possibly produce 1000–100,000 m^2 OSCs per day on a process line and for silicon (Si) solar cells the same area would take around a year for fabrication [2]. It shows the potential of printing technology and makes OSCs very cost effective. Initially, printed OSCs showed poor performance, but advanced research and development has resulted in high-performance solar cells. The performance of printed devices still remains low compared to spin-coated devices, but the difference is small. Development in R2R processing would allow transfer of laboratory polymer solar cells to large-scale industrial production. Before I discuss the R2R fabrication of OSC modules, it is important to understand some of the related coating techniques.

7.2 Printing Techniques

7.2.1 Doctor Blading

Doctor blading is a kind of printing technique that is used for coating of few micro meter thin films of the materials on plane surfaces from their pastes or dense solutions. Figure 7.1a schematically represents formation of a thin film via doctor blading. A sharp blade is placed at a fixed distance from the surface to be coated and the coating solution is dropped on the substrate in front of the blade. The blade is moved at a constant speed parallel to the surface making a thin wet film after the blade. Thickness of the wet film depends upon the gap between the blade and substrate, surface energy of the substrate, surface tension and viscosity of the coating solution, and meniscus formed between the blade and wet film. Once dried, the final thickness of the film can be calculated from [1]

$$d = \frac{1}{2}\left(g\frac{c}{\rho}\right), \tag{7.1}$$

where g is the gap between the substrate and blade, c is the concentration of solution, and ρ is the density of material in the dry film. This method does not cause wastage of any material, however, it is slow and some materials may get aggregated or crystallized. Moreover, the films prepared via doctor blading are not as smooth and uniform as those by spin coating. The solvent evaporation is also very slow compared to spin coating. In solar cells this technique is usually employed for coating of nanoparticle films such as ZnO and TiO_2.

7.2.2 Screen Printing

Screen printing is a more versatile technique and is used for printing of different patterns through screens. Like doctor blading, it does not cause material loss. It requires highly viscous and low volatile solutions, which give large thickness of wet film. Figure 7.1b schematically shows the screen printing method. The screen has openings in a predefined pattern, which is to be transferred on the substrate. The rest of the screen remains impermeable to the solution or ink. During printing, the impermeable areas prevent printing of substrate in those regions. The screen is mounted on a frame under tension and a squeegee filled with the ink is moved across the screen to fill the openings. Subsequently the screen is brought into proximity of the substrate and the squeegee is forced to bring the screen in contact with substrate. Then the squeegee is moved across the screen and this makes the ink filled in the open areas transfer onto the substrate and the screen is pulled back as the

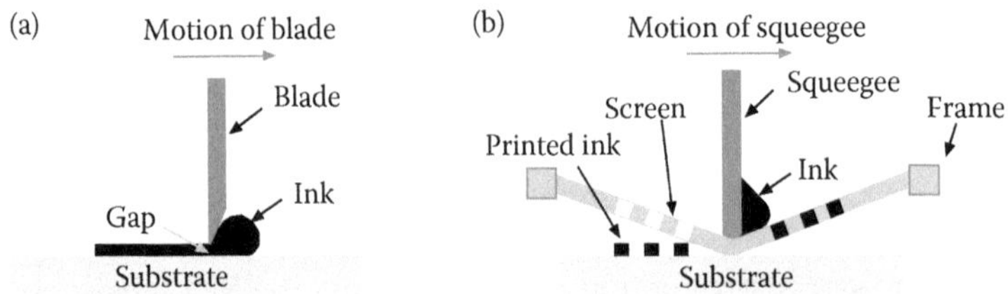

FIGURE 7.1
Schematic representation of ink coatings by (a) doctor blading and (b) screen printing.

squeegee moves on. The thickness of the wet film is determined by the paste volume of the screen per its open area. However, it is influenced by the force applied on the squeegee, the snap-off distance, speed of the squeegee, and viscosity of the solution. The thickness of screen printed dry film could be given by [1]

$$d = V_{\text{screen}} k_p \frac{c}{\rho}, \tag{7.2}$$

where V_{screen} is the volume of screen per open area, c is the concentration of ink, ρ is the density of material in the dry film, and k_p is the pick-out ratio. Sometime screen printing may not be suitable for active layer printing, but it is frequently used for printing of other layers like PEDOT:PSS and Ag electrodes in R2R fabrication [2–5]. The R2R printers possess a rotary screen printer where both the screen and web support are in cylindrical shape and the squeegee is inside the screen cylinder. The squeegee is stationary and both the screen and support roller of the web move at the same speed. Screen printing is an important printing technology for large-scale cost-effective production of OSC modules.

7.2.3 Inkjet Printing

Inkjet printing has high resolution and does not require any screen or mask for printing of a pattern. It is a complex printing technology and is completely computer controlled. The printed image is digital, just like from normal inkjet printers used in offices, but the ink is the solution of organic materials and the cartridge head is solvent resistant. The films are prepared by a to-and-fro motion of the head in transverse direction while the substrate moves in a forward direction. Figure 7.2a schematically shows the printing of a pattern

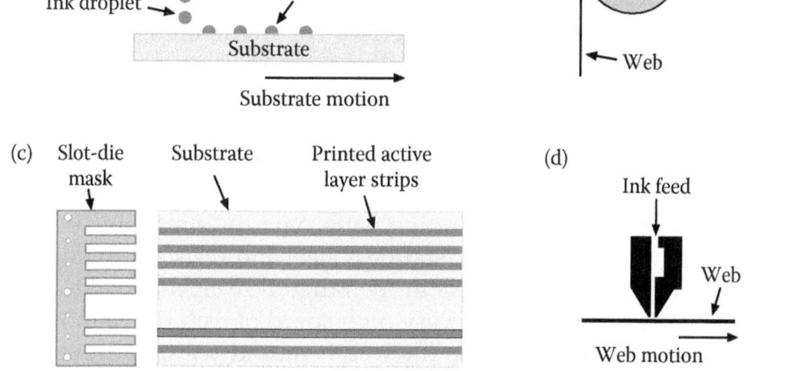

FIGURE 7.2
Schematic representation of (a) inkjet printing, (b) knife-over-edge coating, (c) the slot-die coating mask with corresponding printed substrate, and (d) the slot-die coating head.

on the substrate via inkjet printing. The ink is released on the substrate from cartridge in the form of droplets. The droplets are either formed by mechanical compression of the ink through piezoelectric effect or by heating the ink, which is electrostatically charged by an electric field and pushed toward the substrate for deposition. Thickness of the film printed by inkjet printing is calculated from [1]

$$d = N_d V_d \frac{c}{\rho}, \quad (7.3)$$

where N_d is the number of droplets per unit area, V_d is the volume of the droplet, c is the concentration of the ink, and ρ is the density of material in the dry film. For inkjet printing the ink is required to be of low viscosity and electrostatically chargeable, which put limitations on the ink formation. For the formation of droplets the ink requires high surface tension, therefore the ink possesses a mixture of many solvents. Formation of ink of organic materials for inkjet printing is a challenge, and it is difficult to say if this technique will lead the high-volume production of OSCs. However, this may be an important technique in generating very fine and complex patterns for OSCs.

7.2.4 Knife-Over-Edge Coating

Knife-over-edge coating is similar to doctor blading except the knife is stationary and the substrate moves. Figure 7.2b schematically shows knife-over-edge coating. It possesses an ink bath in front of the knife and the ink is fed in the gap between the knife and substrate. It gives a smooth and uniform film. Knife-to-web distance is important to define the film thickness and the coating area. The edge definition depends upon the knife-to-web gap and speed of the web. The flow of ink to the knife could be controlled with a piston pump or a pressure tank. There is no wastage of ink in this technique and all the material supplied to the knife is coated on the substrate.

7.2.5 Slot-Die Coating

Slot-die coating is a very simple technique that allows the materials to coat in strips and causes no material loss. In multilayer solar cells, different materials could be easily coated one over another in the same pattern. This technique is most suitable for R2R production of solar panels where cells are serially connected. For alignment and coating of different materials, the coating head could be easily aligned by shifting it suitably vertical to the direction of web motion. The ink is fed to the coating head via a piston pump or a pressure tank. The coating head is the most important part of the slot-die coating and it is very complex. The coating head comprises several parts, in which the mask is the most important one. It defines the slots and voids in the coating head. Figure 7.2c schematically shows the slot-die coating mask and the corresponding pattern on the substrate after printing. The mask and the coating head are made of stainless steel and resistant to ink chemicals. For the polymer solutions where viscosity is in the range 1–20 cP, masks 20–50 μm thick are used [1]. It is important to note that for very high speed of web the coating becomes very difficult or almost impossible, therefore the web is moved within a limited speed range to get good-quality film. The schematic picture of the slot-die coating head is shown in Figure 7.2d. Just by changing the mask, the coating patterns can be changed quite easily. Thickness of the film coated by the slot-die technique depends upon the web speed and rate of ink feeding to the coating head. Thickness of a dry film coated from the slot-die technique can be calculated from [1]

$$d = \frac{f}{Sw}\frac{c}{\rho}, \tag{7.4}$$

where f is the flow rate of the ink, S is the web speed, w is the coating width, c is the solid content in the ink, and ρ is the density of the material in dried film.

7.2.6 Gravure and Meniscus Coatings

There are two rollers in gravure coating: one, known as coating roller, has some engraved pattern on it; and the other roller supports and directs the web. The engraved coating roller is filled with the ink by dipping it in an ink bath. Before the coating roller is brought in contact with the web, the excess ink is removed by doctor blading. Upon contact with the web, the ink filled in the gravure of the coating roller is transferred to the web, resulting in formation of a patterned layer on it. The change in pattern by gravure coating requires complete replacement of the coating roller, which makes it very expensive and is the main drawback of this technique. But it has the benefit that even low viscous ink can form good-quality films even at high web speeds.

In meniscus coating too there are two rollers: one that supports and directs the web and the other that does the metering of the ink. Both rollers have a gap in between and the ink is fed between the two rollers. The meniscus of the liquid ink formed in the gap draw a film whose thickness could be controlled by controlling the gap and speed of the web. The roller may be rotating in the same or opposite direction to the web motion. Meniscus coating gives a layer as even as that from knife-over-edge coating. There are several other coating techniques as well, including curtain coating, multilayer slot coating, slide coating, spray coating, flexographic printing, and offset printing, and the information on these techniques can be found in Krebs [1]. I present here a short review on R2R processing and characterization.

7.3 Fabrication of Roll-to-Roll (R2R) Solar Modules

In general, for R2R fabrication of polymer solar cells on indium tin oxide (ITO)-coated PET sheet, the sheet is patterned using a R2R etching machine. The patterning includes printing of a UV-curable photoresist, stripping the photoresist with sodium hydroxide solution, developing and hardening of pattern, etching out undesired ITO in copper chloride ($CuCl_2$) solution, washing with deionized water, and finally drying. Prior to any coating, the PET/ITO substrates are cleaned on both sides using isopropanol. Thin films of transport materials and active layer are printed via slot-die coating, and the silver ink is screen printed for back electrode. Let us take the example of R2R modules, prepared by Krebs in the structure PET/ITO/ZnO/P3HT:PCBM/Ag [6]. First, the ITO-coated PET sheet was patterned in a way that the solar cells could be prepared individually or connected in series together. The roll was mounted on the web roller and marked at edges for reference. The UV-curable photoresist was screen printed and UV cured in the desired striped pattern for ITO patterning. Due to limited printable length of the screen printer, the strips were 250 mm in length along the web direction and 9 mm wide. Then the roll was etched by passing it through the stripping bath (NaOH), etching bath ($CuCl_2$), washing bath, and drying section at an etching speed of 40 m/h. The ZnO film over ITO was prepared by knife-over-edge or slot-die coating. In the knife-over-edge coating the ZnO nanoparticle solution was fed manually in front of the knife and then the coated roll was allowed to pass through the hot air oven for drying. However, in slot-die coating the ZnO solution was fed into the slot-die coating head. Similar to slot-die coating, the knife in knife-over-edge coating remains stationary and the web moves. The ZnO lines were 9 mm wide and 1 mm shifted laterally with respect to the ITO pattern, therefore the overlap was of 8 mm.

For the active layer P3HT and PCBM, inks in 1:1 mixture of 1,2-dichlorobenzene and chloroform were fed to the slot-die coating head and printed on the roll. The solutions of P3HT and PCBM in desired solvents were kept in separate vessels and pumped via separate channels into the mixing head. This is the beauty of R2R processing that the donor and acceptor materials are not mixed until they are used and one can use the same stock of solution with different materials by simply changing the source solution. The active layer lines were 9 mm wide and 1 mm shifted with respect to ITO, therefore the overlap was 7 mm. After coating, the active layer was allowed to dry in a hot air oven, and then PEDOT:PSS and Ag electrodes were screen printed or slot-die coated onto it. Ag coating was done using a silver ink. PEDOT:PSS (Orgacon EL-P 5010, diluted with isopropanol) was printed with the pattern 3 mm shifted with respect to ITO and dried in the hot air oven at 130°C for 15 min. The total overlap of ITO and PEDOT:PSS was 6 mm. Dupont 5007 silver ink was used for the back electrode, printed on the top of PEDOT:PSS, and dried in the hot air oven at 130°C for 3 min. The Ag pattern was shifted slightly with respect to PEDOT:PSS. Figure 7.3 schematically illustrates the different processing steps in fabrication of R2R solar modules, whereas the structure of three serially connected cells along with layer sequence prepared by slot-die coating is shown in Figure 7.4. Figure 7.5 compares R2R processing with the conventional spin-coating method on glass substrate. For encapsulation of R2R-processed modules, a barrier film is employed along with an adhesive or the modules are laminated by sandwiching them between two barrier films via high pressure thermal lamination. The modules are protected from the coating side first, however, for longer life the modules are protected from both sides. The adhesive is applied on the barrier film, which is cut in a size such that some of the Ag contacts from the sides are left exposed for electrical connections and measurements. Silver inks with high Ag content (80%–86% wt/wt) and less solvent show high performance, whereas those with low Ag content do not work well. Low solvent contents cause slow diffusion of solvent into active layer. Some of the inks are solvent free and require UV curing for solidification and such inks show excellent performance [7].

Figure 7.6 shows the picture of an inline coating and printing machine from Solar Coating Machinery GmbH, Germany [1]. A R2R coating machine could be small and compact with a narrow web for slow printing or it could be very large for wide web printing with high printing speed. In general a printing machine possesses an unwinder, edge guide, corona treater, double side web cleaning unit, antistatic bars, slot-die coater/printer, oven, coating roller, screen printer, rewinder, and cutter. One the other hand, a laminator possesses an unwinder, edge guide, splicing table, lamination rollers, and rewinder for laminated cells [8]. Printing of web can be carried out without any support, but mostly it is supported by a roller that provides a solid support for the coating and guides the web.

In the context of operational stability and large-scale production without vacuum coating steps, the normal cell geometry is far from ideal, therefore other device geometries

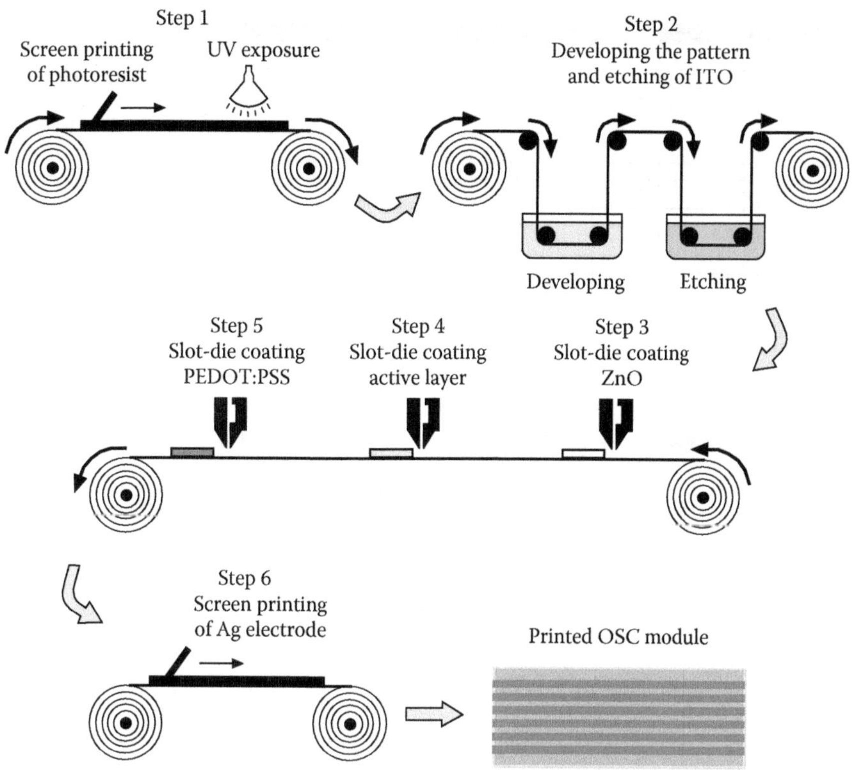

FIGURE 7.3
Schematic illustration of different steps in R2R fabrication of OSC modules.

are worth pursuing. Inverted cell geometry is quite compatible to R2R fabrication. Krebs et al. first fabricated an inverted solar cell of the same PCE as that in normal structure with the same active layer, and then transferred the process to a full R2R fabrication on ITO-coated PET substrate using slot-die coating and screen printing techniques. R2R-processed modules exhibited nearly the same performance as that of a small-area laboratory cell [9]. The normal cell was structured ITO/PEDOT:PSS/P3HT:PCBM and finally evaporated Al as cathode, whereas the inverted geometry employed ITO/ZnO/P3HT:PCBM and finally evaporated Ag. In inverted geometry, Ag was chosen as the top electrode because it possesses similar work function as that of Al, and it could be solution processed, which is prime requirement for full R2R processing. Normal and inverted cells exhibited PCEs of 2.7% and 2.67%, respectively. For R2R modules, ZnO, P3HT:PCBM, and PEDOT:PSS were employed by slot-die, whereas the top Ag grid lines were applied by screen printing of silver ink. Since PEDOT:PSS does not wet P3HT:PCBM, printing of PEDOT:PSS over P3HT:PCBM did not form good-quality film. To overcome this problem, a small amount of isopropanol was mixed in PEDOT:PSS and also the active layer was wet with isopropanol just before the slot-die coating head. Application of isopropanol resulted in a smooth and uniform coating of PEDOT:PSS. The complete process was performed in air without any vacuum coating step. The modules were cut in 1, 2, 3, and 8 strips (25×0.6 cm), and the characterization was done without any encapsulation or lamination with a simple PET foil using an acrylic adhesive. The module consisting of eight serially connected cells gave a PCE of ~2.1% for 120 cm^2 active area and ~2.3% for 4.8 cm^2 active area. In the photovoltaic

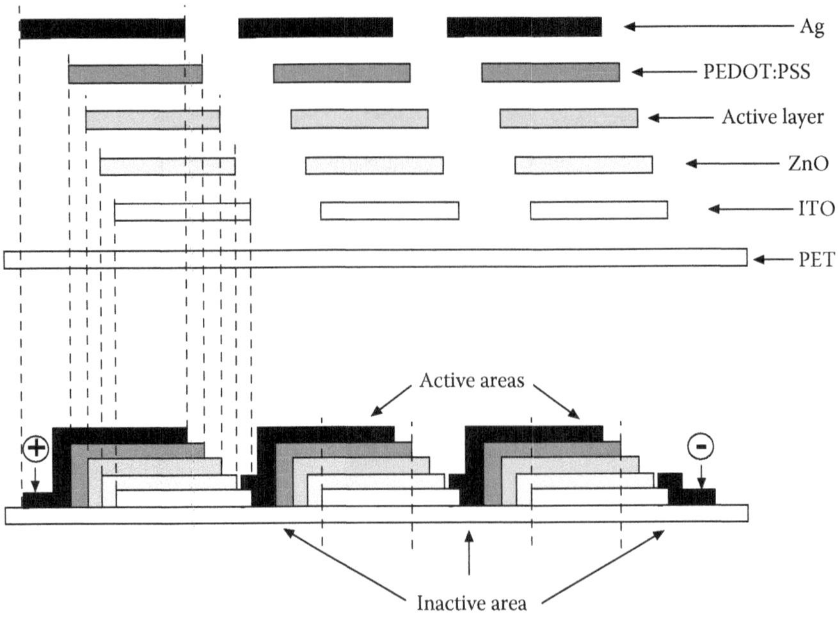

FIGURE 7.4
Schematic structure of serially connected OSCs module (bottom) and the layer sequence (top) prepared by R2R slot-die coating method.

industry, it is commonly reported that manufactured devices typically reach 60%–70% of the performance of laboratory cells. As far as stability is concerned, the normal device exhibited stable operation in dry oxygen but did not gave stable performance in a humid environment, whereas inverted solar cells behaved differently and gave stable operation in humid environment but poor stability in the presence of oxygen.

R2R processing is useful not only for large-area manufacturing but also in optimizing the device performance. Optimization of polymer solar cell performance requires optimization of device thickness, processing solvents, donor–acceptor ratios, and nanoscale morphology. Usually in spin processing these parameters are optimized by performing complex experiments individually, which is not only time-consuming but also requires great effort and causes a significant loss of materials. Additionally, there are various unintentionally introduced variations in processing that are difficult to identify. For a typical material system, Alstrup et al. studied and reported the complexity and limitations in getting an optimized device using spin coating [8]. To overcome the difficulties, R2R processing was employed for optimization of polymer solar cell performance. R2R fabrication is not only a time-saving process, but it also reduces material loss and provides accurate material parameters for efficient power conversion with only few sources of experimental variations. For example, a rough estimate shows that a typical laboratory experiment for different donor–acceptor ratios using 10 sampling points would require ~100 mg of material and ~60 min per sample, and it would enable reasonable accuracy in the determination of the optimum donor–acceptor ratio. On the other hand, in R2R processing, 200 sampling points would require ~60 mg of materials and ~35 s per sample with high accuracy in donor–acceptor ratio.

To demonstrate the application of R2R processing for optimization of device parameters, Alstrup et al. used P3HT and PCBM as basic active materials and the substrate was

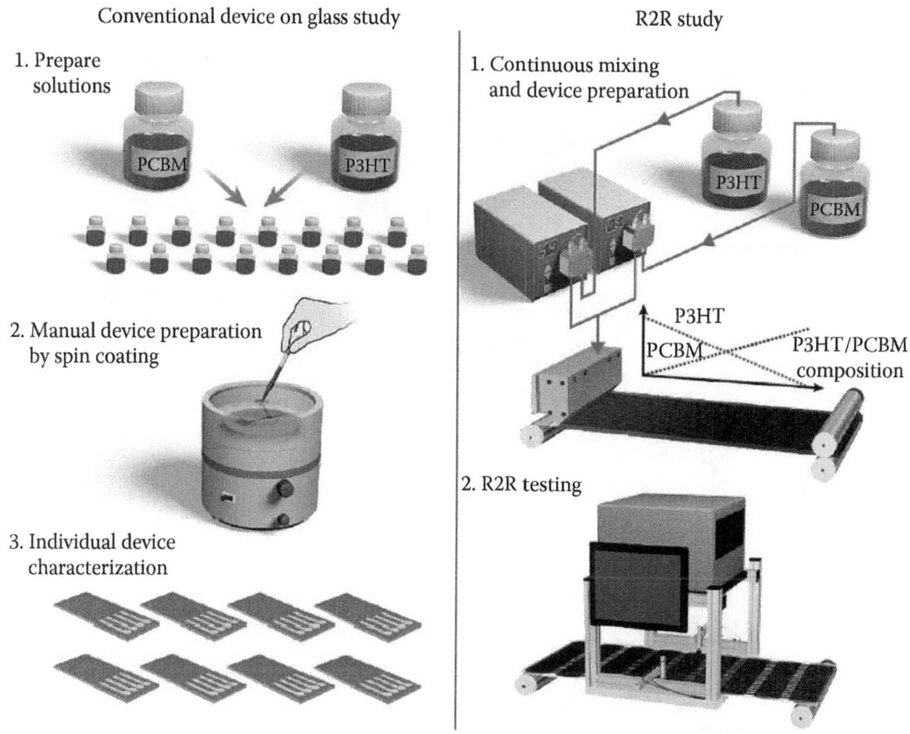

FIGURE 7.5
Comparison of conventional processing of OSCs on glass with that via R2R processing. (Reprinted with permission from J. Alstrup et al., Ultra fast and parsimonious materials screening for polymer solar cells using differentially pumped slot-die coating, *ACS Appl. Mater. Inter.* 2, 2819–2827. Copyright 2010 American Chemical Society.)

ITO-coated PET foil (305 mm width), which was patterned into 16 parallel strips (13 mm width with 2 mm separation) with a length of 250 mm. The solar cells were prepared in ITO/ZnO/P3HT:PCBM/PEDOT:PSS/Ag configuration. The ZnO layer was coated on all the strips using slot-die coating head and dried at 140°C for 10 min. Similarly, P3HT:PCBM (chlorobenzene solution) and PEDOT:PSS were applied via a slot-die coating head and the top Ag back electrode was deposited by screen printing in grid structures. P3HT and PCBM solutions were pumped through the mixing head to the slot-die coating head and their different compositions were applied to the substrate. The experiment started with pumping 100% P3HT and 0% PCBM solutions. The flow of the solutions was changed in steps of 0.5%–1% by decreasing the P3HT flow and increasing the PCBM flow. For the optimization of the P3HT:PCBM ratio, the step size 0.5%–1% resulted in 100–200 cells covering the entire phase diagram. The R2R process typically yields an enormous amount of solar cells (750–1500 in one run). For thickness optimization as well, the step size was kept 0.5%–1%. For characterization, the roll was mounted in an R2R characterization machine and the cells were forwarded to the measuring positions. The devices were masked such that only the active area was illuminated during measurements and electrical contacts were made using pneumatic cylinders. After measurements the contacts were broken and the web was moved to the position of the next cell. Figure 7.7 shows the variation in device performance as a function of composition of the P3HT:PCBM mixture.

FIGURE 7.6
Photographs of the laboratory scale inline printer from Solar Coating Machinery GmbH, Germany. (Adapted with permission from F. C. Krebs, 2009. *Sol. Energy Mater. Sol. Cells* 93: 394–412.)

It can be seen from Figure 7.7 that the optimum performance is obtained in a relatively broad range of composition, which is around 50% of P3HT and 50% of PCBM. There have been numerous publications where the optimum ratio for P3HT and PCBM has been reported to be 1:1. Where the traditional method would fail to identify the high-performance materials as the extreme mixing ratio is not explored, this method is more efficient in identifying the potential materials and their efficient combinations. R2R processing was also used to find the optimum thickness of the active layer, which could be performed by diluting or concentrating the mixture of donor–acceptor materials during coating. The wet layer remained constant, but the dry layer exhibited gradual change in

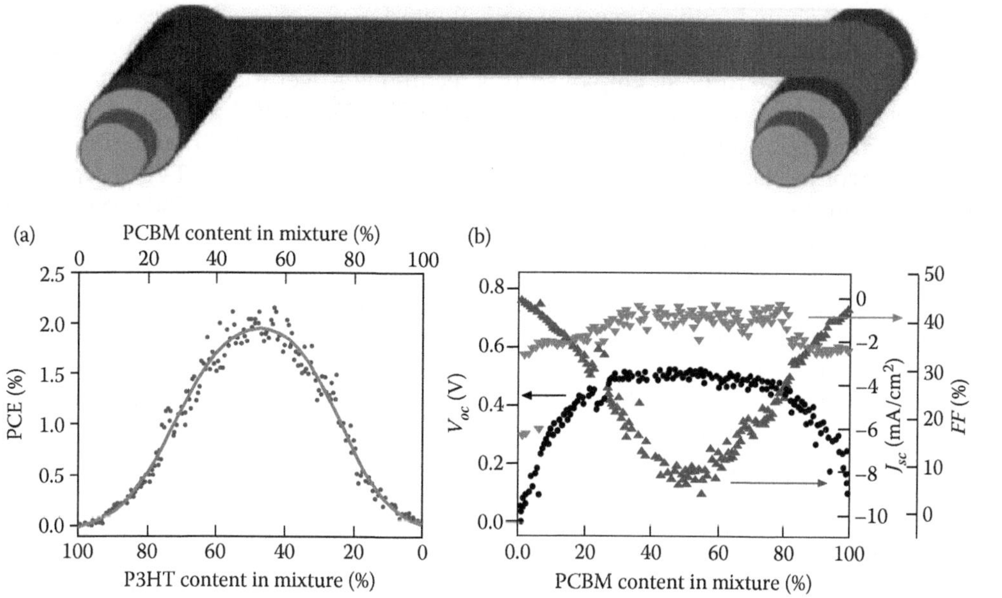

FIGURE 7.7
The top image shows the schematic roll-to-roll patterning of P3HT:PCBM with different mixture compositions. (a) Variation in PCE and (b) variation in J_{sc}, V_{oc}, and FF as a function of mixture compositions. The compositional step size in the experiments was 0.5%. (Reprinted with permission from J. Alstrup et al., Ultra fast and parsimonious materials screening for polymer solar cells using differentially pumped slot-die coating, *ACS Appl. Mater. Inter.* 2, 2819–2827. Copyright 2010 American Chemical Society.)

thickness. Alstrup et al. started with pure solvent and then gradually increased the concentration of active materials to full concentration. Figure 7.8 shows the variation in PCE with the active layer thickness from R2R processing. As expected, the efficiency dropped to zero for very thin films, but increased throughout the thickness range (50–800 nm). With the increase in film thickness, V_{oc} quickly stabilized to its maximum value, but J_{sc} gradually increased because of the increase in light absorption. From the calculations done by Brabec et al. [10], it was expected that J_{sc} would drop for higher thicknesses giving maxima and minima in efficiency as a function of thickness. But it was not so in this case

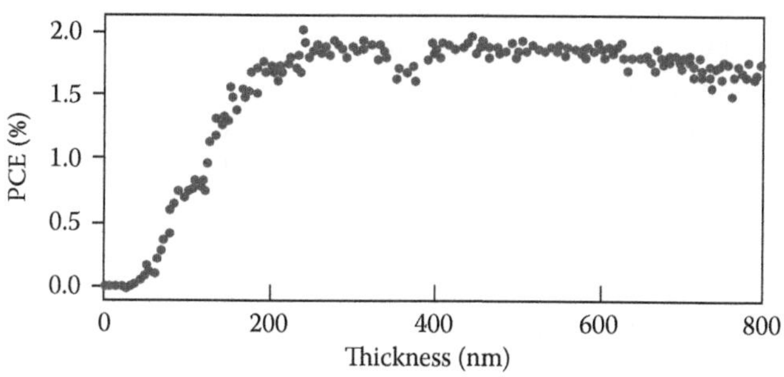

FIGURE 7.8
Variation in the performance of P3HT:PCBM solar cells on varying the active layer thickness and keeping all other parameters constant. (Reprinted with permission from J. Alstrup et al., Ultra fast and parsimonious materials screening for polymer solar cells using differentially pumped slot-die coating, *ACS Appl. Mater. Inter.* 2, 2819–2827. Copyright 2010 American Chemical Society.)

and this was attributed to the absence of optical interference effects in these devices. R2R processing is really useful for very fast screening of new materials with limited quantities. For example, Bundgaard et al. employed this technique to get the optimum parameters for efficient device fabrication using a newly synthesized low band gap copolymer based on dithienothiophene and dialkoxybenzothiadiazole named poly(dithienothiophene-co-dialkoxybenzothiadiazole) (PDTTDABT) (band gap 1.6 eV) [11]. PDTTDABT was mixed with $PC_{61}BM$ in steps of 1% variation in composition for both materials. The optimum mixing ratio for maximum efficiency of ~0.38% was found to be relatively broad around 1:2. Both the J_{sc} and V_{oc} were maximum at the ratio 1:2, whereas *FF* was nearly independent of the polymer:PCBM ratio. Devices were prepared under ambient conditions employing all solution-processing steps. The solar module consisted of 16 serially connected cells with a total module active area of 96 cm^2.

One of the drawbacks in R2R processing is that the freshly prepared devices exhibit poor performance because of the inflection point phenomenon resulting in low *FF* [12]. Lilliedal et al. examined the inflection point behavior in the context of R2R-processed polymer solar cells with a layer structure of PET/ITO/ZnO/P3HT:PCBM/PEDOT:PSS/Ag [12]. Solar cells were prepared using a combination of slot-die coating and screen printing, and encapsulated using a polymer-based barrier material. Freshly prepared devices exhibited consistent inflection point and poor performance. They found that this problem could be overcome by photoannealing under illumination and elevated temperatures to get optimum performance. With photoannealing, the inflection point gradually disappeared and the performance drastically increased. They did further study the stability of photoannealing by subjecting the devices to different temperatures, illumination intensities, and atmospheres, and found that the inflection point was a dynamic phenomenon, which could be removed under specific conditions. Disappearance of the inflection point has been attributed to photoconductivity and degradation of impurities in ZnO and a favorable redistribution of oxygen within the encapsulated cell. Because photoannealing requires considerable energy, it may not be suitable for scale-up of the technology. Therefore, alternate methods should be developed, and that is why the effect and understanding of various parameters on the inflection point behavior became very important. Successful implementation of OSC technology requires reasonably high performance and stability, and it has been discussed earlier that there are some inconsistencies in the reported data from laboratory to laboratory. This problem could have been overcome if the devices were sent to other laboratories for testing and comparison. But sharing and measurement of the cells at different laboratories had been difficult because of degradation. There are now frequent reports on sharing and testing of polymers solar cells, which include the certification of efficiency and the measurement of record devices at a reference laboratory (e.g., National Renewable Energy Laboratory [NREL], National Institute of Advanced Industrial Science and Technology [AIST], Fraunhofer Institute for Solar Energy Systems [ISE]). Earlier round-robin studies (or interlaboratory studies) had been employed to transport the same reference cell to different laboratories for calibration of equipment so that a very high level of agreement on the measurement data could be achieved. Krebs et al. fabricated large-area polymer solar cell modules via R2R processing and reported that it was possible to encapsulate flexible solar cells in a way that round-robin studies were possible. They presented a round-robin study on performance evaluation of flexible polymer solar cells, which involved 18 different laboratories in North America, Europe, and the Middle East [13]. The mode of transportation was air mail and surface mail. The devices were prepared and characterized at Denmark Technical University (DTU) and then transported to participating labs for testing and returned back to the starting lab at DTU for retesting. The layered architecture of the device was PET/ITO/ZnO/P3HT:PCBM/PEDOT:PSS/Ag. The devices were prepared in one, two, three, and eight serially connected strips, and the modules with eight serially connected devices were employed in these studies. The encapsulation was done using a 25 μm thick PET barrier layer incorporating a UV filter, such that degradation played little role in experiments, which were carried out for over 4 months. The devices were encapsulated on both the front and back sides. Immediate after the encapsulation, the devices exhibited a significant drop in performance, which was quickly recovered and even became better than before encapsulation when devices were annealed under a solar simulator (photoannealing) by measuring the current density–voltage (*J–V*) characteristics under illumination. The modules were sent to different laboratories with a list of rough guidelines on how to record measurements. The measurements were carried out using real and simulated sunlight [13]. Such work demonstrated the possibility to share devices and obtain consistent data. However, the technology that made this possible was the barrier material and encapsulation, which gave the devices sufficient stability.

Krebs et al. [7] presented a detailed work on up-scaling the production of polymer solar cells from laboratory to industrial level and identified the instruments for complete manufacturing. They presented an analysis on processing, labor, and material cost to estimate the manufacturing and electricity cost. Search and application of an alternate to ITO in OSCs has been previously emphasized. Finding novel approaches that avoid the need of patterning of some or all of the layers is also being encouraged. From that point of view, Krebs did fabricate a large-area polymer solar cell by R2R process that did not involve ITO [14,15]. He reported R2R devices in normal [14] and inverted [15] geometries. Normal R2R devices consisted of a kapton foil with a layer of copper, which was used as the substrate. The substrate was introduced into the R2R sputtering system to coat titanium metal, which worked as a back electrode (Figure 7.9a). The P3HT:PCBM layer was coated by slot-die and dried at 130°C. PEDOT:PSS was slot-die coated over the P3HT:PCBM layer and cured at 130°C. For smooth and proper coating of PEDOT:PSS, it was necessary to wet the surface of P3HT:PCBM with isopropanol just before PEDOT:PSS coating. Finally, the top Ag electrodes were screen printed in grids using a silver ink. The devices were then cold laminated with a 25 μm thick PET film using an acrylic adhesive. Figure 7.9b shows the R2R lamination of the solar modules prepared by Krebs. The electrical contacts to the copper back electrode were made by mechanical removal of PEDOT:PSS and the active layer, followed by soldering the tinned copper strips onto copper. A typical 7 × 7 cm^2 cell, when illuminated through a 7 × 7 cm^2 mask, exhibited I_{sc} of 42.57 mA, V_{oc} of 0.178 V, *FF* of 25%, and efficiency of 0.061%. This approach did not require the patterning of the back

FIGURE 7.9
(a) Processing of a kapton/Cu/Ti substrate, where dark gray coating band is Ti on the kapton/Cu substrate, (b) roll-to-roll lamination of the modules, and (c) the modules cut into sheets and final cells. (Adapted with permission from F. C. Krebs, 2009. *Sol. Energy Mater. Sol. Cells* 93: 1636–1641.)

electrode and other layers except the final top grid Ag contact. This simplifies the process and increased the possibility for implementation. In the case of inverted geometry, the bottom electrode was a layer of silver nanoparticles on a PEN substrate [15]. Subsequently ZnO, P3HT:PCBM, and PEDOT:PSS layers were applied by slot-die coating, and the final electrode was applied by screen printing in a grid structure. These devices exhibited a typical efficiency of ~0.3%.

On the basis of current performance and manufacturing cost of polymer solar cells, Nielsen et al. presented a review on their business potential and market opportunities [16]. Although the business potential of polymer solar cells is large, the current performance leaves little room for competition with other thin-film technologies. Analysis shows that the annual market could be in the range of $500 million by 2018. However, PCE, operational lifetime, and cost would all need to be much better before a significant market share can be expected. An overview on the patents and an intellectual property rights analysis found that aside from Europe, United States, and the Far East, there are regions including South America, Africa, Middle East, Central Asia, and Australia that are open for polymer solar cells, and there is significant chance that many academic and industry players may create small patent pockets. Polymer solar cell technology has developed remarkably and essentially came from laboratories to several commercial companies including Konarka, Heliatek, Plextronics, Solarmer, and Mitsubishi [17–20]. Konarka, in the United States, was the first company to launch polymer solar modules in market in 2009, where the modules were developed and sold among the primary designers, product developers, and partners of Konarka. Unfortunately, in 2012 Konarka went bankrupt and other companies like Heliatek and Solarmer became the leaders.

Krebs et al. demonstrated the versatility of polymer solar cell technology using flexible substrates, complete solution processing in air using screen printing, and simple mechanical encapsulation [20]. They printed the solar cells in many square meters and demonstrated the processing of inks, processing of devices, processing speed, processing cost, materials consumption, sacrifice in the performance going from laboratory to mass production, module performance, and their application by public demonstration. They demonstrated that the production technique is scalable and well suited for the direct transfer to the printing industry employing existing equipment. They demonstrated the operation of small radio and many other electronic gadgets using R2R polymer solar cells. Krebs et al. fabricated R2R solar modules based on P3HT:PCBM and integrated them into a flexible electronic product. They demonstrated the capacity of polymer solar cells in terms of complete manufacturing of a small lamp using flexible electronic components [17]. The solar modules were used to charge a lithium ion battery through a blocking diode. Krebs et al. [19] manufactured a lamp for illumination purposes using fully R2R-processed semitransparent flexible polymer solar cell modules, which were printed on a silver-based circuit, and on the back side, integrated white LEDs, blocking diodes, a lithium ion battery, vias, and button contacts. When the solar module is illuminated from either side, it charges the battery, and while operating the battery gets discharged through LEDs, illuminating the surface in front of them. They prepared different lamps using the same solar cell and circuitry by employing more LEDs and batteries. They developed prototype lamps and subjected them to field tests in Zambia. The experiences gained in immediate response and spontaneous handling by a layman was also shared in view of further developments [19]. Figure 7.10a shows the solar hat comprising a polymer solar module, an FM radio, ear plug, and neck strap. Figure 7.10b shows a large parasol of 8 segments with each having 30–33 polymer solar modules prepared by Krebs et al. that could be connected in series or parallel [20].

(a) (b)

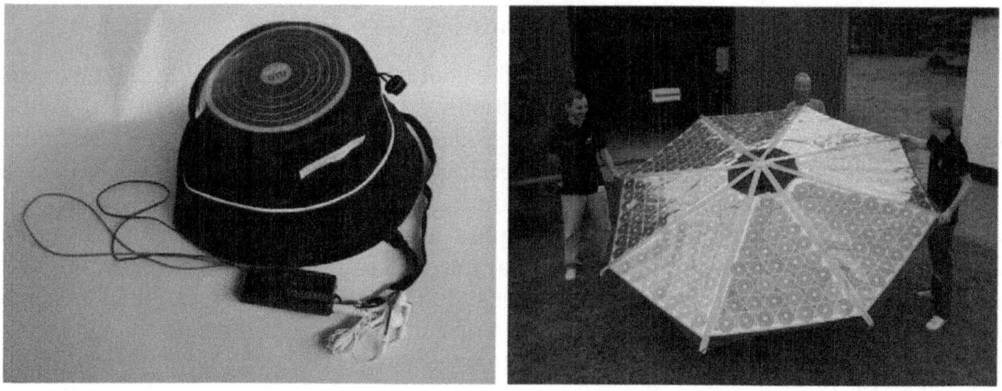

FIGURE 7.10
(a) Solar hat comprising a polymer solar cell, FM radio, ear plugs, and neck strap. (b) A large-area parasol having 8 segments with each having 30–33 flexible polymer solar modules, prepared by Krebs et al. (Adapted with permission from F. C. Krebs et al., 2009. *Sol. Energy Mater. Sol. Cells* 93: 422–441.)

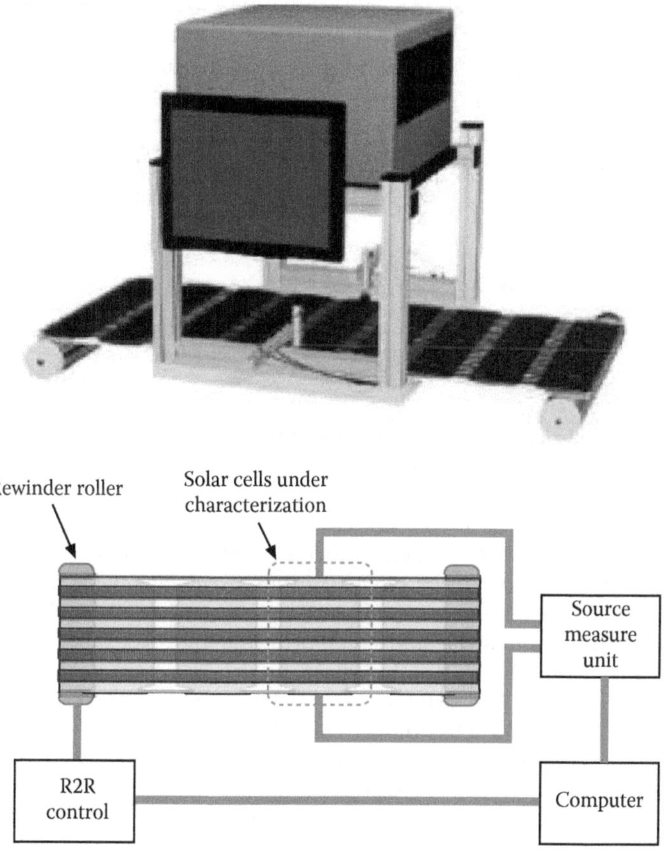

FIGURE 7.11
Schematic of the R2R characterization setup. (Reprinted with permission from Alstrup et al., Ultra fast and parsimonious materials screening for polymer solar cells using differentially pumped slot-die coating, *ACS Appl. Mater. Inter.* 2, 2819–2827. Copyright 2010 American Chemical Society.)

7.4 Characterization of R2R Solar Modules

For characterization the roll is mounted on the characterization system. A R2R characterization system possesses an unwinder, positioning camera, video camera, solar simulator with source meter unit, transport rollers, pneumatic contact pads, vacuum table with connections, tensioning rollers, and rewinder. It is a computer controlled, fully automatic system. The modules are placed on the vacuum table where contact strips on the modules make contacts with connections on the table that are fixed by pneumatic pads. A printed solar cell roll possesses several cells, for example, the roll prepared by Alstrup et al. [8] had 6 strips and each strip comprised 200 cells; this way there was a total of 1200 cells on the roll. Because of the limited illumination area, such a large number of solar cells cannot be characterized simultaneously, instead they are characterized in groups in different segments. While preparation, the roll is marked at several positions and the measuring position of the roll is computer controlled through these marks. During measurements, the cells are forwarded to the illuminated area and the measuring position is computer controlled by the marks on the roll. A R2R characterization system is shown schematically in Figure 7.11. The current–voltage (*I–V*) characteristics are recorded using a solar simulator and source meter unit. It has the provision of annealing the device and can have multiple curves of the device with predefined periods of time. It analyzes all the cells on the module very quickly, and the cells behaving abnormally can be detected easily. To avoid contributions from neighboring cells, the roll is masked and only the active area to be measured is illuminated. After measurement of one set of solar cells, the contacts are broken and the roll is moved to the next position for measurement of the next set of solar cells. As the roll moves it is unwinded by one roller on one side and rewinded by the other on the other side.

References

1. F. C. Krebs, 2009. Fabrication and processing of polymer solar cells: A review of printing and coating techniques, *Sol. Energy Mater. Sol. Cells* 93: 394–412.
2. F. C. Krebs, H. Spanggard, T. Kjær, M. Biancardo, and J. Alstrup, 2007. Large area plastic solar cell modules, *Mater. Sci. Eng. B* 138: 106–111.
3. S. E. Shaheen, R. Radspinner, N. Peyghambarian, and G. E. Jabbour, 2001. Fabrication of bulk heterojunction plastic solar cells by screen printing, *Appl. Phys. Lett.* 79: 2996–2998.
4. F. C. Krebs, J. Alstrup, H. Spanggaard, K. Larsen, and E. Kold, 2004. Production of large-area polymer solar cells by industrial silk screen printing, lifetime considerations and lamination with polyethyleneterephthalate, *Sol. Energy Mater. Sol. Cells* 83: 293–300.
5. T. Aernouts, P. Vanlaeke, J. Poortmans, and P. Heremans, 2004. Polymer solar cells: Screen printing as a novel deposition technique, *Proc. SPIE* 5464: 252.
6. F. C. Krebs, 2009. Polymer solar cell modules prepared using roll-to-roll methods: Knife-over-edge coating, slot-die coating and screen printing, *Sol. Energy Mater. Sol. Cells* 93: 465–475.
7. F. C. Krebs, T. Tromholt, and M. Jørgensen, 2010. Upscaling of polymer solar cell fabrication using full roll-to-roll processing, *Nanoscale* 2: 873–886.
8. J. Alstrup, M. Jørgensen, A. J. Medford, and F. C. Krebs, 2010. Ultra fast and parsimonious materials screening for polymer solar cells using differentially pumped slot-die coating, *ACS Appl. Mater. Inter.* 2: 2819–2827.

9. F. C. Krebs, S. A. Gevorgyan, and J. Alstrup, 2009. A roll-to-roll process to flexible polymer solar cells: Model studies, manufacture and operational stability studies, *J. Mater. Chem.* 19: 5442–5451.
10. G. Dennler, M. C. Scharber, and C. J. Brabec, 2009. Polymer–fullerene bulk-heterojunction solar cells, *Adv. Mater.* 21: 1323.
11. E. Bundgaard, O. Hagemann, M. Manceau, M. Jørgensen, and F. C. Krebs, 2010. Low band gap polymers for roll-to-roll coated polymer solar cells, *Macromolecules* 43: 8115–8120.
12. M. R. Lilliedal, A. J. Medford, M. V. Madsen, K. Norrman, and F. C. Krebs, 2010. The effect of post-processing treatments on inflection points in current–voltage curves of roll-to-roll processed polymer photovoltaics, *Sol. Energy Mater. Sol. Cells* 94: 2018–2031.
13. F. C. Krebs, S. A. Gevorgyan, B. Gholamkhass, S. Holdcroft, C. Schlenker, M. E. Thompson, B. C. Thompson et al., 2009. A round robin study of flexible large-area roll-to-roll processed polymer solar cell modules, *Sol. Energy Mater. Sol. Cells* 93: 1968–1977.
14. F. C. Krebs, 2009. Roll-to-roll fabrication of monolithic large-area polymer solar cells free from indium tin oxide, *Sol. Energy Mater. Sol. Cells* 93: 1636–1641.
15. F. C. Krebs, 2009. All solution roll-to-roll processed polymer solar cells free from indium tin oxide and vacuum coating steps, *Org. Electron.* 10: 761–768.
16. T. D. Nielsen, C. Cruickshank, S. Foged, J. Thorsen, and F. C. Krebs, 2010. Business, market and intellectual property analysis of polymer solar cells, *Sol. Energy Mater. Sol. Cells* 94: 1553–1571.
17. F. C. Krebs, J. Fyenbob, and M. Jørgensen, 2010. Product integration of compact roll-to-roll processed polymer solar cell modules: Methods and manufacture using flexographic printing, slot-die coating and rotary screen printing, *J. Mater. Chem.* 20: 8994–9001.
18. A. J. Medford, M. R. Lilliedal, M. Jørgensen, D. Aarø, H. Pakalski, J. Fyenbo, and F. C. Krebs, 2010. Grid-connected polymer solar panels: Initial considerations of cost, lifetime, and practicality, *Opt. Exp.* 18: A272–A285.
19. F. C. Krebs, T. D. Nielsen, J. Fyenbo, M. Wadstrøm, and M. S. Pedersen, 2010. Manufacture, integration and demonstration of polymer solar cells in a lamp for the "Lighting Africa" initiative, *Energy Environ. Sci.* 3: 512–525.
20. F. C. Krebs, M. Jørgensen, K. Norrman, O. Hagemann, J. Alstrup, T. D. Nielsen, J. Fyenbo, K. Larsen, and J. A. Kristensen, 2009. Complete process for production of flexible large area polymer solar cells entirely using screen printing: First public demonstration, *Sol. Energy Mater. Sol. Cells* 93: 422–441.

8

Cost Analysis, Technological Impact, Challenges, and Outlook

8.1 Introduction

Photovoltaic (PV) technology is quite futuristic for clean and long-term energy production, but unfortunately the present cost of traditional PV technologies is not low enough for their widespread and frequent use for energy generation. The cost of different PV technologies is determined on the basis of their per watt energy generation that comes out to be different for different PV technologies. The manufacturing process and efficiency determine the cost of a PV technology. The manufacturing of PV devices and panels requires lots of energy, which is mostly derived from nonrenewable sources of energy. Therefore, the cost of a PV technology in a way also depends on the cost of fossil fuels. The dependence of PV production on fossil fuels makes them not 100% clean, however, the subsequent use of solar cells will cause no harmful effect on the climate. The overall damage done to the environment by PV technologies is negligible compared to that by fossil fuels. Once the energy production cost of solar cells comes down below that of fossil fuels, it will reduce our dependence on fossil fuels. The energy required for production of PV cells and modules could be drawn directly from solar PV plants that would eliminate any harm done to the environment by production of PV panels.

In general, the module cost is determined by raw material cost, cell processing, manufacturing, and module assembly cost. The economic assessment of a PV module is shown schematically in Figure 8.1. The breakdown of average worldwide PV module price of different PV technologies through cost analysis done by the International Renewable Energy Agency (IRENA) in 2012 is shown in Figure 8.2. The absolute cost of a PV module varies from technology to technology, and the accurate global average price of PV modules is difficult to obtain as the cost depends on various parameters like expenditure by manufacturer, module efficiency, and the market features. The installation cost includes site preparation, module array preparation, inverter, transformer, battery, wiring, and other installation parameters. The total installed cost of PV modules varies place to place because of, for example, variation in module manufacturing cost, local labor charges, and the local market. After multijunction concentrated PV technology, conventional crystalline silicon (c-Si) PV technology is the most expensive but also has the maximum commercial efficiency. The cost of solar power production from conventional c-Si PV devices in the late 1970s was ~\$70/W_p capacity, which has now come down to less than \$1/W_p capacity as a result of accelerated research and development. Although the cost of solar power production has dropped about 100 times, it is still higher than the cost of fossil fuels; therefore PV technologies require government subsidies to be in the market. For PV technology to be

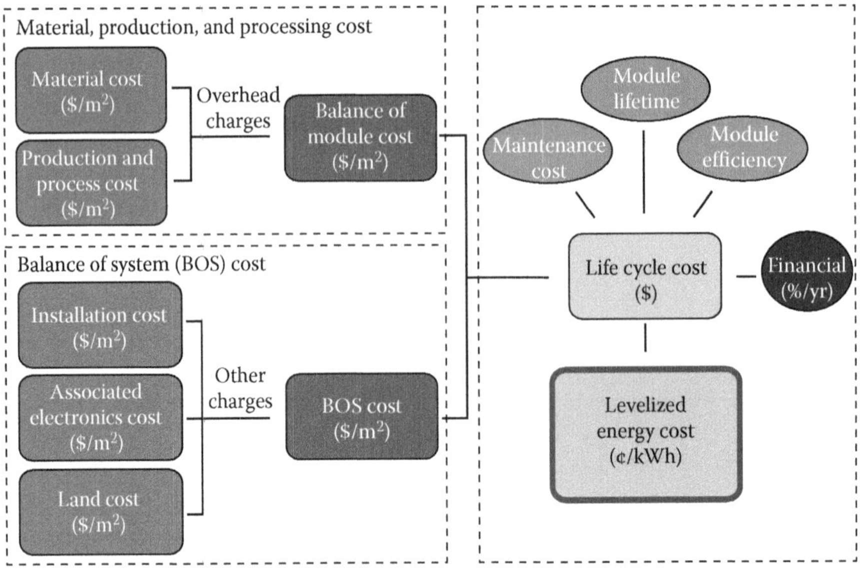

FIGURE 8.1
Economic assessment of electricity cost from PV modules.

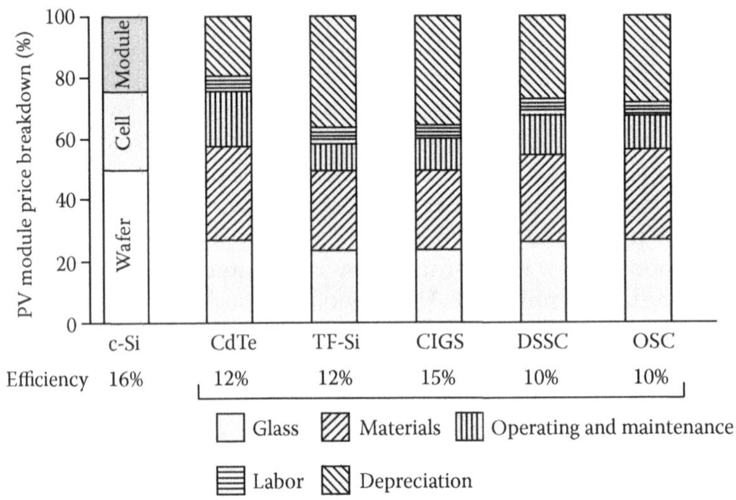

FIGURE 8.2
Average worldwide PV module price breakdown for different PV technologies. (Data from International Renewable Energy Agency (IRENA), June 2012. *Renewable Energy Technologies: Cost Analysis Series*, 1, issue 4/5.)

a free-standing technology in the marketplace without any subsidy, the cost should come down to less than \$0.20/W_p capacity, which is expected to be realized in the near future. As far as organic solar cells (OSCs) are concerned, they incorporate inexpensive materials with great abundance of raw materials. They have lower material and energy consumption and can be processed at high speed via roll-to-roll (R2R) processing, which makes them much more cost effective than conventional PV technologies.

8.2 Estimation of Manufacturing Costs of Organic Solar Cells (OSCs)

Because OSC technology is quite young and still in the research and development (R&D) phase, estimation of its correct cost is very difficult. However, I discuss here some of the methods adopted to calculate the approximate cost of small molecular and polymeric solar cell modules. In 2009, Kalowekamo and Baker made an estimation on the manufacturing cost of OSCs based on the small molecular active materials CuPc, SnPc, and C_{60} [1]. Thin films of small molecular materials are prepared by thermal evaporation in a vacuum chamber. Compared to large-area R2R printing, thermal evaporation limits the area of solar cells and is comparatively more expensive. Yet considering 5% efficiency and 5 years lifetime, the module cost was calculated to be between \$1.00 and \$2.83/W_p with the levelized electricity cost (LEC) between \$0.49 and \$0.85/kWh [1]. However, for the solar cells having 15% efficiency and 20 years of lifetime, the LEC was calculated to be between \$0.07 and \$0.13/kWh. The substrate cost and cell efficiency were found to be the most sensitive parameters in deciding the module cost. The module cost is calculated by dividing the manufacturing cost per square meter by the output power from 1 m^2 of the module. For calculation of module cost, Kalowekamo and Baker calculated the materials cost, processing, and other overhead charges separately [1].

Depending upon the material sources, the estimated cost would vary place to place, company to company, and time to time. For material cost, Kalowekamo and Baker considered that the film thickness was 1 μ, and in 1 m^2 area the volume of material would be 1 cm^3, and the film had about 1 g/cm^3 material. Considering only 75% of material is deposited as film and the remaining 25% is wasted, ~1.3 g of material would be required for a 1 m^2 solar module [1]. Considering CuPc and SnPc costs \$1.00/g and that of C_{60} to be \$0.50/g, the material cost was found to be \$3.3/m^2 and the maximum about \$5.00/m^2. The cost of electrical contacts and interconnections was estimated to be \$3.42/m^2 and the maximum about \$5.00/m^2. The substrate was considered to be flexible plastic with \$5.00/m^2 coated with ITO (\$2.40/g). Considering 1 g of ITO is coated on 1 m^2 of plastic with a coating rate of \$0.50/m^2, the final rate of ITO-coated plastic came out to be \$7.90/m^2, with a maximum \$13.68/m^2. Based on previous estimates for dye-sensitized solar cells (DSSCs) [2], the cost of flexible protective cover and sealant was considered to be from \$2.90 to \$4.40/m^2. Similarly the packing cost was estimated from \$2.00 to \$3.00/m^2. The cost of specialty chemicals to prevent degradation was estimated to be from \$1.00 to \$2.00/m^2. The total material cost was estimated as \$23.40 to \$37.50/m^2, where substrate cost had the largest share. For processing cost estimation, the cost of capital equipment was assumed to be in the range of \$14.5 million to \$43.5 million with 10 MW of annual production capacity of 6.5% efficient solar cells. For 1000 W/m^2 modules, the annual production came out to be ~150,000 m^2. Considering the life of a production plant to be 7 years, with 10% annual interest the initial capital investment came out to be \$2.98M to \$8.94M/year, which gave the equipment cost per square meter of module production between \$19.86 and \$59.57/m^2. Based on previous calculations for minimum equipment cost for DSSCs [2–4], the minimum equipment cost for OSCs was \$3.30/m^2. Considering 8 h per day, 350 working days per year, a wage rate \$17/h per person, the labor cost for 30 people involved in production was calculated to be \$10/m^2. The minimum labor cost was considered to be \$6.00/m^2. Estimates of overhead charges for facilities, utilities, equipment maintenance, customer warranty and miscellaneous were taken as \$13.78 to \$25.21/m^2. The total manufacturing cost was calculated to be from \$46.48 to \$132.29/m^2. Considering 95% cell yield and 98% module yield, the module cost was estimated from

$$\text{Module cost } (\$/W_p) = \frac{\text{Manufacturing cost} \times 1.05 \times 1.02}{\text{Output power per unit area}}. \tag{8.1}$$

With 5% module efficiency, the output power per unit area came out to be 1000 $W_p/m^2 \times 5\%$. The calculation gives the module cost from \$1.00 to \$2.83/W_p [1]. The cost sensitivity calculations showed that the module cost was very sensitive to cell efficiency and substrate cost, whereas machine lifetime and interest rates were not that important.

Kalowekamo and Baker [1] also estimated the LEC, which along with the module cost also depends upon the module lifetime and balance of systems (BOS) cost. The BOS cost depends upon a number of parameters, including land cost, support structure (ground base, rooftop, building integrated), transportation, wiring, installation, and lighting conditions. It is, therefore, difficult to estimate accurate value. Considering the BOS cost to be \$75.00/m^2 and adding it to the manufacturing cost, the installed capital cost (ICC) was found to be between \$2.48 and \$4.28/W_p and the LEC was calculated from the formula

$$\text{LEC} = \frac{(\text{ICC} \times 1000 \times \text{CRF})}{(\text{CF} \times 8760)} + \text{Operating and maintenance cost}, \tag{8.2}$$

where capital recovery factor (CRF) = $\{i \times (i+1)^n\}/\{(i+1)^n - 1\}$, where i is the discount rate and n is the lifetime of PV module. The capital factor (CF) = (1700 × 0.8)/8760, for a place where solar illumination is 1700 kWh/m^2-year, and total output is 80% of peak power rating due to temperature, resistance, and power conditions. Here 8760 accounts for the total hours in a year. For the module of 5% efficiency and 5 years of lifetime, the LEC was calculated to be between \$0.49 and \$0.85/kWh; whereas for 15% efficiency and 20 years of lifetime, the LEC was calculated to be between \$0.07 and \$0.13/kWh [1], which is about 4 times lower than the conventional Si PV technology. Figure 8.3 shows the variation in LEC as a function of lifetime and efficiency of small molecular OSCs, calculated by Kalowekamo and Baker [1].

Azzopardi et al. estimated the manufacturing cost of R2R-processed OSCs modules from ProcessOne, and for a 5% efficient module the cost was estimated to be in the range €63–192/m^2 with electricity cost of €1.26–3.84/W_p [5]. The ProcessOne modules were made

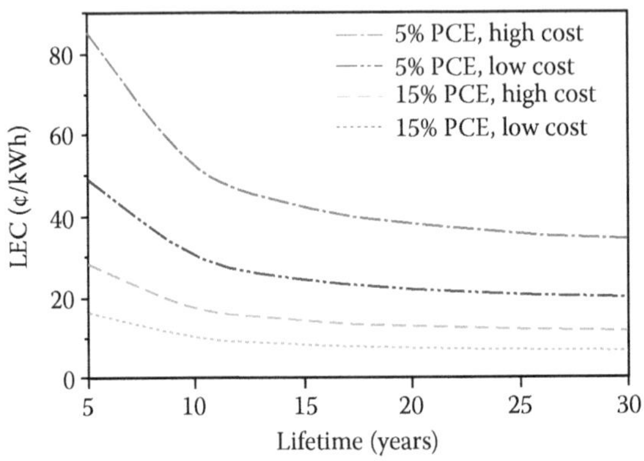

FIGURE 8.3
Variation in low and high LECs as a function of lifetime for small molecular OSCs modules calculated by Kalowekamo and Baker for 5% and 15% efficiencies. (Redrawn from J. Kalowekamo and E. Baker, 2009. *Sol. Energy* 83: 1224–1231.)

of ITO-coated PET substrates with sequentially deposited solution-processable ZnO, P3HT:PCBM, PEDOT:PSS thin films, and Ag ink on the top as back electrode. For estimation of module cost, first the materials cost (per m^2 of module area) was calculated considering the material price from three different material sources. The patterned ITO-coated PET accounted for €15.32–79.24/m^2, the materials required in coating of ZnO thin films accounted for €0.07–1.08/m^2, the materials required in coating of P3HT:PCBM thin film accounted for €8.36–42.06/m^2, the materials required in coating of PEDOT:PSS thin film accounted for €4.3–12.57/m^2, the materials required in coating of Ag ink accounted for €7.67/m^2, and finally the encapsulation with 3M 467 MPF and PET accounted for €5–12.08/m^2, resulting in the total material cost of €40.72–154.7/m^2. PET/ITO had the maximum share in total cost and was followed by active layer materials and then by Ag ink. The production and process cost was calculated that included the cost of electricity used in manufacturing, equipment and infrastructure cost, labor cost, and overhead charges. The cost of electricity was calculated by the energy consumed in each process multiplied by the minimum and maximum electricity cost and was calculated to be between €0.49 and €1.08/m^2. The total investment and infrastructure cost including a R2R screen printer, R2R slot-die coater, R2R etching machine, R2R laminator, R2R characterization, R2R sheeter, and contactor was calculated to be €530,250. Such a manufacturing plant has around 20,000 m^2/year production capacity with 95% of module yield. Now considering the machine lifetime of 10 years, CRF of 0.14 with 7% interest rate, levelized capital cost of €75,495.67/year and maintenance cost of €3019.83/year, the levelized investment cost per m^2 was calculated to be €3.93/m^2. For estimation of labor cost, the labor time was multiplied with labor cost per unit time for each step and the total labor cost was estimated to be €5.32/m^2. The overhead cost for facilities, customer warranties, and miscellaneous was estimated to be €6.97–9.43/m^2. As a result, the total module cost was estimated as €57.42–174.45/m^2, and with 10% of overhead margins, the balance of module (BOM) cost was estimated as €63.16–191.89/m^2, which corresponded to €1.26–3.84/W_p for 5% of module efficiency. For comparison with conventional PV systems the BOS cost was calculated for a 1 kWp, grid connected, roof-mounted system with a lifetime of 25 years. The BOS cost included associated electronics cost (e.g., inverter, metering, and switchgear), the installation cost, and other costs such as maintenance and the total BOS cost ranged between €1281.81 and €2484.76/kW_p.

For estimation of electricity cost for OSCs modules, Azzopardi et al. estimated the life cycle cost (LCC) first [5]. LCC is actually the cost of a PV system per unit electricity yield over its lifetime. It is the sum of initial investment plus future investment with possible discounts minus the value of unused module components at the end of its life. Module efficiency, module lifetime, interest rate, and the active area on the module control final cost of the system. For 5% module efficiency and 5-year module life, the module cost had a major share for system cost, which could be reduced by increasing the module efficiency and module active area. The energy output of the OSC system was calculated from [5]

$$E_{PV} = \sum_{t=0}^{T-1} H_{yr}\eta_{PV}\delta_{avg}A_m\alpha PR, \tag{8.3}$$

where H_{yr} is annual radiation, η_{PV} is the efficiency of active area, δ_{avg} is the average linear degradation, A_m is the module area, α is geometric fill factor (GFF) defined as the ratio of active area on the module over the module area, PR is the performance ratio, t is the time, and T is the system lifetime. Considering H_{yr} to be 1700 kWh/m^2, δ_{avg} to be 80%, α to be

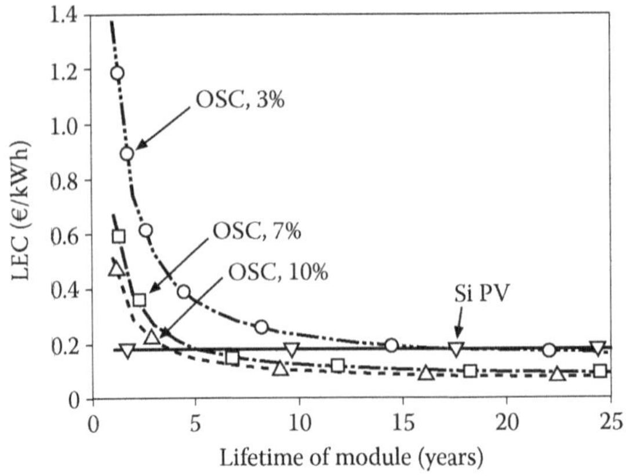

FIGURE 8.4
Compared to Si PV technology, the LEC of OSC technology as a function of module lifetime for different efficiencies. (Redrawn from B. Azzopardi et al., 2011. *Energy Environ. Sci.* 4: 3741–3753.)

67%, and PR to be 85%, the total system output power over a 25-year system lifetime for 7% efficiency was calculated to be 71,114 kWh. The LEC could be calculated from the LCC and energy output of the system. For 5 years of module lifetime and 7% of cell efficiency, the LEC was estimated as €0.19–0.50/kWh. LEC from OSCs modules strongly depends on module efficiency and module lifetime. Azzopardi et al. calculated LEC for OSCs systems for different module efficiency and module lifetime, and the results are shown in Figure 8.4, which also shows the LEC for wafer-based Si PV technology. An organic PV module of 3% efficiency would require 15 years of lifetime to compete with Si PV technology. However, the competitiveness with Si PV could be achieved for module lifetime of less than 4 years with 10% of module efficiency [5]. Figure 8.5 compares the minimum and maximum estimated LEC's of OSCs with some of other sources of energy.

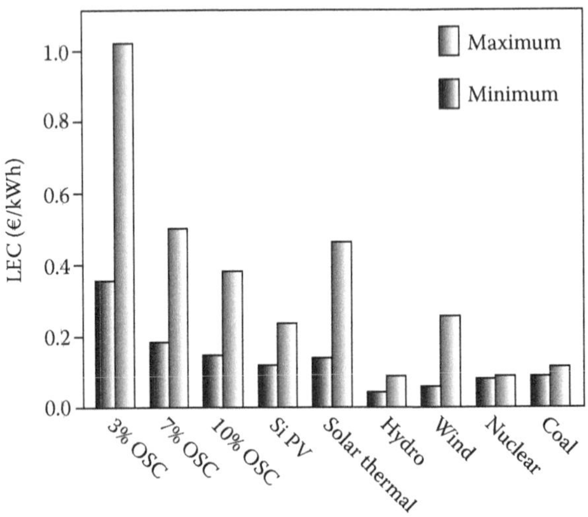

FIGURE 8.5
Comparison of electricity cost from different sources. (Redrawn from B. Azzopardi et al., 2011. *Energy Environ. Sci.* 4: 3741–3753.)

TABLE 8.1

Cost Estimates of Organic Photovoltaic Modules Based on Different Materials and Processing Techniques

Cell Structure	Processing	Module Cost	Electricity Cost	Reference
PET/ITO/CuPc/C_{60}/Al	Thermal evaporation	\$50–140/m²	\$1.0–\$2.83/W_p (@5% PCE)	1
Glass/ITO/PEDOT:PSS/P3HT:PCBM/LiF/Al	Inkjet printing and thermal evaporation	€180/m² (~\$204/m²)	€4.18/W_p (~\$4.7/W_p) (@5% PCE)	14
PET/ITO/ZnO/P3HT:PCBM/PEDOT:PSS/Ag	Roll-to-roll, using slot-die coating and screen printing	€89/m² (~\$100/m²)	€8.0/W_p (~\$9/W_p)	15
PET/ITO/ZnO/P3HT:PCBM/PEDOT:PSS/Ag	Roll-to-roll, using slot-die coating and screen printing	€63–192/m² (~\$71–217/m²)	€1.26–3.84/W_p (\$1.43–4.35/W_p) (@5% PCE)	5
PET/Ag/PEDOT:PSS/ZnO/P3HT:PCBM/PEDOT:PSS/Ag	Inkjet printing and screen printing	€28.63/m² (~\$33/m²)	€4.28/W_p (\$4.8/W_p) (@1.82% PCE)	10
PET/Ag/PEDOT:PSS/P3HT:PCBM/Al	Inkjet printing and thermal evaporation	\$5.5–10.2/m²	—	6

There have been a number of other reviews on estimation of cost of OSCs modules and a brief comparison of different estimates is shown in Table 8.1. These are rough estimates and future development in technology, and scale-up in production would affect the LEC of OSC systems. As the PET/ITO substrate has maximum share on the total cost, Dastoor et al. recently performed analysis to determine the projected cost of mass-manufactured ITO-free OSC modules, and their calculations showed that the OSC module cost would be considerably lower than the earlier costs estimations [6]. They calculated that the cost of OSCs with lifetimes as low as 3 years and efficiency of just 2% can produce electricity at ~27 Australian cents/kWh. Therefore, efforts are being devoted to develop ITO-free OSCs [7–9]. For this purpose, the Flextrode substrates, which are completely R2R-processed low-cost transparent electrodes, are used as a convincing alternate of expensive ITO. The concept of Flextrode substrate has been employed for both single bulk-heterojunction (BHJ) and tandem OSCs, and the large-area ITO-free flexible single BHJ solar cells have shown PCE of 3.5%, whereas tandem solar cells have demonstrated PCE up to 2.2% [7–9]. Machui et al. performed the cost analysis of such fully encapsulated, ITO-free and vacuum-free R2R-processed single BHJ and tandem organic solar modules [10]. They presented the analysis for three cases—current status, up-scaling, and industrial and based on real data of R2R processing and scale-up effects, it was shown that with optimized processes and projections to medium- and large-scale scenarios costs below \$0.05/W_p are achievable. The cost calculation for current status was done in kW production regime, for up-scaling the cost calculation was done in MW production regime, and for the industrial scenario the calculations were made in GW production regime. For single BHJ solar modules P3HT:PCBM (low cost and low efficiency) and MH301:PCBM (high cost and high efficiency) were used as active layers, whereas for low-cost tandem modules both subcells were made of P3HT:PCBM and high-cost tandem modules were made of MH301:PCBM and MH306:PCBM active layers [11,12]. Figure 8.6 shows the present cost structure of an ITO-free organic solar module and the inset shows the schematic structure of the fully R2R-processed OSCs. Barrier foils and the Ag electrode were observed to have the biggest share of total cost. The analysis was performed based on the real process data and the

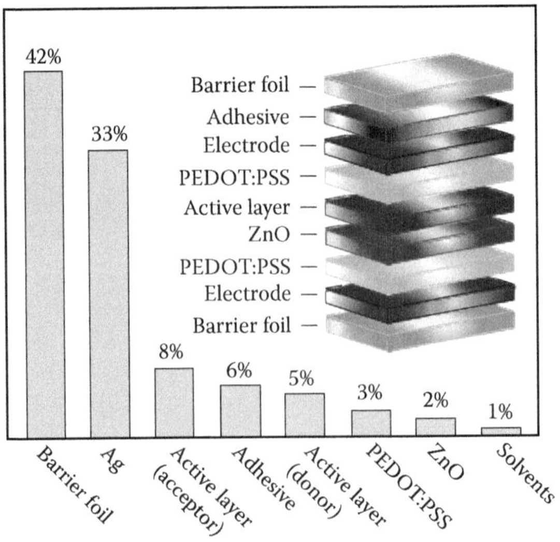

FIGURE 8.6
Cost structure of ITO-free fully-R2R processed OSC module. Inset shows the layer structure of R2R-compatible single BHJ OSC module analyzed. (Data from F. Machui et al., 2014. *Energy Environ. Sci.* 7: 2792.)

estimated cost was nearly exact. The material cost was calculated to be €28.63/m^2, and for 37% of GFF and 1.82% of module efficiency, the electricity cost for a single BHJ ITO-free OSC module was calculated to be €4.28/W$_p$ at 1000 W/m^2 illumination [10,13]. At laboratory scale, however, material costs dominate the overall cost but processing costs had a significant share. However, upon up-scaling, the processing cost decreases to insignificant values. The overall cost of a P3HT:PCBM single BHJ and tandem module came out to be €28.63 and €34.56/m^2, respectively, whereas for the MH polymer-based single BHJ and tandem module, the overall cost ended up as €42.81 and €61.71/m^2, respectively [10]. The electricity cost with this range of module costs came out to be €0.8–8.4/W$_p$ for PCE ranging from 2% to 10%. To be competitive with the mature PV technologies, the cost of OSC modules should be much lower than these values, and it could be achieved at up-scaling and industrial levels.

GFF has a crucial effect on overall cost of the module, and a dramatic cost reduction is expected with an increase in GFF. Figure 8.7 shows the effect of GFF on the single BHJ ITO-free OSC module costing €28.63/m^2. For up-scaling and industrial level, the GFF would be more than that at laboratory scale. Machui et al. considered the GFF to be 75% and 98% for up-scaling and industrial scenarios where production capabilities were 100 MW and 100 GW, respectively. Due to higher production capabilities, the material costs and material consumptions were considered to be much lower than that in the current scenario. For the MH polymer-based tandem module, the overall cost at the up-scaling scenario was calculated to be €12.81/m^2 with Ag electrodes. Replacement of Ag with Al was proposed to reduce the cost up to 77%, whereas use of carbon in place of Ag is expected to reduce the cost up to 90% [10]. The inset in Figure 8.7 shows the effect of carbon electrode compared to Ag electrode on the cost of single BHJ OSC modules at up-scaling and industrial scenarios. The electricity cost on up-scaling the low-cost and high-cost single BHJ and tandem OSC modules for 2%–10% efficiency ranges between €0.08 and €0.9/W$_p$. With process optimization at industrial scale, the electricity cost is

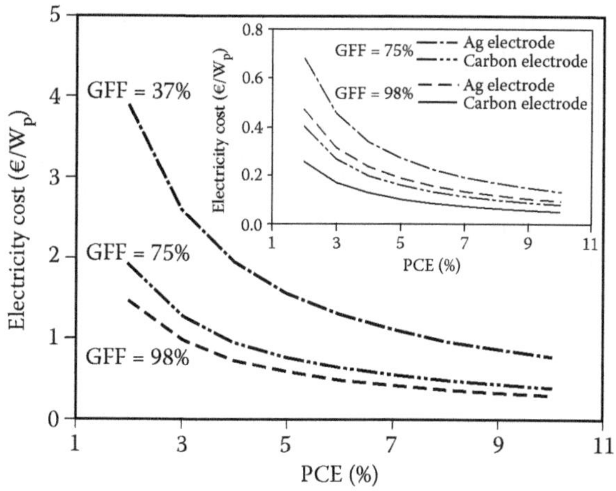

FIGURE 8.7
Effect of GFF, corresponding to current status, up-scaling, and industrial scenarios on the overall cost of single BHJ P3HT:PCBM-based OSC modules. Inset shows the effect of carbon electrode on the overall cost at up-scaling (GFF = 75%) and industrial (GFF = 98%) scenarios. (Data from F. Machui et al., 2014. *Energy Environ. Sci.* 7: 2792.)

expected to be in the range €0.05–0.6/Wp. Most important for all scenarios, the 10% efficient modules are shown to produce electricity at below €0.15/W_p, which shows that the OSC is a competitive technology that would outperform conventional PV and other renewable energy technologies. It gives a boost to rapid development in OSC technology.

8.3 Technological Impact and Outlook

The Si PV modules are quite robust and the manufacturer warranty covers a lifetime of over 20 years. A typical warranty includes 90% of the rated output for the first 10 years and then 80% of the rated output for the next 10 years. The efficiency of commercially available c-Si solar modules ranges from 14% to 19%, whereas the module efficiency of poly-c Si solar modules ranges from 13% to 15%. The module efficiency of a-Si PV ranges from 5% to 8%, whereas that of CdTe and CIGS solar modules ranges from 7% to 11%. On the other hand, the DSSC and OSC module efficiencies range between 4% and 5%. c-Si and poly-Si PV technologies are quite mature and are produced on a large scale, whereas a-Si, CIGS, and CdTe PV technologies are cheaper but are in early development and are produced on small scale. Due to the high cost of conventional PV technologies, the interest is shifting towards DSSCs, OSCs, and perovskite solar cells, which are expected to be highly cost effective when processed R2R on plastic base. DSSCs, OSCs, and perovskite solar cells are still in the R&D phase, and efforts are being made to make them commercially viable.

The potential uses of OSC modules include building integration, mobile phones, laptops, radio, power umbrella, power tent, power jacket, flashlights, toys, and several other hand-held electronic devices. OSC modules can be affixed anywhere to anything. The performance of OSCs has improved remarkably in the last decade but still is not high enough for commercialization. To compete with existing and other emerging PV technologies, OSC panels must demonstrate efficiency exceeding 10% with a lifetime more than 10 years and a manufacturing cost <$0.50/W$_p$. Where the efficiency of OSCs at lab scale has reached ~12%, efficiency in small modules with serially connected thin strips lies between 5% and 9%. For achieving higher efficiency on bigger modules with high stability, a number of technical challenges need to be overcome. Europe has set a target to achieve OSCs modules of <$0.50/W$_p$ by 2020 with a 60 GW global production capacity.

There are some areas that need significant development before we expect OSCs in the market. Substrates, metal electrodes, and encapsulants are the primary cost drivers for OSCs modules. For cost-effective production of OSCs modules, some of the areas where high priority research must be carried out are development of large-area monolithic modules, replacement of costly components like ITO with suitable low-cost alternatives, cost-effective encapsulant and barrier materials with high barrier for long-term stability, and mass production with minimum density defects. The manufacturing speed should be increased further with inline coatings. The operational lifetime needs to be improved significantly along with high efficiencies. The OSC market is expected to generate revenue from portable DC power, building integrated PV systems and conventional power. For commercial expansion the inverters and power electronics for grid connections are other areas that require research attention for timely implementation and full utilization of this technology. Multijunction OSCs have shown very high efficiency due to enhanced absorption and photon harvesting, but they are very complex and costly. Therefore, for cost-effective, efficient solar cells, some alternate strategies like donor–acceptor materials with complementary absorption spectra should be adopted. OSC modules have monolithic interconnection of cells and the area covering interconnections does not contribute to power conversion. Therefore, for higher output of the module GFF should be as high as possible and for this purpose gap between patterned ITO strips on the substrate should be as small as possible and also the gap between slot-die coated strips should also be as small as possible. Some of the area is taken up by the serial connections of the strips, which causes loss in module output. Therefore, the strip width and gap between them should be optimized by compromising the active area, device performance, and technical yield. OSC technology is highly scalable with high feasibility of manufacturing and up-scaling. Very large-area flexible modules have been successfully fabricated and demonstrated.

8.4 State of the Art, Challenges, and Opportunities

Despite OSCs being quite fascinating, there have been several questions raised about their success and feasibility for commercial viability. Will OSCs ever go beyond 10% PCE? Will they ever be as stable as their inorganic counterparts? Will they ever be able to give competition to conventional solar cells? Around one decade ago a P3HT and PCBM combination in BHJ structure resulted in around 5% PCE and it remained almost the same for many years. But now the situation has changed. There has been remarkable progress in the performance of OSCs in the last couple of years. There are several reports in the literature that now claim more than 10% PCE in both single and tandem OSCs [16–19]. Companies such as Mitsubishi Chemicals Corporation, Japan; Solermer Energy, United States; Heliatek, Germany; and Plextronics, United States, are the leaders in producing high-efficiency OSCs from both small molecular and polymeric materials. On April 27,

2012, Heliatek announced 10.7% PCE from tandem OSCs based on thermally evaporable small molecules. Then on January 16, 2013, Heliatek announced a record 12% PCE from small-molecular-based tandem OSCs [19].

Achievement of high efficiency in OSCs has been possible because of new donor–acceptor materials with better optical and electrical properties. Distributions of interpenetrating bulk heterojunctions and their morphologies had been the key in achieving high-efficiency solar cells. A lot of research in physical and chemical areas has been helpful in identifying the weak areas and overcoming them. Use of suitable low and high work function metal electrodes, suitable donor–acceptor combinations for improved photon harvesting, optical spacers, thin active layers, inverted structures, metal oxides, understanding the parameters affecting J_{sc}, V_{oc}, and *FF*, and controlling the nanoscale morphology are some of the examples that have led to improved performance in OSCs. Inverted structure of polymer solar cells has shown record-breaking PCE and very high stability in air without encapsulation. These solar cells are fully capable of high throughput R2R production via printing with completely vacuum-free processes, and show practical approach for mass production in future.

Printing is a highly automated, low-energy consuming, and low production-cost technique. Recent development of new processing techniques like inkjet printing, soft lithography and microcontact printing have increased the cost effectiveness of OSCs by allowing the fabrication of large solar cells on rigid and flexible substrates. New device architectures are being developed to overcome the problems related to scale-up of production. Control of nanoscale morphology of donor:acceptor blends toward best performance is an important fundamental issue. It is observed that there could be more than one morphology of active layers giving nearly the best performance. Recently, Liu et al. presented an approach to control the aggregates and morphology of polymer:fullerene systems [16]. They found that the 2-octyldodecyl alkyl chains on quaterthiophene in a polymer make the polymer aggregation highly temperature dependent, and it allows the polymers to be processed at elevated temperatures leading to control aggregation and high crystallization. For optimum aggregation behavior, the position and size of the alkyl chains play a very crucial role. This approach can be applied to multiple polymer:fullerene systems for optimization, and it would accelerate the materials and process development toward enhanced performance.

Device modeling is an important tool to achieve high-performance solar cells. For example, Yang et al. performed optical simulation and transfer matrix formalism (TMF) modeling toward solving the current matching problem between subcells in tandem devices [18]. The simulated results showed that the problem could be solved by fine-tuning the thickness of each subcell and interconnecting layers. The experimental data showed good agreement with simulated results, and a PCE of 11.55% was achieved in an optimizing triple-junction solar cell using P3HT, PTB7, and PDTP-DFBT polymers. There is still room for further improvement in the performance of OSCs, but it includes lots of challenges. Development of new donor and acceptor materials, their compatibility and processing toward maximum photon harvesting with energy levels having minimum energy loss, and excellent morphology with high charge carrier mobilities are the real challenges for achieving higher efficiency. Long-term stability and large-scale manufacturing are other challenges that need to be addressed to make OSCs commercially viable. Morphology of a film can be controlled by varying the processing conditions, changing the solvents and solution concentrations, changing the donor–acceptor combinations and their ratio, and by changing the deposition methods. The PCE of solar cells and modules on commercial scale are lower than those reported on lab level, and this difference is maximum and very large for OSCs [20]. Therefore, manufacturing of high-efficiency organic solar modules is very challenging. OSC modules are expected to be manufactured by printing techniques and the film morphology would depend on a number of parameters, including printing method, solvents used, drying conditions, morphology of different interfaces, and deposition of the top electrode. Encapsulation of R2R modules toward high stability is also very challenging.

OSC modules have opportunities of integration in a number of objects for energy generation and power supply. Some of such applications are toys, mobile phones, laptops, transportation, building integration, facades, windows, power plants, and off-grid remote sectors. OSCs have a unique feature in that they can efficiently harvest solar energy even at high temperatures and low light. OSCs offer truly transparent solar cells with tunable transparency for various applications. Heliatek recently developed OSCs having transparency of up to 40% with a PCE of over 7% [19]. By changing the color, thickness, shape, and size, the appearance of OSC modules can be made more esthetic. OSCs are expected to have a small energy payback time and low carbon footprint. Where energy payback time of a c-Si solar module is ~3 years, the energy payback time of organic solar modules would be around 4 months only. Similarly the carbon emission is expected to be ~9 g eq/kWh for organic modules, which is much lower than that for c-Si solar modules (~36 g eq/kWh). With their current performance, OSCs can still compete with some photovoltaic applications, however, market analysis shows that different photovoltaic technologies could be used together to meet energy demand. Toward commercial success, the issue of long-term stability in OSCs can be compensated by lower cost.

References

1. J. Kalowekamo and E. Baker, 2009. Estimating the manufacturing cost of purely organic solar cells, *Sol. Energy* 83: 1224–1231.
2. T. Meyer, 1996. Solid state nanocrystalline titanium oxide photovoltaic cells, Thesis, Ecole Polytechnique Federale de Lausanne.
3. R. G. Little and M. J. Nowlan, 1997. Crystalline silicon photovoltaics: The hurdle for thin films, *Prog. Photovolt.: Res. Appl.* 5: 309–315.
4. G. Smestad, 1994. Testing of dye-sensitized TiO_2 solar cells I & II, *Sol. Energy Mater. Sol. Cells* 32: 259–272.
5. B. Azzopardi, C. J. M. Emmott, A. Urbina, F. C. Krebs, J. Mutalea, and J. Nelson, 2011. Economic assessment of solar electricity production from organic-based photovoltaic modules in a domestic environment, *Energy Environ. Sci.* 4: 3741–3753.
6. C. J. Mulligan, M. Wilson, G. Bryant, B. Vaughan, X. Zhou, W. J. Belcher, and P. C. Dastoor, 2014. A projection of commercial-scale organic photovoltaic module costs, *Sol. Energy Mater. Sol. Cells* 120: 9–17.
7. J. E. Carle and F. C. Krebs, 2013. Technological status of organic photovoltaics (OPV), *Sol. Energy Mater. Sol. Cells,* 119: 309–310.
8. M. Helgesen, J. E. Carle, and F. C. Krebs, 2013. Slot-die coating of a high performance copolymer in a readily scalable roll process for polymer solar cells, *Adv. Energy Mater.* 3: 1664–1669.
9. T. R. Andersen, H. F. Dam, B. Andreasen, M. Hosel, M. V. Madsen, S. A Gevorgyan, R. R. Søndergaard, M. Jørgensen, and F. C. Krebs, 2014. A rational method for developing and testing stable flexible indium- and vacuum-free multilayer tandem polymer solar cells comprising up to twelve roll processed layers, *Sol. Energy Mater. Sol. Cells* 120: 735–743.

10. F. Machui, M. Hosel, N. Li, G. D. Spyropoulos, T. Ameri, R. R. Søndergaard, M. Jørgensen et al., 2014. Cost analysis of roll-to-roll fabricated ITO free single and tandem organic solar modules based on data from manufacture, *Energy Environ. Sci.* 7: 2792.
11. T. R. Andersen, H. F. Dam, M. Hosel, M. Helgesen, J. E. Carle, T. T. Larsen-Olsen, S. A. Gevorgyan et al., 2014. Scalable ambient atmosphere roll-to-roll manufacture of encapsulated large area, flexible organic tandem solar cell modules, *Energy Environ. Sci.* 7: 2925–2933.
12. J. E. Carle, M. Helgesen, M. V. Madsen, E. Bundgaard, and F. C. Krebs, 2014. Upscaling from single cells to modules-fabrication of vacuum- and ITO-free polymer solar cells on flexible substrates with long lifetime, *J. Mater. Chem. C* 2: 1290–1297.
13. F. C. Krebs, M. Hosel, M. Corazza, B. Roth, M. V. Madsen, S. A. Gevorgyan, R. R. Søndergaard, D. Karg, and M. Jørgensen, 2013. Freely available OPV: The fast way to progress, *Energy Technol.* 1: 378–381.
14. A. L. Roes, E. A. Alsema, K. Blok, and M. K. Patel, 2009. Ex-ante environmental and economic evaluation of polymer photovoltaics, *Prog. Photovoltaic.: Res. Appl.* 17: 372–393.
15. F. C. Krebs, T. Tromholt, and M. Jørgensen, 2010. Upscaling of polymer solar cell fabrication using full roll-to-roll processing, *Nanoscale* 2: 873–886.
16. Y. Liu, J. Zhao, Z. Li, C. Mu, W. Ma, H. Hu, K. Jiang, H. Lin, H. Ade, and H. Yan, 2014. Aggregation and morphology control enables multiple cases of high-efficiency polymer solar cells, *Nat. Commun.* 5: 5293.
17. J. You, L. Dou, K. Yoshimura, T. Kato, K. Ohya, T. Moriarty, K. Emery et al., 2013. A polymer tandem solar cell with 10.6% PCE, *Nat. Commun.* 4: 1446.
18. C. C. Chen, W. H. Chang, K. Yoshimura, K. Ohya, J. You, J. Gao, Z. Hong, and Y. Yang, 2014. An efficient triple-junction polymer solar cell having a power conversion efficiency exceeding 11%, *Adv. Mater.* 26: 5670–5677.
19. Heliatek, http://www.heliatek.com.
20. M. A. Green, K. Emery, Y. Hishikawa, W. Warta, and E. Dunlops, 2012. Solar cell efficiency tables (version 39), *Prog. Photovolt.: Res. Appl.* 20: 12–20.
21. International Renewable Energy Agency (IRENA), June 2012. *Renewable Energy Technologies: Cost Analysis Series*, 1, issue 4/5.

Index

F

J

K

L

M

N

P

S

An environmentally friendly book printed and bound in England by www.printondemand-worldwide.com

PEFC Certified

This product is from sustainably managed forests and controlled sources

www.pefc.org

PEFC/16-33-415

This book is made of chain-of-custody materials; FSC materials for the cover and PEFC materials for the text pages.

$108
I/D